Energy: Insights from Physics

Energy: Insights from Physics

Philip DiLavore

Indiana State University

John Wiley & Sons

New York ● *Chichester* ● *Brisbane* ● *Toronto* ● *Singapore*

Library of Congress Cataloging in Publication Data:

DiLavore, Philip, 1931–
  Energy, insights from physics.

  Bibliography: p.
  Includes index.
  1. Power resources.   2. Power (Mechanics)
3. Physics.   I. Title.

TJ163.2.D54   1984   621.042   83-19840
  ISBN 0-471-89683-7

Printed in the United States of America

10 9 8 7 6 5 4 3 2 1

To Carmela

Whose yesterdays encouraged intellectual growth.

To Santa, Pam, Phil, and Teresa

Whose tomorrows are what this book is all about.

And, most of all, to Evelyn

Who makes every today worth the effort.

Preface

To the Teacher:

This book is intended for a one-semester course for students who have avoided science and mathematics and who are taking a physics course primarily to satisfy a general education requirement. The topic is frankly energy, and I have made no attempt to survey physics. Thus, many of the topics normally found in an introductory physics text are omitted. However, the techniques of physics and, especially, the experimental aspects of physics are presented in an elementary manner; students are thereby introduced to different ways of approaching a topic that is at least vaguely familiar to all of them and that certainly will affect their lives. Students need the opportunity, along with the motivation, to develop an analytical approach to important problems and to practice grappling with approximate numbers for insights that remain otherwise elusive. With this text, and the guidance of a skillful teacher, they will be given such an opportunity.

You might find that there is more material contained here than can be considered in depth in the usual one-semester course. As the teacher, you will therefore have to make judgments about what material to omit, cover lightly, or assign for out-of-class reading. I have included some material that might have been omitted without disrupting the flow of the text, but that many teachers will want to include to some degree. My philosophy has been that, in general, it is easier for a teacher to omit material than to add and face the plaintive cry, "it is not in the book." The physics presented provides a tool with which to tackle the applications at hand. For those teachers who are inclined to use it, the Teacher's Guide, available from Wiley, will assist you in developing a teaching strategy.

Decisions affecting the future use of energy by humankind will be crucial. If you and I can help the coming generations, objectively and rationally, to approach those decisions then a worthwhile purpose will have been served.

Philip DiLavore

Dear Student:

Energy may well turn out to be one of the most, if not the most, significant problems facing our society in the years to come. In your lifetime you will undoubtedly see enormous change, much of it caused by the availability, or lack, of energy sources. Careful planning for the future will require knowledge of the energy situation, past and present, and an understanding of future prospects.

As the title suggests, this book examines the present energy situation, using the methods of physics developed over a period of hundreds of years, and considers some possible avenues for the near future (say, the next 50 years). No text, unless it is a many volume encyclopedia, could examine every aspect of energy, so this book is limited to a few important ones. There is some emphasis on small-scale energy uses, such as home heating and solar energy, because each of us can have a very immediate effect on those uses. However, large-scale energy production and consumption, including nuclear power, are also considered, since those uses are certain to be of great consequence to us as well. More important, the methods of examination and analysis in this book may be extended to a wide range of problems not discussed here. In fact, you will soon forget the facts and figures presented (undoubtedly, immediately following the final exam). However, the problem-solving, inquisitive approach, once properly developed, endures: It will serve you well so I urge you to develop it as best as you can.

An extensive mathematics background will not be necessary in order to study this text; basic algebra at the high school or junior high school level is sufficient. Often, though, energy-related numerical calculations will be done, some involving very large numbers. If your algebra, computation, or graphing, is rusty the appendices at the back of the book will help you to review for the mathematics in this book. An inexpensive calculator will also assist you with the calculations.

The many "activities" scattered throughout the text could be the most beneficial aspect of this text for you. I urge you to complete as many of the activities as you possibly can—in the lab, at home, in your dorm room, wherever and whenever you have the opportunity. Don't be afraid to "get your hands dirty." You will find that you will better understand the physics principles and applications after you have seen them in action for yourself. The activities here may also suggest other activities that will help you gain insight into the workings of the surrounding world. You may be inspired, I hope, to invent some activities of your own.

Where will you be if you diligently study the material in this book? Certainly, you will be more informed about energy than you were at the outset. You will have had some experience with the techniques of problem solving and therefore be better able to evaluate statements about energy. You will more effectively interact with or respond to heating contractors, solar installers, wood stove salespeople, local utility representatives, newspapers, television, public service commissions, and even your congressmen, the state legislature, or local government. Above all, you may acquire an inquisitive attitude, be less likely to accept authoritative statements, and be prepared to objectively analyze any situation. If this book helps you to develop such an ability, then my efforts will have been worthwhile.

P.D.

Acknowledgments

I am grateful to Indiana State University, which generously permitted me a sabbatical and additional time for the preparation of this book.

I would like to express my thanks to many former teachers and mentors for guiding me into the field of physics and physics teaching. Foremost among them were Professors Peter A. Franken and Richard H. Sands, both talented physicists, dedicated teachers, and fine human beings.

During the preparation of this text, Professor Uwe J. Hansen was of invaluable help. He was the first to use an early version of the manuscript for a class, and his insights have been most helpful. In addition, as head of my department, he has provided assistance and encouragement in ways too numerous to mention. Professors L. Eugene Poorman and Vincent A. DiNoto have also used the manuscript on a trial teaching basis and have provided valuable suggestions. Norman L. Cooprider contributed technical assistance for many of the photographs and drew some of the original maps. Robert Lavette was the first to read the manuscript in an early draft, and he made numerous useful suggestions. Other colleagues who have contributed by reading and commenting on portions of the text are Professors John A. Swez, Carl O. Sartain, and Walter H. Carnahan. All of these people, and the several anonymous reviewers who so carefully and thoughtfully reviewed early manuscripts, have my gratitude. Additionally, I am grateful to Kathy Bendo, of John Wiley and Sons, who was quite helpful in locating hard-to-find photographs. By far, most constructive was the work of Ruth Greif, production supervisor for this text. The quality of the book and the sanity of the author owe much to her careful, knowledgeable, and ever-cheeful self. Finally, I would like to thank my good wife, who has suffered nearly three years of unmowed lawns, poor home maintenance, and shameful neglect on my part, but with not one word of complaint and with ever-ready encouragement when the going got tough. I'll try to fix those back steps now.

P.D.

Contents

1

Introduction

FAR FUTURE NUMBER ONE

Ralf Worrom's return from deep sleep to full awareness was gradual and slow. Sensing his increased brain activity, his sleeping area also slowly increased its activity: The lighting gradually replaced the darkness with full artificial sunlight at a comfortable level, soft music filled the background, and the video walls began their day-long three-dimensional portrayal of scenes corresponding to Ralf's moods and desires. (Currently, Ralf appeared to be in the middle of a small clearing in a wooded area.) As he became fully awake, the computer-controlled air bed, which was suspending him on supremely comfortable jets of warm air, slowly levered him to his feet on the plastic surface. Ralf, still young at 83, stepped into the bathroom, pushed a button, and was cleansed by jets of water. They came from all directions and were programmed to wash, then rinse, and follow by warm air drying. He then had facial hair removed by the depilator, a necessary weekly nuisance in an age when fashion demanded clean faces, and his teeth and gums attended to by an ultrasound cleaner. A total of 0.1 ergons had ticked off on the ergonmeter he wore on his wrist.

Walking back to the sleeping area, Ralf dialed the color combinations and material types for his clothes for the day and, after a 30-second wait for them to be fabricated, put them on. The cost was 0.5 ergons.

In the dining area, Worrom dialed in his favorite breakfast of hamneggs. While it was being synthesized from the basic ingredients—essential amino acids, glucose, vitamins, minerals—he somewhat ruefully pondered the cost of one-half ergon, but decided that it was well worth it. The meal was perfectly balanced, with no cholesterol or unnecessary ingredients, and it would keep him comfortable until his second and final meal of the day. And, besides, it was perfectly delicious. As he ate, he considered his activities for the day.

While the atomic trash receptacle was reducing his plate and eating utensils to their basic constituents, for future use in fabricating needed articles, he dialed up Nada Tomo, who instantly appeared in all her loveliness, full size and in three dimensions, in the video wall. His wife, whose main place of residence was on another continent, brightened and said, ''Ready for another walloping in spacebol, darling?'' Chortling, he answered, ''That will be the day—you barely won yesterday after several strokes of sheer luck." Sweetly, she said, "We shall see, my dear, we shall see. I'll be there in 30 minutes; it's your turn to drive." While he waited for her, he indulged in a bit of reading, a hobby practiced by very few persons these days.

Spacebol, a type of handball played in the weightlessness of an orbit around the Earth, was an excellent way to stay in condition, but quite expensive. The energy cost for a few hours in outer space was 20 ergons, and the only way they could afford it out of their daily allotment of 25 ergons each was by sharing the cost.

Nada and Ralf lived in an energy economy, with the costs of goods and services measured in terms of the energy required to produce them. The basic unit, the ergon, was equivalent to the ancient, twentieth-century measure of 1000 kilowatt-hours. With the Earth's population stable at 1 billion persons, and with plentiful energy from deuterium fusion and the sun, each person was allowed 25 ergons per day, with metering provided by the wrist-worn ergonmeters. Some overage was allowed in any given day, but it had to even out later. About half of the 25 ergons went for the needs of daily living, the rest being available for use as desired.

When Nada arrived, they went immediately to Ralf's spacebol rocket for the trip to orbit. . . .

REALITY

Is the future world portrayed in this fanciful bit of fiction a possibility? Well—discounting such items as the instantaneous fabrication of ham and eggs from basic constituents—perhaps. That is, perhaps there will be an affluent, comfortable society, with enormous amounts of energy available. That will depend critically upon what we do right now, and in the next few years, particularly in the United States. The future of all humankind has never before depended so sensitively upon the actions of the present generation. The possibility of almost limitless energy from fusion and from the sun made available to humankind may exist, but its beginnings are probably 30, 40, or 50 years in the future. What we do between now and then could destroy the chance of future generations to have reasonable comfort and could, perhaps, even destroy life itself. The purpose of this book is to start you on the road to being able to make intelligent and thoughtful decisions about the course of the immediate energy future. Before we go on, let's take a look at another conceivable future.

FAR FUTURE NUMBER TWO

Ralf awoke with a lurch, heart pounding, and although he was cold as always, he was covered with sweat. It was still dark in the cave. Was that a noise? A wild dog, or a wolf, perhaps? Or was it a headhunter from the neighboring tribe? He grasped his club and crept toward the mouth of the cave. Toward the east, there was the glow of the approaching sunrise, and in the opposite direction there was a glow of one of the many radioactive deserts left by the Great Energy War centuries ago. This one was in the location of the former city of St. Louis. Ralf didn't know what it was, and just accepted it as part of his environment. Seeing no immediate threats, he relaxed a little.

Aroused by the noise, his mate Nada crept out, shivering. She grunted a greeting. Their first thoughts were to find food. Hunting was increasingly difficult for Ralf, who, at 29, was the oldest member of his tribe, and they had not eaten anything other than roots and berries for three days. Although they had no notion that the glowing deserts were responsible, Ralf and Nada, like most of their tribe, were childless. The only child born to them had been horribly deformed and died soon after birth, so they had no help with the hunting. (The population of the entire world was now under 2 million and declining.) Anyway, even if meat were caught, Ralf and Nada could barely eat it—most of their teeth were gone. Cooking it would have helped, but they knew nothing of fire, other than the wildfires they sometimes saw in the forest. They both knew they were near the end of their long lives.

Nonetheless, and with little hope of success, they set out to find meat. . . .

REALITY

This is a grim picture, indeed. Is it a possible future? Although we sincerely hope not, it is conceivable. We have seen in the past few years how changes in fuel production patterns, with no real shortages for Americans as yet, have contributed to widespread economic distress and increasing world tensions. We Americans, being the world's energy gluttons, have become vulnerable to the rules of the Organization of Petroleum Exporting Countries (OPEC) and other forces over which we have little or no control. Because we have greatly increased our consumption of energy and have thus far found no reasonable substitutes for petroleum, we are in a situation we must get out of—or else. The situation is far from hopeless, but we must approach it sensibly. The aim of this book is to give you some of the background necessary for intelligent decision making. Although it covers a broad range of topics, the emphasis is on those aspects of energy use you, as an individual, can influence most directly.

THE ENERGY CRISIS

In recent years, it seems we have heard continually about the "The Energy Crisis." In a way, *crisis* is the correct word, for it implies that we are on the edge of drastic change, possibly for the worse. Yet, as with the little boy who cried *wolf* too often, *crisis* can be repeated so frequently that it loses its mean-

ing, particularly if no obvious disaster or other dramatic change follows soon after the cry. True, we have had rapidly increasing prices at the gas pump, but that doesn't seem to stop us from buying gasoline. True, we have had runaway inflation and high unemployment, but their relationship to the energy crisis is not at all clear. True, we have a government that seems unable even to begin coping with the situation, but we always think the next administration will do better. Sometimes we think that the whole energy crisis thing is phony.

However, we should stop deluding ourselves. Call it what we may, we have an energy problem. Actually, it sorts itself out into two related problems: the problem of providing energy for the near future, and the problem of ensuring that future generations have sufficient energy resources for their needs. Certainly, we could continue to gobble up resources, particularly petroleum, just as fast as they can be pumped or dug out of the ground, but if we do, surely our children and grandchildren, and even the younger ones of us now living, will pay a price that will cause them to curse our present thoughtlessness. As pointed out by Figure 1.1, the resources of the Earth are limited. One of the things that made us a great nation was the consideration for future generations shown by the founding fathers in the establishment of our form of government. Another was a great abundance of natural resources. Now that the resources are shrinking, we need to show some of the same kind of consideration for future generations. To start, let's take a look at how we got to this point of energy use.

ENERGY ADDICTION

Yes, *addiction* is precisely the right word. Without realizing what was happening, we have become a nation of energy junkies. Although it might be desirable to kick the habit, to do so is not easy and, done in the wrong

Figure 1.1 The Earth, as viewed from outer space. Except for the energy coming to us from the sun, the Earth has only the resources already here. (Photo courtesy NASA.)

way, kicking the habit might do more harm than good.

The Ralf Worrom of future number one was an incredible energy hog, consuming 25,000 kilowatt-hours (kWh) of energy each day. However, the Joe Doaks of the 1980s does pretty well too, with the consumption of one-hundredth of that amount. That is, if the overall rate of energy consumption in the United States, including industrial and commercial use, transportation, and heating, is divided equally among the population, it amounts to over 250 kWh per day per person. This number may not mean much to you yet, so to put it in perspective, compare it to these rates: about 125 kWh per day per person in highly industrialized nations like West Germany and England; less than 90 in Japan; and about 75 in Italy. In the other extreme, consider that our near neighbor Mexico consumes under 30 kWh per person per day; Ecuador and Egypt consume about 7; Guinea about 2.5, and Ethiopia less than 1!

It would not be correct to equate quality of life to the amount of energy consumed, but there certainly is some relationship between the two. The underdeveloped nations are beginning to want—and should have—a fairer share of the world's energy resources. Ironically, some nations with great energy resources, being used up by others, are among the lowest consumers. One can understand that they may wish to reserve some of their resources for their own future use.

THE ROAD TO ADDICTION (EXPONENTIAL GROWTH)

Figure 1.2 is a graph showing the growth of energy use in the United States for this century. Except for the period of the Great Depression in the 1930s and—on the scale of this graph—some nearly unnoticeable effects following the Arab oil embargo of 1973, the picture is one of continuous growth. Further, it is growth of a particular kind.

The kind of growth exhibited is called *exponential* growth. To demonstrate the characteristics of exponential growth and the trouble it can get us into, consider the following example. Suppose two ambitious joggers go on self-improvement programs. Both are presently doing a mile a day, and both are determined to increase the distance. They place a bet on whether they will be able to stick to their respective schedules. Doris, being very ambitious, vows to add a half mile per week, so that at the end of the year she will be doing the marathon distance each day. Walter is not nearly as ambitious, so he states he will simply increase his distance by 2 percent each day. That seems reasonable, doesn't it? Well, try the numbers, and you will see otherwise. Just multiply 1 by 1.02, then the result by 1.02, and so on for 365 multiplications. At first, there is no problem. One mile the first day, 1.02 the second, 1.04 the third, 1.06 the fourth,. . .,and about 2 miles at the end of 35 days. Not so bad. By this time, Doris, being a determined person, is up to 3.5 miles a day. Continuing on in the same manner, one finds that Walter's distance doubles again to about 4 miles at the end of another 35 days. Still not too bad. Poor Doris is now struggling with 6 miles per day. In another 35 days Walter is doing 8 miles and Doris is at 8.5. Every 35 days Walter's distance doubles, whereas Doris's is increased by 2.5 miles. The 35-day period for Walter is called the *doubling time*, and every 35 days his distance will double again. Now that we know that, let's not do the multiplication all 365 times. Rather, we know that there are about ten 35-day periods in a year, and that Walter's distance doubles every 35 days, so let's do an approximate calculation of the distance. It goes 1 mile, 2 miles, 4 miles, 8, 16, 32, 64, 128, 256, 512, 1024 miles each day before the year is out! Ludicrous! A graph of Doris's progress and Walter's struggles is

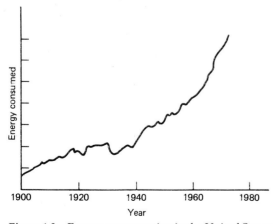

Figure 1.2   Energy consumption in the United States for this century.

shown in Figure 1.3. The straight-line graph represents a *linear* relationship, and the upward curving one is the exponential curve. (Incidentally, if the growth of 2 percent per day were figured exactly for a year, the final figure would be about 1377 miles per day, but our approximation is close enough to make the point.)

This somewhat silly example of Walter's jogging is, unfortunately, nearly matched in silliness by reality. For example, during the period from the middle 1950s to the early 1970s, electric power generation in the United States grew exponentially at a rate of about 7.5 percent per year, or with a doubling time of 10 years. Many people who should have known better assumed that such a growth pattern would continue. (We seem to think that "bigger is better" in our society.) I recall seeing television commercials in the early 1970s by large companies involved in the production of power describing the problem of a 10-year doubling time and their own contributions to the solution of the problem—helping to double the generating capacity in the coming 10 years! This kind of reasoning, carried to logical ends, leads to some crazy results. For ex-

ample, in 1970 approximately 0.3 percent of the area of the continental United States was covered by electric transmission lines. Making the assumption that the growth pattern would continue unabated, with a doubling time of 10 years, and the reasonable assumption that the area covered by transmission lines would follow a similar pattern, then, before the year 2055 the entire continent would be totally covered by a shroud of wires! That can't happen, and neither can continued exponential growth—of anything—at any rate whatever.

Activity 1.1

To get some notion of the energy use around your home, try checking your electrical appliances. Most, if not all, of them will indicate somewhere the amount of power consumed (although it may sometimes be difficult to find). Make a table listing the appliances and then estimate the amount of use each gets, say in a month, and find the total energy consumed by that appliance for the month. For example, a 900 watt (0.9 kilowatt) toaster may be used for 10 min a day, or about 5 hr per month. The energy consumption for the month is then

$$\text{energy} = 0.9 \text{ kW} \times 5 \text{ hr}$$
$$= 4.5 \text{ kWh}$$

The unit kilowatt-hours (kWh), is a measure of energy. If the cost of electric energy is $0.05 per kWh, then it would cost about $0.23 to run the toaster for a month. Make a table like the one here.

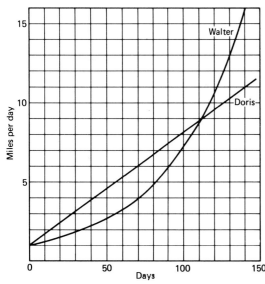

Appliance	Power required	Monthly use	Energy used	Cost
Toaster	0.9 kW	5 hr	4.5 kWh	0.23
Electric range	6.0 kW	30 hr	180 kWh	9.00

Figure 1.3 Doris and Walter's progress.

When you have made your best estimates, add up the last column and see how close it comes to an actual electric bill. (Of course, you will need to use the actual rate charged in your area.) The power to run your appliances comes from an electric power generating station like the one shown in Figure 1.4.

Activity 1.2

Take a close look at your electric meter. The turning disk is attached to the shaft of a little motor that turns at a rate that is proportional to the amount of power being consumed. If possible, have someone in the house turn various appliances on or off, and see what effect it has. If you can get everything shut down except, say, one electric clock, then the disk barely creeps around. If an electric furnace or all the elements of an electric range are turned on, it fairly flies! If you wish, you can time how long it takes the disk to make a complete revolution with different appliances running. If the disk goes around in one-half the time, twice as much electric power is being used.

The dials on the meter show how much energy, in kilowatt-hours, has been consumed since the meter read zero. Some of them read "forward" and some "backward," whichever makes the gearing easier. In Figure 1.5, the meter reads 45,743 kWh. It is interesting to keep a daily log of the meter readings. On some days, more energy will be consumed than on others. (You can also check on the electric company this way, to see if they billed you correctly.) Try this and see if you can account for differences from one day to another. Just be sure to take a reading at the same time every day and subtract the previous day's reading from the present reading each time.

Figure 1.4 The familiar stacks of a coal-burning power plant reach toward the sky.

There is not a great deal of mathematics used in this book. However, physics is a quantitative science, and you will need to be reasonably handy with high school algebra, arithmetic, and graphing. If you have any problems with any of these, refer to Appendices A through C in the back of the book. They are designed to help you. Read them, as needed. Your teacher can provide additional help, if necessary.

ADDITIONAL QUESTIONS AND PROBLEMS

1.   Which of the following situations represent exponential growth, and which are other kinds of growth?

   a.   John Farmboy lifts his calf every day to build his muscles. The calf gains 3 lb a day, so John gets stronger every day.

   b.   Sarah Brainhead started a new business selling soap. In her first year, the profit was $1000. In the second year, she made $1100, and in the third and fourth years, $1210 and $1331, respectively. She expects this pattern of growth to continue.

   c.   Joe Workerman belongs to a union that gets him a 5 percent raise each and every year.

   d.   Irma Bombast makes her living by giving public speeches once a week. Her usual speech lasts 30 min. For the last five weeks, however, she has been getting more and more wordy, and each speech has lasted 10 min longer than the speech of the previous week.

2.   Make a graph of each situation described in the preceding question. Were your previous answers correct?

3.   Do the following arithmetic operations. If you have any trouble with them, Appendix B will help.

Figure 1.5   An electric meter (this one reads 45,743 kWh).

$6.3 \times 10^5 \times 2.1 \times 10^4 =$

$7.21 \times 10^{-4} \times 3.44 \times 10^3 =$

$25 \times 10^5 - 3.1 \times 10^6 =$

$5 \times 10^{-4} / 2.2 \times 10^{-5} =$

$2.36517 \times 10^{-4} + 3.36 \times 10^{-5} =$

4.   One of the numbers in the following table was copied incorrectly. Which one is it? Approximately what should it have been? (*Hint*: You may be able to pick it out of the table, but making a graph would make the problem much easier. If you have any difficulty in making the graph, consult Appendix C.)

x	y
0	5.3
1	6.8
2	11.4
3	19.0
4	24.7
5	43.4
6	60.2

2 Motion

An understanding of energy requires some knowledge of the behavior of physical objects, their motions, and the forces they exert upon one another. A first step toward this understanding can be made by carefully observing the motions of some familiar objects and the forces that cause motions to change. The objective of this chapter is to help you make these observations.

UNIFORM MOTION

Activity 2.1

Put a book or a block of wood on the floor and give it a good push, letting it go to slide across the floor. Carefully try to describe its motion after you let it go. Now try exactly the same thing with a toy truck or car or with a roller skate. Again describe the motion carefully. What are the differences in the two cases? What similarities are there? In each case, why does the object keep moving after you stop pushing? Why does it finally stop moving?

From your practical experience, you have probably answered that "friction"—whatever that is—was responsible for slowing and stopping the objects. That is quite correct. In fact, as you observed, if you reduce the friction on an object, (for example, by putting wheels on it) it takes a lot longer to stop. If you used a hard smooth floor, the roller skate probably kept going until it bumped into something, because its wheels are made to be very low-friction devices.

The other question, why does the object keep going after the push has stopped, is much more difficult, conceptually. It is such a difficult concept that for centuries people tried to resolve it with theories such as the one that supposed that the air pushed aside by the front of the moving object rushed around and pushed the object from behind. In this age of space travel, the concept is somewhat easier to grasp. Imagine that you are an astronaut doing a space walk, and you have a wrench in your hand. Give it a push and let it go. What happens?

There is no friction at all out in space, and the wrench keeps moving forever, unless it runs into a planet, or something. Further, since there is nothing to deflect it from its path, it moves forever in a straight line.

Coming back to Earth while the wrench

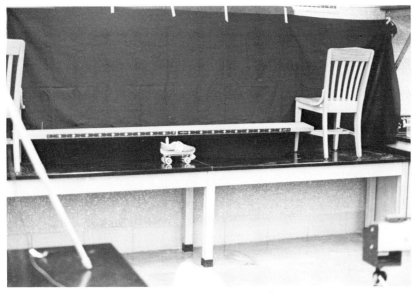

Figure 2.1 The setup for taking a strobe photo of a moving skate.

continues on its way, let's see if we can approximate the same effect and measure the results. Figure 2.1 shows an arrangement for doing so with a roller skate. The skate is given a push on a level, hard surface and allowed to coast. To get a measure of the motion, we use a strobe light and a camera. The stroboscopic light produces short, bright flashes at equal time intervals and serves as a clock. For these trials the strobe is set to flash five times each second. The room is darkened, the camera's shutter is opened and held open, the strobe is started, and the skate is given a push. Since the room is dark, the camera can "see" the skate only at each instant that the strobe is flashing. The result is Figure 2.2, a picture showing successive positions of the skate

spaced 0.2 s apart in time. Instead of looking at the images of the skate, focus your attention on the white pointer attached to it. The point to which it is cut is perfect for making measurements.

In Figure 2.2 the important thing to note is that the successive positions of the skate are equally spaced. That is, in each 0.2 s interval, the skate moves about the same distance, as we can see in the picture. This is called *uniform motion*. The skate executes uniform motion only approximately, because there is some friction present. Thus it slows down a little. The wrench in outer space performs uniform motion exactly and forever, as long as it does not get close to any other bodies in space.

Figure 2.2 A strobe photo of uniform motion.

ANALYZING THE MOTION

Figure 2.3 shows measurements made on the photo of Figure 2.2, including some unavoidable errors in measurement. The measurements above the skate show that it moves about the same distance between each of the flashes. They were made by taking measurements from the picture itself and converting to the proper scale by measuring the size of the meter stick in the picture. One could also read positions directly from the meter stick in the photo and make the appropriate subtractions.

The measurements below the skate show the *total* distance moved to the right for each position, starting with an arbitrarily selected zero position. Here we choose the zero point to be the first position showing in the photo, and we also choose the starting time at that point to be zero. The time that has passed since the $t = 0$ time is also shown below the skate. Note that any other position could have been called the zero point.

Table 2.1 shows the relationship between the distance moved and the time elapsed, as taken from the strobe photo.

A graph of distance moved (x) versus time (t) is shown in Figure 2.4. Notice that the points do not all lie exactly on the straight line. Since the data came from an actual experiment, there is some unavoidable error in measurement. However, the straight-line trend of the graph is quite obvious, and the line that is drawn is a good average, with some of the measured points lying above the line and some below.

Because the graph is a straight line, the relationship between x and t is expressed in the equation:

$$x = B + At$$

Here B is the point on the distance axis where the straight line starts. (If you have any problems with this, see Appendix C at the end of the book.) Putting $x = 0$ and $t = 0$ into the equation shows that $B = 0$ in this case. Understand that this is simply the result of how we picked the starting point. For example, we could have chosen to use the readings from the meter sticks. Then, when $t = 0$, $x = 0.32$ m, so B would be 0.32 m, also. In that case, the graph would look like Figure 2.5, with the same *slope* as Figure 2.4, but with the whole thing raised up by 0.32 m on the x axis.

The slope is a measure of the degree of tilting of the straight line, and it is equal to A in the equation. Figure 2.6 shows how to find

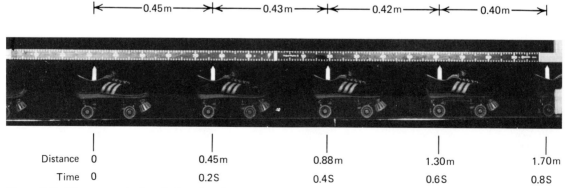

Figure 2.3 The same strobe photo, showing measurements of successive positions of the skate.

TABLE 2.1 *A Skate Rolling Along a Level Surface*

Time elapsed, t (s)	Total distance traveled, x (m)
0.0	0.00
0.2	0.45
0.4	0.88
0.6	1.30
0.8	1.70

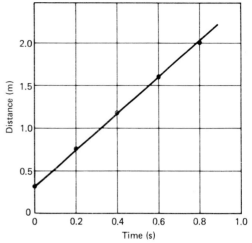

Figure 2.5 The same graph as Figure 2.4, but with the starting point chosen as 0.32 m.

the slope. It is the same as Figure 2.4, but with a right triangle drawn in. Any two points on the line can be used to make the triangle. Here the two points chosen were those for x = 0.5 m and x = 1.5 m. Then the length of the vertical leg of the triangle is Δx = 1.5 m − 0.5 m = 1.0 m (Δ is the Greek letter *delta*. It is often used to denote a *difference* in two quantities or a *change* in a quantity.) The length of the horizontal leg of the triangle, Δt, is found by simply measuring from the graph, and it is about 0.47 s. The slope is defined as Δx divided by Δt.

Note that both quantities have dimensions associated with them; in the one case the unit is meters, in the other, seconds. When the slope is found by dividing one by the other, it also has dimensions, with the units being meters per second (m/s). That is:

$$A = \frac{\Delta x}{\Delta t} = \frac{1 \text{ m}}{0.47 \text{ s}} = 2.1 \text{ m/s}$$

Since we now know both A and B, we can write the equation above as:

$$x = (2.1 \text{ m/s})t.$$

Think for a moment about the significance of the slope, which we have called A. It has the units of meters per second, so it is

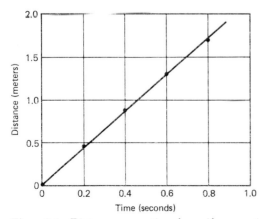

Figure 2.4 Distance versus time for uniform motion.

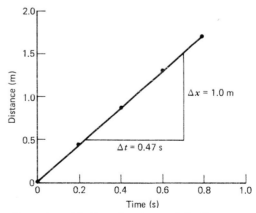

Figure 2.6 Finding the slope of the line.

related to the distance traveled in a second. If the skate is pushed more softly so that it goes slower, the slope decreases, as in Figure 2.7, which is taken from another experiment. If the skate moves faster, the slope increases. Thus the slope is a measure of the *speed*, or *velocity*, of the skate. To generalize our earlier equation, let us replace A with v for velocity. Also, since B is just the starting point along the x axis, we shall use x_0 for it, so our equation now becomes

$$x = x_0 + vt \qquad (2.1)$$

This is a very general equation. It says that the distance traveled (x) is equal to the starting point (x_0) plus velocity multiplied by time (vt). In the particular case we are examining, x_0 has been chosen to be 0 m, and $v \approx 2.1$ m/s. ("$\approx$" means approximately equal to.)

Questions

Using the equation $x = (2.1 \text{ m/s})t$, one can make a table for x versus t, as shown in Table 2.2. This is done by choosing values for t—for example $t = 0$ s, 0.2 s, 0.4 s, 0.6 s—and substituting these values into the equation.

TABLE 2.2   *Distance versus Time*

t (s)	x (m)
0.0	0.00
0.2	0.42
0.4	0.84
0.6	1.26
0.8	1.68
1.0	2.10
1.2	2.52

Now make a graph like that of Figure 2.4. How does the "theoretical" graph you just made compare with the "experimental" graph of Figure 2.4?

What is the slope of the graph of Figure 2.7?

Before we go any farther, let's take a look at the units for Equation 2.1. As everyone knows, you cannot add apples and oranges. Likewise, you cannot add meters (m) and meters per second (m/s), so we need to be careful. If we take the case where $v = 0$, there is no problem, for then

$$x = x_0$$

Since both x and x_0 are in meters, they can be added or equated without difficulty. Taking a closer look at vt, let us assign values, say, of $v = 5$ m/s and $t = 3$ s. Then vt becomes

$$vt = 5\,\frac{\text{m}}{\text{s}} \times 3\,\text{s} = 15\,\frac{\text{m} \times \text{s}}{\text{s}} = 15\,\text{m}$$

What we have done here is to divide seconds in the numerator by seconds in the denominator. Any quantity divided by itself is 1, so the end result, in this case, is meters. Thus, we see that vt is just the distance traveled in a given time; it has the units of meters, and we may add it to x_0 without difficulty.

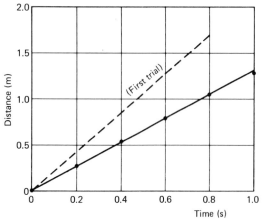

Figure 2.7   The same type of graph for a slower moving skate.

Questions

By choosing $t = 0$ s, $t = 1$ s, $t = 2$ s, and so forth, make a table of distance traveled versus time for each of the following cases. Then, by using your table for each case, make a graph of x versus t.

$x_0 = 0$ m and $v = 5$ m/s

$x_0 = 1.5$ m and $v = 5$ m/s

$x_0 = 50$ m and $v = -5$ m/s

$x_0 = -10$ m and $v = 5$ m/s

ACCELERATION

Activity 2.2

Find or make a long, smooth incline. A "two by eight" 8 ft long is ideal. Get a pair of roller skates (or "hot wheels" cars), a meterstick or a ruler, and a stopwatch.

Prop up one end of the board about a foot or so higher than the other. The idea is to let a skate roll down the incline without being pushed. Try it a few times to get practice aiming it down the board. Describe the motion that takes place.

Mark the board off, from a line close to the top end, in equal intervals of convenient length, perhaps 0.5 m each. With the nose of the skate held at the top line, get the stopwatch ready. Let the skate go and start the watch, simultaneously. Be careful not to push the skate either up or down the incline as you let it go. When the front of the skate reaches the first mark, stop the stopwatch and record the time. When you have had a few practice runs, take readings at least an additional five times. Take the average of the five recorded times and use this as the time it takes the skate to go that distance. (Remember, the average is the sum of the five values divided by five.)

Now repeat the run to get the time it takes the skate to reach the second mark, the

third, and so on. If the skate goes off the side of the board discard that trial. Be patient and aim well.

When you have completed the data taking and averaging, make a table and a graph of distance traveled (x) in meters versus time (t) in seconds. Interesting, isn't it?

If you have the time, it would be instructive to tie the other skate securely on top of the first one to double the weight and repeat the experiment. If time is short, just try it for one or two distances. What is the result?

Analysis of the Experiment

You have undoubtedly made the observation that, as the skate goes down the incline, it goes faster and faster. That is, the velocity increases. When the velocity of an object changes, we say that it has an *acceleration*. Table 2.3 and Figure 2.8 show the results of a similar experiment.

We certainly know that there is error in the measurements just by looking at the third column in the table. There, each Δt, the difference in times the skate takes to go past two successive points, should get smaller and smaller as the skate goes down the ramp. Instead, Δt seems to jump around in size. This is because the measurements are not exact.

However, you can tell that the graph of distance traveled versus time is definitely not a straight line this time. It is a curve. It has a slope, but that slope increases as time goes by. As you learned earlier, the slope of a graph of x versus t is just the velocity. Thus, since the slope keeps increasing, this graph is another way of showing that the skate goes faster and faster as time passes.

More precise measurements may be made by, again, using a strobe photo, like the one shown in Figure 2.9. There a skate rides down a tilted board while the strobe light "blinks" every 0.2 s. Measurements taken from

TABLE 2.3 *A Skate Rolling Down an Incline*

Distance traveled (m)	Time elapsed (s)	Difference in times (s)
0.0	0.00	
		0.74
0.4	0.74	
		0.42
0.8	1.16	
		0.26
1.2	1.42	
		0.24
1.6	1.66	
		0.26
2.0	1.92	
		0.24
2.4	2.16	
		0.14
2.8	2.30	
		0.22
3.2	2.52	
		0.16
3.6	2.68	

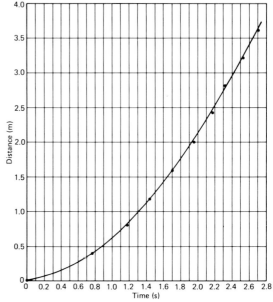

Figure 2.8 The graph for a skate rolling down a sloped ramp.

this photo are displayed in Table 2.4 and Figure 2.10.

This, once again, is accelerated motion, with the skate going faster and faster as it moves down the slope. The increasingly larger distances between blinks is the proof. To see just how the velocity increases, look at the distance traveled in each 0.2 s time interval. For example, for $t = 0$ s to $t = 0.2$ s, the skate goes a distance $\Delta x = 0.01 - 0.00$ m $= 0.01$ m. Its average velocity in this time interval is then the distance traveled divided by the time interval, or:

$$v = \frac{\Delta x}{\Delta t} = \frac{0.01 \text{ m}}{0.2 \text{ s}} = 0.05 \text{ m/s}$$

Since the velocity is increasing during the 0.2

Figure 2.9 Strobe photo of a skate riding down an incline.

TABLE 2.4 *Measurements from Strobe Photo of Cart Rolling Down Incline*

Time elapsed (s)	Distance from start (m)
0.0	0.00
0.2	0.01
0.4	0.05
0.6	0.12
0.8	0.23
1.0	0.36
1.2	0.52
1.4	0.72
1.6	0.94
1.8	1.20
2.0	1.48
2.2	1.81

TABLE 2.5 *Velocity versus Time for Cart Rolling Down Incline*

Time elapsed (s)	Velocity (m/s)
0.1	0.05
0.3	0.20
0.5	0.35
0.7	0.55
0.9	0.65
1.1	0.80
1.3	1.00
1.5	1.10
1.7	1.30
1.9	1.40
2.1	1.65

s interval, this average velocity is neither the velocity at the beginning of the interval ($t = 0$ s) nor at the end ($t = 0.2$ s). It is the velocity in the middle of the time interval, that is, at $t = 0.1$ s. The approximate velocity at $t = 0.3$ s can be found in the same way by dividing 0.05 m − 0.01 m = 0.04 m by 0.2 s, giving a result of 0.2 m/s. Now it is just a matter of doing the arithmetic for each time interval to produce Table 2.5 and the graph of Figure 2.11 for velocity versus time.

Because of the subtractions and divisions, I'm not very confident of the final digit

in the velocity shown in Table 2.5, and it could have been rounded, but wasn't. However, it is clear that the graph of velocity versus time is a straight line. When the graph of v versus t is a straight line, we can write the relationship between the two as a linear equation:

$$v = v_0 + at \tag{2.2}$$

In words, the velocity equals the initial velocity plus a constant multiplied by time. The constant a is the slope of the line. This constant is called the *acceleration* of the skate, and it has the units of m/s divided by s or m/s^2.

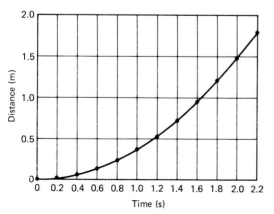

Figure 2.10 A graph of distance versus time for the skate going down the ramp.

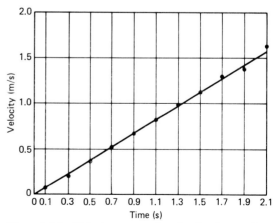

Figure 2.11 Velocity versus time for the skate going down the ramp.

(This is read as "meters per second per second" or "meters per second squared.") When the initial velocity is zero, as it is for our skate, Equation 2.2 simplifies to:

$$v = at \qquad (2.2a)$$

Questions

Make tables and graphs of v versus t for the following:

$v_0 = 0$ m/s and $a = 1$ m/s^2

$v_0 = 5$ m/s and $a = 4$ m/s^2

$v_0 = 10$ m/s and $a = -2$ m/s^2

How do you interpret this last graph?

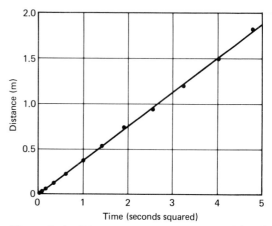

Figure 2.12 Distance versus time *squared* for the accelerating skate.

We have now found the relationship between velocity and time for an object undergoing *uniform acceleration*. Finding the relationship between distance traveled and time is a little more difficult and, rather than go through a long argument, let's try a little trick to get some understanding. The trick consists of looking at distance versus time *squared*. To do this, simply square each time in Table 2.4 to produce Table 2.6 and Figure 2.12. Lo! It is a straight line. (Of course, I

probably would not think of this trick if I didn't already know the answer!)

Since the graph is a straight line, the relationship between x and t^2 may be written as:

$$x = A + Bt^2$$

A complete analysis would show that the constant A is the starting point, x_0, and the constant B is just one-half the acceleration of the skate. That is:

$$x = x_0 + \tfrac{1}{2} at^2$$

If x_0 is taken to be zero, this becomes:

$$x = \tfrac{1}{2} at^2 \qquad (2.3)$$

This simpler form of the equation is the one we will usually use, so it is given an equation number.

One final point: If the skate is given a shove down the incline, then it has an initial velocity, v_0. As a result, it will go farther in a given amount of time than if it starts from rest. To take this into account, the equation

TABLE 2.6 *Distance versus Time Squared*

Square of the time (s²)	Distance traveled (m)
0.00	0.00
0.04	0.01
0.16	0.05
0.36	0.12
0.64	0.23
1.00	0.36
1.44	0.52
1.96	0.72
2.56	0.94
3.24	1.20
4.00	1.48
4.84	1.81

has to be modified slightly. The complete
equation is:

$$x = x_0 + v_0 t + \tfrac{1}{2}at^2$$

Questions

Does each term in the last equation have the
same units? What are the units?

Graph Equation 2.3, for $a = 1$ m/s^2.

The kind of acceleration we have been
studying is called *uniform acceleration*. That is,
the acceleration is constant. It is possible, of
course, to have situations in which the acceleration is not constant, but we will have no
need of studying those.

FALLING BODIES

Galileo is reputed to have dropped objects of
different weights from the Leaning Tower of
Pisa late in the sixteenth century to prove that
they fell at the same rate. Back in those days
this was a difficult concept to swallow and,
despite careful experiments, Galileo had trouble in trying to convince people that he was
right.

Today, it is a great deal easier to see.
Figure 2.13 shows a strobe photo of two balls
being dropped. The two balls fall side by side,
with motions that match all the way down.
Of course, this should not surprise you, because you have already duplicated Galileo's
experiments. Because he did not have strobe
lights and cameras available, he had to "di-
lute" the effect of gravity. He did so by rolling
balls of different weights down a ramp. When
you rolled the two roller skates tied together—double the weight—down your own
ramp, didn't you find that the motion was the
same as for one skate? I did.

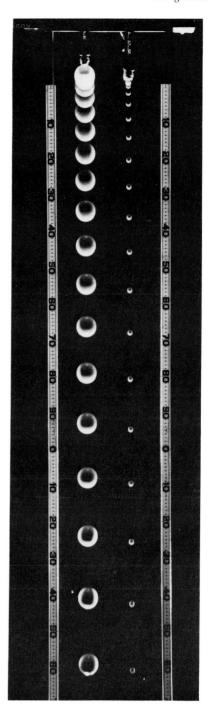

Figure 2.13 A strobe photo of two objects of dif-
ferent masses falling side by side. (Courtesy of
Educational Development Center, Newton,
Massachusetts.)

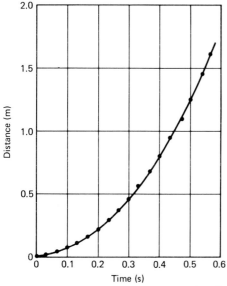

Figure 2.14 Distance versus time for the falling balls.

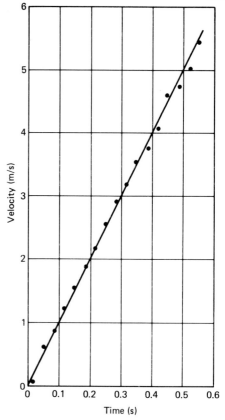

Figure 2.15 Velocity versus time for the falling balls.

Figures 2.14 and 2.15 show graphs of distance versus time and velocity versus time, respectively, for the falling balls (distance is measured from the top downward). These graphs look exactly like Figures 2.10 and 2.11, respectively. Thus they represent motion with uniform acceleration. In other words, when a ball, or other object, falls, it undergoes a constant acceleration. If very careful experiments are done, using a lot of different objects, that acceleration is measured to be 9.8 m/s². It is usually given the symbol g:

$$g = 9.8 \text{ m/s}^2$$

Using g for the acceleration, the equations of motion for a falling body can be written as:

$$v = v_0 + gt \tag{2.4}$$
$$x = x_0 + v_0 t + \tfrac{1}{2} g t^2 \tag{2.5}$$

Questions

Drop two objects simultaneously. Use a sheet of paper and a wadded-up sheet of paper as

the two objects. Why don't they fall at the same rate? Why do the balls of Figure 2.13 seem to fall at the same rate? How should the simple statement of Galileo's discovery be modified for objects falling in the air?

For the case of dropping a ball (initial velocity = 0) and taking the starting point (x_0) to be zero, what simplifications can be made to Equations 2.4 and 2.5?

FORCE

Question

Remember the wrench out in space? It is still undergoing uniform motion, at a constant speed and in a straight line. On the other hand, if we drop a ball here on the Earth, it undergoes an acceleration. What is it that makes the difference in these two cases?

Activity 2.3

I'm sure that, in practical terms, you have some idea of what a force is. For now, call it a push or a pull, if you wish. Try pulling a skate on a smooth, level surface with a long rubber band. (You can make a long rubber band by tying several short ones together.) Describe the motion.

Now try to quantify this a little by keeping the rubber band stretched to the same length throughout the pull. This will take a little practice, but it can be done by pulling as shown in Figure 2.16. Loop the rubber band over the end of a ruler and, as you pull, take care to keep the same point on the ruler over the end of the skate. What is the motion like now?

Do the same thing several times with the rubber band stretched to a different length

each time. How does the motion compare for the different trials?

Finally, load the skate down with some extra weight: the other skate, a big rock, or anything else that is heavy. Now pull with the rubber band stretched the same amount as for a previous trial. How do the motions compare?

As long as you pull—exert a *net* force—on the skate, it accelerates. (The net force is the force you exert along the table minus the frictional forces in the opposite direction.) You were probably surprised by how difficult it was to keep up with it, exerting the same force throughout the motion. Careful measurements of this motion show that the acceleration is constant as long as the force is constant. In the trials in which you increased the force by stretching the rubber band more, the acceleration was greater, and when a smaller force was used, the acceleration was smaller. In general, if the force on an object is doubled, the acceleration also doubles; if force is tri-

Figure 2.16 Pulling a skate with a rubber band.

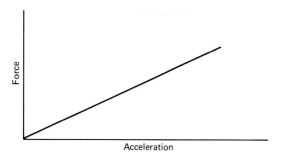

Figure 2.17 A graph of acceleration versus force. For a given object, a greater force produces a greater acceleration.

pled, then the acceleration triples, and so on. In other words, the graph of acceleration versus force is a straight line, as shown in Figure 2.17. The line passes through the origin ($F = 0$ and $a = 0$) because, when no force acts on the object, there is no acceleration.

The relationship between force and acceleration could be written as

$$F = Ba$$

where B is the slope of the line on the graph. But every symbol in a physics equation represents some actual quantity. In this case, the slope is a new quantity, called *mass*, and it is given the symbol "m." Thus, the equation above becomes:

$$F = ma \tag{2.6}$$

This deceptively simple equation is exceedingly important, for it describes the motion of every ordinary object we can observe, such as roller skates, balls, automobiles, and space shuttles.

The mass in this equation is a measure of what is referred to as *inertia*. Inertia is the resistance of a body to a change of motion—resistance to acceleration. Thus, the greater the mass, the more force is required to produce a given acceleration. You saw this when you put the extra mass on the skate, for you then increased the mass of the thing being pulled and more force was needed to produce the

same effect. Putting it in more down-to-earth terms, the mass of an object is a measure of the amount of "stuff" it has. A large truck has a much greater mass than does a small automobile. Thus, the truck requires much more force to speed it up (or slow it down) at the same rate as the car.

Mass is *not* the same thing as weight, although the two are related. The mass of any given object is the same, wherever it is, but its weight depends on its location. The weight of an astronaut may be 180 lb on Earth, 30 lb on the moon, and nothing at all in outer space. His weight changes because it depends on the pull of gravity, which is different in the three places. His mass is *the same* in all three places; it takes the same force to give him the same acceleration. Think again about the wrench in outer space; it is weightless, but it has mass. To change its motion—to give it an acceleration—we must exert a force on it.

The unit of mass is the *kilogram* (kg), which originally was defined as the mass of a cube of water 0.1 m on each side. (That is, 0.001 cubic meters, or 1000 cubic centimeters.) The kilogram, along with the meter and the second, is a basic unit of the SI system of units. (The SI system—*Le Système Internationale d'Unités*—is described in Appendix B.)

The units for force, on the other hand, are *derived* from Equation 2.6. To see how, just substitute into the equation the conditions of $m = 1$ kg and $a = 1$ m/s^2:

$$F = ma$$
$$= 1 \text{ kg} \times 1 \text{ m/s}^2$$
$$= 1 \text{ kg m/s}^2$$

So the unit of force is the kg m/s^2. This is such an important unit that it is given its own name, the *newton* (N). That is:

$$1 \text{ N} = 1 \text{ kg m/s}^2$$

Thus a force of 1 N is required to give an acceleration of 1 m/s^2 to a mass of 1 kg.

EXAMPLE

An object with a mass of 5 kg (about 11 lb) is given a constant push so that it accelerates with $a = 2$ m/s². What force does this require? If it is pushed with this force for 7 s, how fast is it traveling at the end of that time? How far does it go?

Solution

To find the force, just use the equation

$$F = ma$$
$$= 5 \text{ kg} \times 2 \text{ m/s}^2 = 10 \text{ N}$$

The speed at the end of 7 s is found from:

$$v = at \qquad \text{(since } v_0 = 0)$$
$$= 2 \text{ m/s}^2 \times 7 \text{ s} = 14 \text{ m/s}$$

The distance traveled is found from:

$$x = \tfrac{1}{2}at^2 \qquad (x_0 = 0, \text{ also})$$
$$= \tfrac{1}{2} \times 2 \, \frac{\text{m}}{\text{s}^2} \times (7\text{s})^2 = 49 \text{ m}$$

Questions

Assume that your loaded-down roller skate has a mass of about 2 kg and that you can exert a force of 5 N on it for a period of 3 s. What is its acceleration? What speed does it have at the end of the 3 s period and how far has it traveled?

NEWTON'S LAWS OF MOTION

A *law* of physics is essentially a description of what happens about us, which seems always to hold true. That is, in any observations we can make, in any experiments we can do, the *law* seems to be valid. We shall see several such laws in this book; three such descrip-

tions are called Newton's laws of motion. You already know two of them. They follow.

1. **Newton's first law.** Unless some net force is exerted upon an object, it will continue forever in motion at a constant speed and in a straight line. This is sometimes called the law of inertia. Note that the constant speed can be zero. That is, if the object is at rest, it will stay at rest unless some force acts upon it.

2. **Newton's second law.** This is just Equation 2.6 relating force, mass, and acceleration:

    $$F = ma$$

3. **Newton's third law.** Forces always come in pairs. If your hand pushes a book along the table, simultaneously the book pushes back on your hand with a force of equal size. This is indicated in Figure 2.18. The important things to note are that the two equal forces *always* act on two different objects and always in opposite directions. Sometimes the force applied is referred to as the *action* force and the force that pushes back in return is called the *reaction* force. Then, a concise way of stating Newton's third law is. For every action, there is an equal and opposite reaction. Remember, the two forces always act on two different objects and they are always in opposite directions.

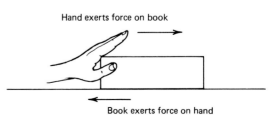

Hand exerts force on book

Book exerts force on hand

Figure 2.18 Action and reaction forces.

EXAMPLE

A book lies on a table. What are the action and reaction forces involved?

Solution

To answer such questions, one can practically set up a "recipe," consisting of two sentences:

a. The action force is the force exerted on A by B.
b. The reaction force is the equal and opposite force exerted on B by A.

In this case, the two sentences become:

a. The action force is the force exerted on the table by the book.

b. The reaction force is the equal and opposite force exerted on the book by the table.

Note that it makes absolutely no difference which force is called action and which reaction. For example, the above question could equally well be answered:

a. The action force is the force exerted on the book by the table.
b. The reaction force is the equal and opposite force exerted on the table by the book.

Questions

See if you can identify the action and reaction forces in the drawings of Figure 2.19.

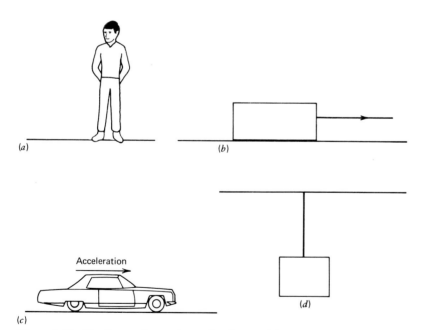

(a) (b)

Acceleration

(c) (d)

Figure 2.19 Finding action and reaction forces. (a) There are forces between this person's feet and the floor. (b) String pulling a block. (c) There are forces between an accelerating car's tires and the road. (d) There are forces between the string and block *and* between the string and ceiling.

MORE ABOUT FALLING BODIES

Weight

Let us return to a consideration of the ball falling near the surface of the Earth. We have already seen that as it falls it accelerates, with a constant acceleration of about 9.8 m/s². Why does it accelerate? It must be because there is a net force acting on it. Where does that force come from? You know that the force comes from the gravitational attraction of the Earth. It is called the *weight* of the object. Figure 2.20 shows the force acting on the ball. Note that the weight is the *only* force acting on the ball (if we ignore air resistance), and it produces the acceleration. (Incidentally, we don't know the *cause* of gravity, but we do know that any two bodies with mass attract each other, and we can measure the forces involved.)

Now, you know from Galileo's experiments that if you drop a ball with double the mass, its acceleration will also be 9.8 m/s². In order for this to happen, the force must also have doubled. But the force producing the acceleration is just the weight of the ball. If the mass is doubled, then the weight must also double. With this understanding, we can rewrite Newton's second law, $F = ma$, as:

$$Wt = mg \qquad (2.7)$$

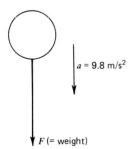

Figure 2.20 The only force acting on a falling ball is its weight (the force of gravity).

Here, Wt is the weight of the object (a force) and g is the acceleration due to gravity near the surface of the Earth—that is, 9.8 m/s².

EXAMPLE

Using $Wt = mg$, find the weight of a person whose mass is 60 kg.

Solution

$$Wt = mg$$
$$= 60 \text{ kg} \times 9.8 \text{ m/s}^2 \approx 590 \text{ kg m/s}^2 \approx 590 \text{ N}$$

To put this into more familiar terms, 1 lb is about 4.5 N, so this person weighs about 130 lb.

Questions

What is your own weight in newtons? What is your mass? The acceleration due to gravity on the surface of the moon is about 1.6 m/s². What would your weight be on the moon in newtons? In pounds?

SUMMARY

In this chapter you have learned how to analyze and graph simple motions. The two motions of interest are uniform motion (where velocity is constant) and uniform acceleration (where acceleration is constant and velocity changes.) More important, to change the velocity of an object, a force or forces must act on it. Newton's laws describe the effects of forces and apply to all ordinary objects. Mass is a measure of the amount of "stuff" in an object (its inertia). But weight is a force, due to the gravitational pull of another object with mass, such as the Earth or the moon. In outer space, objects and persons are weightless.

With the understanding you have gained here, you are now ready to tackle the main concept of this book: energy.

ADDITIONAL QUESTIONS AND PROBLEMS

1. A skate is given a push on a level floor and, for a time, it travels at a constant speed of 1.5 m/s. Write the equations for the distance traveled and the velocity of the skate.

2. Make tables and draw graphs of distance and velocity versus time for the skate of the preceding problem. What is the acceleration of the skate?

3. An automobile coasts down a long hill, starting from rest. Assuming that the friction is negligible, sketch two graphs, one showing distance traveled versus time and the other showing velocity versus time.

4. By making measurements on Figure 2.13, reproduce the graphs of Figure 2.14 and Figure 2.15. (The time between the "blinks" of the strobe is about 1/30 s.)

5. When an automobile is moving at a constant speed on a level, straight road, the *net* force acting on it must be zero. Why? Obviously the engine, through the drive train and the wheels, continuously exerts a force on the car; if that were not the case, no gasoline would be needed to keep the car moving. What other forces could be acting on the car, and what is the direction in which each acts?

6. When a skydiver jumps out of an airplane, she accelerates for a time but soon reaches a "terminal velocity." This is the maximum speed reached, and it is a constant until the parachute opens. What are the forces acting on the skydiver? What is the net force? After the parachute opens, the skydiver descends at a lower, but constant, speed. What forces act now? What is the net force?

7. The space shuttle *Columbia* began its fourth flight on June 27, 1982. During the first two minutes of its flight, it went from a standstill to a speed of about 3500 mph (1565 m/s). During that period, what was the average acceleration of Columbia? How many "*g's*" is this? (1 "*g*" = 9.8 m/s².) To give the astronauts, Ken Mattingly and Henry Hartsfield, this acceleration, what force must their seats have exerted on each of them? (Estimate that each has a mass of about 80 kg, wt = 176 lb.) What is this in pounds?

8. Figure 2.21 shows the take-off of an *INTELSAT* spacecraft. The satellite itself weighs about 10,250 N (2300 lb). What is the mass of the satellite? If the spacecraft were just hovering above the ground in this picture (no acceleration), what upward force would it have to exert on the satellite to hold it there against the pull of gravity? If the spacecraft is accelerating at a rate of 20 m/s², what additional force does it exert on the satellite to accelerate it? If that is called the "action" force, what is the corresponding "reaction" force?

Figure 2.21 This *INTELSAT V* spacecraft was launched from Cape Canaveral on March 4, 1982. The satellite circling the globe can relay 12,000 two- way telephone calls and two color TV channels simultaneously. (Photo courtesy NASA.)

3

Energy

In this chapter we start to examine the main theme of this book: energy. Although you will find here little that seems to have a direct relation to the world energy situation, there are some basic concepts which are simple but very important to further understanding. Without a solid foundation, a beautiful building may soon crumble into rubble; likewise, without a solid foundation in the basic ideas of physics, the understanding you hope to gain about the energy situation may crumble into a tangle of half-understood facts and figures. Build a good foundation now, and the rest of the structure will come nicely into place. We start by learning about work and its relationship to energy.

WORK

Activity 3.1

Get out the roller skate and long rubber band again. Load the skate down with as much extra mass as you conveniently can. Pull the skate horizontally across a table or the floor, using the stretched rubber band as you did in Activity 2.3, with a constant force and for some given distance, say 1 to 2 m. This is indicated in Figure 3.1*a*. Use a distance for which you can continue to pull the skate with a constant force without having it get away from you. Do this two or three times with the same force to get an idea of how fast the skate is moving at the end of the measured distance. Try this with several different forces but over the same distance.

What can you conclude about the velocity the skate reaches each time as the pulling force is increased?

Now pull straight upward, stretching the rubber band about the same amount as before, as shown in Figure 3.1*b*. Does the skate accelerate?

Now pull the skate with the same force as in the first trial and through the same distance, but with the force directed at various angles to the horizontal, as indicated in Figure 3.1*c*. After several trials this way, what can you say about the velocity at the end of the vertical pull compared to the velocity when the skate was pulled horizontally?

Finally, using the same horizontal force as in the first trials, try pulling the skate over shorter distances, say three-fourths of the original distance, then one-half, then one-

Figure 3.1 Pulling a roller skate with a rubber band at various angles. (*a*) Pulling horizontally. (*b*) Pulling straight up. (*c*) Pulling at an angle.

fourth. What do you conclude about final velocity for each case?

Undoubtedly, you concluded that the greater the horizontal pulling force, the faster the skate is going at the end of the pull. Also, it seems that, for the same force, the velocity increases as the distance over which the skate is pulled increases.

It would be quite reasonable to assume now that the final velocity is proportional to both the force exerted and the distance traveled. However, careful measurements of the final velocity in each trial would prove this assumption to be wrong. On the other hand, it is advantageous to *define* a quantity that is proportional to both force exerted and distance, and that quantity is given the name *work*. That is:

$$W = Fx \tag{3.1}$$

Where W is, by definition, the work done on the object, F is the constant force exerted, and x is the distance traveled.

Wait just one minute! This can't be right. In your experiments you saw that the final speed of the skate depends not only upon the size of the force and the distance traveled but upon the direction of the force, as well. If the direction of the force is the same as the direction of the motion, then Equation 3.1 is all right. However, if the direction of the force is at right angles (90°) to the direction of the motion, no work is done at all. You saw one example of this when you pulled straight up on the skate and it did not move. The definition of Equation 3.1 needs a carefully worded change to take this into account.

In general, we can say that the work done on the object is the distance moved times that portion of the force (the *component* of the force) that is in the same direction as the motion. Equation 3.1 is fine if we remember that the F used in it is the component of the force in the direction of the motion. In this book we shall use only forces that are in the direction of the motion or in the direction opposite to the motion. Some examples follow.

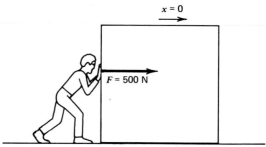

Figure 3.2 The crate must move if work is to be done.

EXAMPLES

Suppose that when pulling the skate with a horizontal force, you are able to exert a force of 5 N for a distance of 2 m. How much work is done?

Solution

Since the force and the motion are in the same direction, we can just make a simple substitution:

$$W = Fx = 5\ N \times 2\ m = 10\ N{\cdot}m$$

This brings up the interesting question of what units are appropriate for work. The unit that seems to fall out of this is the newton–meter, N·m. This unit is so frequently used that it is given a name of its own, the joule (J). Like all other SI units, the joule can be expressed in terms of the basic units. To see this, just substitute kg m/s² for N:

$$1\ J = 1\ N{\cdot}m = 1\ (kg\ m/s^2)m$$

or

$$1\ J = 1\ \frac{kg\ m^2}{s^2}$$

As another example, suppose you are employed in a factory. Your first day on the job, the foreman tells you to move those heavy crates from over here to over there, and he leaves. You push on one of the crates, and it does not move. You push harder and harder. Nothing! You finally give it all you've got, say 500 N, and still it does not budge (Figure 3.2). Suppose you do this all day, first pushing on one crate, then another, to no avail; they all remain where they were. At the end of the day, you will be very tired, but how much pay will you get? Probably, you will get nothing, for you have frittered the whole day away without doing any work. That is:

$$W = Fx = 500\ N \times 0\ m = 0\ J$$

Well, you are not that dumb, so, rather than push without results all day, you call the foreman. He shows you how to load the crate on a dolly and he starts to push it to the other side of the room. Being a sincere person, you rush up and, not wanting to crowd the foreman, you push on the side of the crate—in a direction that is perpendicular to the motion! (See Figure 3.3.) How much work are you doing now? If the two of you push the crate this way for a distance of 10 m, the work you (not the foreman) do is

$$W = Fx = 0\ N \times 10\ m = 0\ J$$

And you still will not get any pay! I used 0 N for the force in this case because none of the force you are exerting contributes to the work done; it is all perpendicular to the motion. You realize this and rush to the back of the crate to help push. If you exert a force of 5 N for a distance of 10 m, you do 50 J of work, and you finally start to earn some pay. Just don't push too hard, or you will get the crate moving so fast you will have trouble stopping it before it crashes into something. Remember inertia?

Questions

In Table 3.1 provide the missing quantity in each row. In each case, the force is either in the direction of the motion or in the opposite direction:

TABLE 3.1

Work (J)	Force (N)	Distance (x)
	100 N	10 m
1000 J		20 m
50 J	25 N	
	− 50 N	5 m
− 10,000 J	400 N	
− 10,000 J		25 m

Suppose that by pushing with a constant force of 500 N you are just able to move one of the big crates at a slow, constant speed across the room. Now you are doing some work, of the amount Fx, but the crate is not accelerating. What is the *net* force acting on the crate? Sketch a drawing showing all of the forces acting on the crate. (Don't forget its weight or the fact that the floor is holding it up.)

KINETIC ENERGY

Let's return to skate pulling for a moment to see what further insights we can get.

When you pulled the skate with a constant horizontal force F for a distance x, the work done was defined as $W = Fx$. However, the force is always given by $F = ma$, so we can write this as

$$W = Fx = (ma)x$$

Since the force is constant, the acceleration is

Figure 3.3 To do work, the force must be in the same direction as the motion.

also constant and, at the end of the pull

$$x = \tfrac{1}{2} at^2$$

where we choose the starting point as $x_0 = 0$ and the initial velocity as $v_0 = 0$, and t is the time elapsed while you were pulling. Substituting this into the equation for work yields

$$W = (ma)x = (ma)(\tfrac{1}{2}at^2) = \tfrac{1}{2}m(at)^2$$

But, for the kind of motion we are discussing—one with a constant acceleration—$v = at$. Therefore, we can write this equation as

$$W = \tfrac{1}{2}mv^2$$

The quantity $\tfrac{1}{2} mv^2$ is called *kinetic energy*, and it is just one of the aspects of energy we shall examine. Using *KE* for kinetic energy:

$$KE = \tfrac{1}{2} mv^2 \qquad (3.2)$$

Whenever an object has a velocity, then, it is said to have kinetic energy. The units for kinetic energy can be found by looking at the equation for an object with a mass of 1 kg moving with a velocity of 1 m/s:

$$KE = \tfrac{1}{2}mv^2 = \tfrac{1}{2}(1 \text{ kg})(1 \text{ m/s})^2$$
$$= \tfrac{1}{2} \text{ kg m}^2/\text{s}^2 = 0.5 \text{ J}$$

So the unit for energy is the joule, the same as the unit for work.

Questions

Suppose that your loaded skate has a mass of 2 kg and that you do work on it by exerting a force of 5 N for a distance of 3 m. How much work did you do?

   What is the final kinetic energy of the skate? What is the final velocity of the skate?

   The preceding questions indicate one of the advantages of using work and energy: Without even mentioning the *time* elapsed, you are able to find the final velocity of the skate. However, there can be difficulties with this, if one is not careful. For example, suppose the skate were already moving with a velocity of 5 m/s and you were able to push it in the direction in which it was moving for 3 m with a force of 5 N. This will make it go even faster and, although you put in 15 N of work as before, the final velocity will not be 3.9 m/s. After all, it was moving faster than that to start with. In fact, the final velocity will not even be 5 m/s + 3.9 m/s. What is important here is the *change* in kinetic energy, ΔKE. That is,

$$W = \Delta KE = KE_2 - KE_1$$

where KE_2 is the final kinetic energy and KE_1 is the kinetic energy at the start. By substituting $\tfrac{1}{2}mv^2$ for KE, this becomes

$$W = \tfrac{1}{2}mv_2^2 - \tfrac{1}{2}mv_1^2 \qquad (3.3)$$

   This result follows from a line of reasoning exactly like that we used for finding the

kinetic energy, but with a starting velocity that is not zero. It says that the net work put into an object is equal to its change in kinetic energy. It is a very powerful tool for solving many problems, but it must be used with care. The following examples will help to show you how it can be used.

EXAMPLES

Suppose you push a 100-kg box across the floor with a force of 400 N for a distance of 2 m, and you find that the velocity goes from 0 to 2 m/s in the process. Then the work you have done on the box is

$$W = Fx = 400 \text{ N} \times 2 \text{ m} = 800 \text{ J}$$

But the final kinetic energy of the box is only

$$KE = 1/2 \ mv^2$$
$$= 1/2 \times 100 \text{ kg} \times (2 \text{ m/s})^2$$
$$= 200 \text{ J}$$

What happened to the other 600 J?

Solution

The extra work done had to be used to overcome friction. In fact, using this information, we can find the force of friction during the slide:

$$W \text{ (done by friction)} = F \text{ (friction) } x$$

Putting in the numbers that are known:

$$-600 \text{ J} = F \text{ (friction)} \times 2 \text{ m}$$

or

$$F \text{ (friction)} = -300 \text{ N}$$

   The negative sign means that the frictional force was in the direction opposite to the motion. Now we see that the *net* force acting on the box was just 100 N (400 N −

300 N). Thus, the work done on the box was:

$$W = F \text{ (net) } x = 100 \text{ N} \times 2 \text{ m}$$
$$= 200 \text{ J}$$

So we see that the *net* force is the source of the 200 J that were added to the kinetic energy.

Let's try another one. A 1000 kg car moving along the highway (at 55 mph, of course) experiences a force of wind resistance and a force of friction between the tires and the road. Both of these forces tend to slow it down. The total force from wind plus friction is about 500 N. How much work must be provided by the engine per kilometer to keep the car moving at this speed? How much work was needed to get it from a standstill up to this speed?

Solution

The work required is just force times distance. What is the force needed to keep the car moving? Remember, if the car is not accelerating, the *net* force acting on it is zero. Thus, the force of the road pushing the car forward (did you catch that?) is 500 N, and the work done is:

$$W = \text{Fx} \approx 500 \text{ N} \times 1000 \text{ m}$$
$$\approx 5 \times 10^5 \text{ J}$$

To get the car to this speed, one must do work on it to change its kinetic energy from zero to the final value. That is:

$$W = \Delta KE = KE_2 - KE_1$$
$$= 1/2 \ mv^2 - 0$$
$$\approx 1/2 \times 1000 \text{ kg} \times (25 \text{ m/s})^2$$
$$(55 \text{ mph} \approx 25 \text{ m/s})$$
$$\approx 3 \times 10^5 \text{ J}$$

So the work required to get the car up to speed is in the same "ballpark" as that needed to move it a kilometer against wind and friction. Of course, if the work done against friction

and wind resistance were included during the time the car is accelerating, the work needed would be even greater.

Questions

Suppose that you are pushing a 250-kg crate on a dolly with a force of 400 N. Your playful friend, just to be contrary, hides on the opposite side and pushes in the opposite direction with a force of 200 N. The frictional forces add up to 50 N. Make a drawing showing all of the horizontal forces acting on the crate. After you have pushed the crate 9 m, how fast is it moving? How fast would it have been moving without your friend's "help"?

POTENTIAL ENERGY

Activity 3.2

Drop a ball, as in Figure 3.4. Describe its kinetic energy from the point from which it is dropped to a point just before it hits the floor. Where does the kinetic energy come from?

You probably answered that as that ball falls through some distance x, the force of gravity—its weight—does work on it, thus giving it kinetic energy. This is correct, and it can be quite helpful in solving problems. For example, suppose a 2-kg object falls through a distance of 2 m. During the fall, the weight does a certain amount of work:

$$W = \text{Wt } x = mgx = 2 \text{ kg} \times 9.8 \text{ m/s}^2 \times 2 \text{ m}$$
$$= 39.2 \text{ J}$$

Velocity can be found easily by equating this work to the change in kinetic energy which, in this case, is just equal to the final *KE*:

$$W = \Delta KE = KE_2 - KE_1$$
$$= 1/2 \ mv^2 - 0 = 39.2 \text{ J}$$

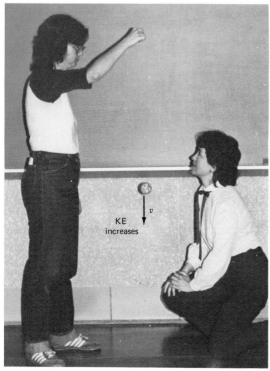

Figure 3.4   When the ball falls it gains kinetic energy.

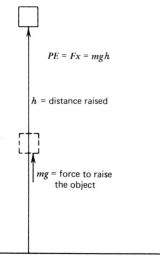

$PE = Fx = mgh$

h = distance raised

mg = force to raise the object

Figure 3.5   The potential energy of an object near the surface of the Earth.

Thus

$$v^2 = 39.2 \text{ m}^2/\text{s}^2$$

and

$$v \approx 6.3 \text{ m/s}$$

Another correct, and often very useful, way of describing this is to say that the object has *potential energy*, often called the *energy of position*. That is, the object has energy because of where it is in space. In the case of gravitational potential energy, the higher the object is, the more potential energy it has. This potential energy (*PE*) then is converted to kinetic energy as the object falls. The potential energy of an object at any given height is just equal to the work required to raise the object to that height. For example, as illustrated in Figure 3.5, suppose the object is slowly raised a distance *h* above an arbitrary starting point. We do it slowly so that it does not gain or lose any significant kinetic energy to complicate the process. Then in order to raise the object to that level a force equal to its weight, *mg*, must be exerted and the distance through which it is raised is *h*. Thus the work done is just *mgh* which, by definition, is also the potential energy of the object after it is raised up.

That is, the gravitational potential energy of an object is

$$PE = mgh \tag{3.4}$$

where *m* is the mass of the object, *g* is the acceleration due to gravity, and *h* is its height above some reference level.

It is important to note that the starting point—the level at which $PE = 0$—is quite arbitrary; we can choose it anywhere that is convenient. In Figure 3.6, for example, we choose the zero point to be at tabletop height, rather than at the floor. Using that zero point,

Figure 3.7 then shows the *PE* of a 1-kg block at various heights. Above the table, the *PE* is positive; below tabletop height, the *PE* is *negative*. This follows directly from the definition: If you lower the block slowly from the zero level—the tabletop—the force exerted by your hand on the block is upward, and the motion is downward. Thus the work you do *on the block* is negative. An equivalent way of saying this is that the block does positive work *on your hand*.

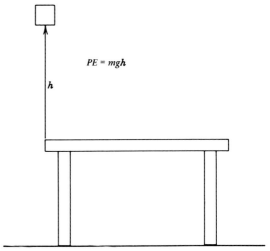

Figure 3.6 We can choose the potential energy to be zero at tabletop height, or anywhere else.

EXAMPLE

Suppose we drop the 1-kg block of Figure 3.7, but first give it a downward push so that, as it passes the point 2 m above the table top, it has a velocity of 2 m/s, and it then falls to the point that is 1 m below the table top. What will its velocity be there?

Solution

The problem is indicated in Figure 3.8. The *change* in potential energy from the starting to the ending points will be

$$\Delta PE = PE_2 - PE_1$$
$$= -9.8 \text{ J} - 19.6 \text{ J} = -29.4 \text{ J}$$

with the negative sign telling us simply that the potential energy has decreased in the process. This *loss* in potential energy is just equal numerically to the *gain* in kinetic energy. That is:

$$\Delta KE = -\Delta PE$$

The negative sign here tells us that what is gained by one is lost by the other. Now substitute for ΔKE:

$$\Delta KE = KE_2 - KE_1$$
$$= \tfrac{1}{2} mv_2^2 - \tfrac{1}{2} mv_1^2$$

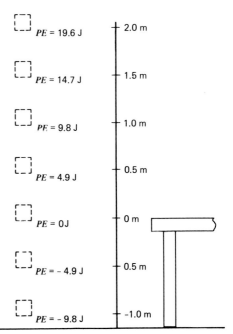

Figure 3.7 The potential energy of a 1-kg block at various heights, with the tabletop chosen to be zero potential energy.

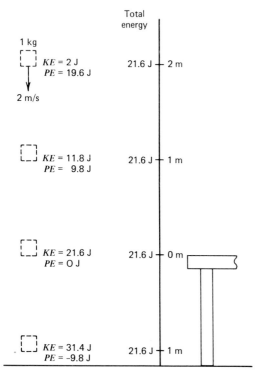

Total
energy

1 kg

$KE = 2$ J
$PE = 19.6$ J   21.6 J + 2 m

2 m/s

$KE = 11.8$ J
$PE = \ 9.8$ J   21.6 J + 1 m

$KE = 21.6$ J
$PE = 0$ J   21.6 J + 0 m

$KE = 31.4$ J
$PE = -9.8$ J   21.6 J + 1 m

Figure 3.8 Potential energy changes to kinetic energy as the block falls, and total energy stays the same.

Since the block has a mass of 1 kg and it started with a velocity $v_1 = 2$ m/s, this becomes

$$\Delta KE = (0.5 \text{ kg})v_2^2 - (0.5 \text{ kg})(2 \text{ m/s})^2$$

But this is just equal to $-\Delta PE$:

$$-\Delta PE = -(-29.4 \text{ J}) = +29.4 \text{ J}$$

so

$$(0.5 \text{ kg})v_2^2 - 2 \text{ J} = 29.4 \text{ J}$$

Solving for v_2:

$$v_2^2 = 2 \times 31.4 \text{ J/kg} = 62.8 \text{ m}^2/\text{s}^2$$
$$v_2 \approx 8 \text{ m/s}$$

Questions

If the previous example is repeated with a 2-kg block or a 3-kg block, what is the resulting final velocity? Is this surprising? (Remember, if you double the mass, you also double the weight.)

Going back to the 1-kg block, suppose you throw it upward in such a way that as it passes the point that is 2 m above the tabletop it is moving with a velocity of 2 m/s upward. What will its velocity be when it gets to the level 1 m below the table? (To get some insight into this, consider the following: As the block is on its way up at the 2-m point, what are its potential and kinetic energies? At the same point on the way down what are the potential and kinetic energies of the block?)

You pick up a box weighing 200 N and carry it up several flights of stairs to a floor 20 m above the starting point. How much work do you do?

Now you raise that same box to a level of 1 m above the new floor, carry it 20 m over to a different location, and put it back on the floor. How much work do you do? (If you don't permit your natural prejudices to becloud your mind, this is an easy one to answer.)

Did you get zero for the answer to the last question? That is the correct answer and a simple, but important, result. To be sure you know why it is zero, let's do an analysis. First, separate the process into the lifting and putting down of the box as one part and the carrying of it horizontally as a second part. For the first part you raise the box 1 m, doing some work on it in the process; then you lower it 1 m, doing the same amount of negative work. The net work for the process is zero.

When you carry the box horizontally, you also do no work on it, for its motion is horizontal and the force you exert is upward, at right angles to the motion. That's right, you may stand there holding the box or moving it in a straight, horizontal line forever; you may faint from exhaustion in the process, but you do no work! This may be a difficult notion for you to swallow, but you should do so, for it is correct.

If you are a good nit-picker you may say, Aha! But what about the work you have to do, after you have picked up the box, to start it moving horizontally—to accelerate it? This is an exceptionally intelligent question and it deserves a serious answer, but the answer is still that you end up doing no work. It is true that you have to do some work on the box to get the horizontal motion started, but once you have done so, it will continue to move in a straight line and with a constant speed; all you have to do is keep up with it and hold it up. When you get to the place where you want to put it down, then you must do some negative work on it to stop its horizontal motion. And guess what? The positive work and the negative work are exactly the same size and cancel each other. If you do not believe this and have a desire to do a little algebra, pick a horizontal velocity and try the calculations. In future considerations we will ignore this fine point, since the work involved is small as long as we keep the velocity small; besides, we get the work back at the end of the trip.

Perhaps you'd better read the preceding paragraph again. It's a bit subtle.

This example makes a very important point! What we may call work in the everyday world may not be work in the strict sense required by our present definition of it. If you move a pile of boxes from one spot to another at the same level, you may get very tired and you may even get paid for the "work" you do. One may say that there is a kind of biological work done by your body and a kind of practical work done in transferring the boxes, but no physical (i.e., physics) work is done.

ENERGY DOES WORK

Some dictionaries define energy as "the ability to do work." This is not a bad definition for, in many instances, objects that possess potential or kinetic energy may be made to do work, either usefully or not. As a trivial example, when you lower an object, it does work on your hand as a result of the gravitational potential energy it possessed. Or, if you catch a ball, your hand is pushed back during the catch; the ball has done work on your hand because of the kinetic energy it had. If you don't believe this, try catching a hard-thrown ball without permitting your hand to move back at all; it can't be done.

More practically, an old-fashioned water wheel, such as the one in Figure 3.9, converts the potential energy of the water into work to run a grist mill. The weight of the descending water turns the wheel. In a modern hydroelectric plant, the water near the top of the dam has potential energy. As water is allowed to flow through nozzles at a level near the bottom of the dam, potential energy is converted to kinetic energy. The rapidly moving water then goes through turbines which convert much of the kinetic energy to work to run the generators.

Elevators are "counterweighted" so that, as the elevator goes up, the counterweight comes down. The counterweight does a good deal of the work in raising the elevator and, as a result, the motor does not have to work so hard (Figure 3.10). Perhaps in the near future, we will store kinetic energy in a rapidly turning and massive flywheel, and then convert the energy of the flywheel into work to accelerate an automobile.

So far, this discussion has been restricted to kinetic energy and gravitational potential energy, two aspects of mechanical energy. There are, of course, other forms of energy, all of which can be converted in some degree to work or to each other. For example, if you

Figure 3.9 A water wheel converts the potential energy of the water into useful work. (The Bettmann Archive.)

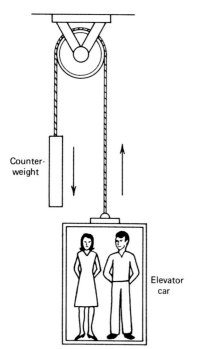

Figure 3.10 A counterweighted elevator.

wind a spring you do work on it, storing potential energy which may then be converted to work to run a clock or a toy car. Other forms of energy are electrical, nuclear, chemical, light, and thermal (heat), some of which you will see in more detail later in this book.

CONSERVATION OF MECHANICAL ENERGY

Look again at the falling block of Figure 3.8. As it falls, it loses potential energy and gains kinetic energy, but the *total* energy remains the same throughout. We say that energy is *conserved*. Conservation of energy in this sense has a very different meaning from that in the recently very popular phrase *conservation of energy*. In the popular sense, conservation of energy means an attempt to save energy, to use as little as possible of energy resources, such as oil and natural gas. In the physical sense, it means that for some collection of objects the total energy remains constant. The collection of objects must be carefully defined and is called *the system*. The system we define for the falling block may surprise you; it consists of the block *and the Earth*. The Earth is included in the system because without it, the block would have no potential energy.

We are creeping up here to a limited wording of the *law of conservation of energy*: In an isolated system with no dissipative forces, the total mechanical energy is conserved.

Hold on here! There are three new words in that definition, and we should look at their meanings. First, by "mechanical" energy we mean just the kinds of kinetic and potential energy we have been talking about. This excludes electric energy, thermal energy, and the like. An "isolated" system means a system upon which no outside forces act. You might imagine drawing a box around such a system. The system is isolated if no forces act between an object inside the box and one outside. A system consisting of only the falling block would not be an isolated system because the gravitational force of the Earth would be acting on it from outside the system. That is why we must include the Earth in the system. Finally, a "dissipative" force can best be defined by describing one. If you slide a book down a ramp it starts out with some kinetic energy and potential energy. As it comes to a rest at the bottom of the ramp, it has lost both kinetic and potential energy. This loss of energy is the result of the frictional force, a prime example of a dissipative force. When such forces are present, total mechanical energy is not conserved. Where does the lost energy go? Well, mostly into heat, but that is a story we shall leave until later.

We can now define a *conservative system* as a system in which the energy does not change. That is, it is an isolated system—one on which no outside forces act—experiencing no dissipative forces, such as friction or air resistance. The restricted law of conservation of energy may then be reworded as follows:

> *In a conservative system, the total mechanical energy remains constant.*

EXAMPLE

A 5-kg ball swings at the end of a string, as shown in Figure 3.11. At the top of its swing, 0.5 m above its lowest position, its velocity is zero. (It stops and turns around.) What is its velocity at the bottom of its swing?

Solution

Using forces and acceleration, this would be a difficult problem to solve. There are two forces acting on the ball: one is its weight, the force of gravity, and the other is the force exerted by the string. The latter force keeps changing direction and size as the ball swings. However, using the fact that potential energy changes to kinetic energy makes it very easy.

First, consider the ball at the top of its swing. Taking the bottom of the swing to be

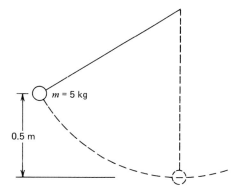

Figure 3.11 A weight swinging on the end of a string (a pendulum).

a convenient level for $PE = 0$, the potential energy at the top is then

$$PE_1 = mgh = 5 \text{ kg} \times 9.8 \text{ m/s}^2 \times 0.5 \text{ m}$$
$$= 24.5 \text{ J}$$

At the bottom of the swing, it is

$$PE_2 = mgh = 5 \text{ kg} \times 9.8 \text{ m/s}^2 \times 0 \text{ m}$$
$$= 0 \text{ J}$$

Thus, the change in potential energy is

$$\Delta PE = PE_2 - PE_1 = -24.5 \text{ J}$$

However, at the top of the swing, the kinetic energy is zero, and at the bottom, it is

$$KE_2 = \tfrac{1}{2}mv^2$$

where v is the velocity we want to find. The change in kinetic energy from top to bottom is

$$\Delta KE = KE_2 - KE_1 = \tfrac{1}{2}mv^2 - 0$$

Thus, using conservation of energy

$$\Delta KE = -\Delta PE$$
$$\tfrac{1}{2}mv^2 = -(-24.5 \text{ J}) = 24.5 \text{ J}$$
$$v^2 = (2 \times 24.5 \text{ J})/5 \text{ kg} = 9.8 \text{ m}^2/\text{s}^2$$
$$v \approx 3.1 \text{ m/s}$$

Questions

The system consisting of the ramp, the roller skate, and the Earth, which you used for earlier experimentation, is approximately a conservative system. The frictional and air resistance forces are small enough to be ignored. As the skate rolls down the ramp, the potential energy decreases and the kinetic energy increases, with total energy being conserved.

Suppose the ramp is 3 m long and its upper end is 1 m higher than its lower end. How fast is the skate moving at the halfway point? At the bottom? Do these answers depend on the mass of the skate? If the ramp is made 5 m long but the upper end is still 1 m above the lower end, what answers do you get? What if the ramp were 10 m long? Trick question: What if the ramp were 1 m long with its upper end 1 m above its lower end?

Imagine a frictionless roller coaster, with no air resistance, moving on a track, as shown in Figure 3.12. At point A it is stationary and just starting its run down the slope. What is its velocity at points B and C? Does it reach point D? If not, describe its motion after it passes C.

One final word about conservation of energy. If we do work on a system from the outside, then the mechanical energy of the system is not conserved. If the outside work is positive, the energy increases; if negative it decreases.

We can state, however that the work input is equal to the *change* in energy of the system. Thus, the limited law of conservation of mechanical energy can be expanded. In equation form, this says

$$W = \Delta PE + \Delta KE \tag{3.5}$$

This states that the work input is equal to the change in potential energy plus the change in kinetic energy. Any, or all three, of these quantities may be negative. If $W = 0$, then we have the original statement of the law:

$$\Delta PE + \Delta KE = 0$$

This is just another way of saying that the total mechanical energy stays constant.

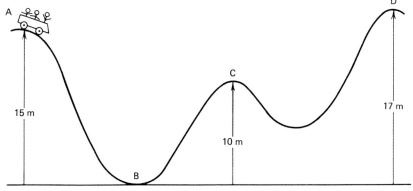

Figure 3.12 A roller coaster.

EFFICIENCY

Activity 3.3

Tie small weights to the two ends of a string and drape the string over a fairly smooth support, as shown in Figure 3.13. (Or, a small pulley could be used, if one is available.) You should have some extra weights to add to one end. Also, leave extra string at that end to tie to the extra weights.

With equal weights, the whole thing just sits there, so add a small amount of weight to one end. What happens? If nothing happens, add some more weight and keep doing so until, with a small push to start it, the greater weight pulls the smaller one upward at about a constant velocity. By weighing them, or just guessing, estimate the two weights in newtons. Make approximate measurements of how far the greater weight goes down and, simultaneously, how far the smaller one goes up.

How much work is done *by* the greater weight as it goes down? How much work is done *on* the smaller weight as it goes up? How do you account for the difference?

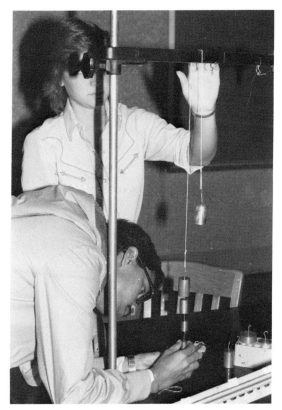

Figure 3.13 Hanging weights on a string.

You can see that the greater weight falling through a given distance does more work than it takes to raise the smaller weight through the same distance. The lost work goes into overcoming friction. In trials with the weights and the string, the students in Figure 3.13 found that it took a weight of about 2 N to raise a weight of 1 N, so half of the work was lost. This process is then said to be *50 percent efficient*. There are many ways to define efficiency in any given situation, and we shall try to be clear as we do so later on. In general, we can define efficiency as

$$Eff = \frac{\text{useful output}}{\text{input}} \times 100\% \qquad (3.6)$$

In this little experiment, we can consider the "useful" output to be the work done in raising the smaller weight a height h, thus increasing its potential energy by an amount mgh. The "input" in this case was the work done by the falling weight on the string, which I shall call Mgh, since the distance moved was the same and only the masses differed. Then, for this case

$$\begin{aligned}
Eff &= \frac{mgh}{Mgh} \times 100\% \\
&= \frac{1\text{ N} \times h}{2\text{ N} \times h} \times 100\% = 50\%
\end{aligned}$$

You do not even need to know how far they moved because the h appears in both numerator and denominator and "cancels out."

In virtually every process one can think of in which work is done to produce energy or energy is converted from one form to another, there are losses involved, and the process is less than 100 percent efficient. In fact, in many cases the efficiency is quite low, indeed. Further, as we shall see later, when thermal energy—heat—is converted into work, there are *always* inevitable losses, even if all of the machinery is perfect and without friction. These losses are very important in the generation and distribution of electric power, in the running of automobiles, and in many other practical applications.

POWER

Suppose you start an automobile from rest and, with steady acceleration, get it up to 55 mph in 15 s. The engine has done a certain amount of work, and the car has a certain amount of kinetic energy. You then stop the car and your friend, who always did have a heavy foot, gets into the driver's seat and accelerates from zero to 55 in half the time. The end result is the same: the same amount of work accomplished and the same final kinetic energy as before. But what a difference in how you got there! (Incidentally, there is also going to be a big difference in how much gas it cost to get there.)

In many applications, the amount of work done or the amount of energy consumed is important, but equally important is the *rate* at which this occurs. The *rate of doing work* or of consuming energy is called *power*. That is, power is work done per unit time:

$$P = \frac{W}{t} \qquad (3.7)$$

To find the units for power, look at the case where 1 J of work is done in 1 s:

$$P = \frac{W}{t} = \frac{1\text{ J}}{1\text{ s}} = 1\text{ J/s}$$

So the unit for power in the SI system is the joule per second (J/s), also called a *watt* (W). We will often be involved with kilowatts (kW):

$$1\text{ kW} = 1000\text{ W} = 1000\text{ J/s}$$

We will also use megawatts (MW):

$$1\text{ MW} = 1{,}000{,}000\text{ W} = 10^6\text{ W} = 10^6\text{ J/s}$$

An interesting unit mentioned earlier is the kilowatt-hour (kWh). This is a unit of *energy* because it is power (kW) times time (hr). That is:

$$1 \text{ kWh} = 1 \text{ kW} \times 1 \text{ hr}$$
$$= 1000 \text{ J/s} \times 3600 \text{ s}$$
$$= 3.6 \times 10^6 \text{ J}$$

So 1 kWh is a rather large amount of energy, 3.6 million joules. But every American is responsible for the consumption of about 250 kWh or nearly a billion joules of energy daily!

SUMMARY

In this chapter, you have learned that work is force times distance. More important, you now know that work can change the kinetic and/or potential energy of a system. Kinetic energy is the energy of motion and potential energy is the energy of position. In an isolated system, the total energy remains the same. In a system that is not isolated, the work put into the system is equal to the total change of energy. Dissipative forces within a system, such as friction and air resistance, may change mechanical energy into other forms, such as heat. The efficiency of a machine is the ratio of useful work or energy output to work or energy input. To put efficiency in terms of a percentage, this ratio must be multiplied by 100 percent. Power is the rate at which work is done or energy is consumed.

ADDITIONAL QUESTIONS AND PROBLEMS

1. Joe Apprentice carries a 40-kg load of shingles up a ladder 3 m to the boss on the roof. How much work does he do? When he gets there, he discovers that the boss doesn't need any more shingles, so he carries them back down. How much work does he do this time? What is the total work done?

2. a. Suppose that you pull a 1-kg skate a distance of 2 m with a constant force of 10 N. How much work do you do on the skate? If the skate starts from rest, how much kinetic energy does it have at the end of the 2-m pull? What is its final velocity?

b. Now pull a 2-kg skate the same distance with the same constant force and answer the same questions.

c. Now start the 2-kg skate with an initial velocity of 5 m/s, pull it with the same force through the same distance, and answer the questions.

3. Suppose you push a heavy box 10 m across the floor at a constant speed by exerting a force of 200 N. (Of course, it will take more force to start the box sliding but, once it is moving, the 200 N will keep it going.) What are all of the forces acting on the box as it slides? How much work do you do in pushing the box 10 m? How much net work is done on the box (considering all of the forces)?

4. In the preceding problem, what happened to the work you did on the box? Is this a conservative system?

5. A 100-kg crate must be lifted from street level to a loading dock 1 m above the street. How much work must be done? Since the crate is so heavy, it is raised to the level of the loading dock by pushing it on a dolly, up an incline 10 m long. Ignoring friction, how much work is done? What is the minimum force needed to push the crate up the incline? Describe what happens if it is pushed with a force of 1000 N.

6. I can throw a 0.5-kg ball straight up with an initial velocity of about 15 m/s. How high will it go? If you would like to outdo me by throwing a 1.0-kg ball to the same height, how fast will it have to be moving as it leaves your hand?

7.   This chapter mentions that, someday, it may be practical to store energy in large wheels turning at high speeds. Give whatever evidence you can think of that such a spinning wheel does have energy.

8.   An elevator and its occupants have a mass of 1500 kg. If they are to be raised a distance of 100 m to the top of a building, how much work must be done by the electric motor? When the elevator is lowered to the ground floor, how much work is done on the motor?

9.   The elevator of the preceding problem is then counterweighted, as in Figure 3.10, with a weight whose mass is 1500 kg. Disregarding friction, how much work must be done by the motor to raise the elevator to the top of the building? Where does the needed work come from? How much work is done on the motor as the elevator is lowered to the ground floor? Why is a motor necessary at all?

10.   Suppose that the total force required to raise an elevator is 10,000 N. This will include the forces required to overcome friction and any weight of the elevator in excess of that of the counterweight. If the elevator is raised at the rate of 2 m/s, how long does it take to go the 100 m from the ground floor to the top floor? How much work is done by the electric motor in the process? What power is expended by the motor? If the upward velocity of the elevator is 5 m/s, what are the answers to the same questions?

11.   A large power plant produces 1000 megawatts (MW) of electric power. This is the same as 1 million kilowatts (kW). If it runs continuously at peak output, how much energy does the plant produce in one day? In one year? Answer both in kilowatt-hours and joules.

12.   If left on continuously, how much energy does a 100-W light bulb consume in a day?

4 *Mechanical Energy Sources*

The story of society for the past 200 years has been most strongly influenced by an ever-increasing replacement of muscle power by machine power. The development of the steam engine and other inventions made the industrial revolution possible, allowing machines to take over much labor previously performed by humans. Water and wind power were among the earliest of the large-scale power sources, and they are still being used today. This chapter takes a look at some of the modern uses of "mechanical energy sources," sources that produce electrical power without the use of steam. Steam-powered sources, including both fossil fuels and nuclear power, are discussed in later chapters.

WATER POWER

Water power was one of the first ways in which energy other than that produced by animals was harnessed. Its use goes back to early civilizations in Egypt, China, and Persia. For many centuries, water power was used for running mills to grind grain, saw lumber, and for many other uses. A visit to an old restored mill is still quite instructive and interesting.

In these old mills, the wheels were either "undershot" or "overshot," and they used principles that are still important today. The overshot wheel, shown in Figure 4.1, was more common, and it turned the potential energy possessed by water into useful work. The wheel was placed at a spot where it could be turned by water that fell down some distance, either at a natural waterfall or at a dam (as indicated in Figure 4.2). Water coming off the *flume*, usually a wooden or stone channel, fell into troughs in the rim of the wheel, carrying the wheel around and thus doing work on the wheel. Then gears were used to transmit the power to the grinding wheels or other machinery.

EXAMPLE

Back in those days, the mill builder must have had to do some kind of "seat of the pants" calculation to see if the stream flow and height of fall were sufficient to supply the needed power. What would that calculation be like?

Solution

We can easily do such a calculation. Suppose a miller needed an average power of one horsepower—about 750 W. (A horsepower was

Figure 4.1 An overshot water wheel. (Bruce Roberts/Photo Researchers.)

supposed to be the power a good horse could provide for an extended period.) Suppose further, that the drop in water level from the flume to the stream bed was 5 m and that the

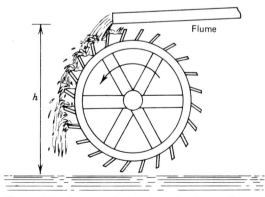

Figure 4.2 An overshot water wheel uses the potential energy of the water which starts at a height h above the stream bed.

overall efficiency of the mill was 10 percent. This estimate of efficiency may be high. It has to account for the fact that much of the water spilled out of the wheel well above the stream bed and for the frictional forces in the gears, which must have been substantial. However, it is convenient to use. The rate of potential energy needed was about 7500 J/s, or about 4.5×10^5 J/min. In order to have this much potential energy available in one minute, the mass of the water flowing by in that time would be given by

$$PE = mgh = 4.5 \times 10^5 \text{ J}$$

or

$$m = \frac{PE}{gh}$$

$$= \frac{4.5 \times 10^5 \text{ kg m}^2/\text{s}^2}{9.8 \text{ m}^2/\text{s}^2 \times 5 \text{ m}}$$

$$\approx 9000 \text{ kg (per min)}$$

Putting this into more familiar terms: Since the mass of 1 gal of water is about 3.8 kg, it is about 2400 gal/min. This seems like a great deal of water but, if the flume is 1 m wide and the water flowing in it is 0.15 m (6 in.) deep, this represents a flow with a speed of only about 1 m/s, or 2 mph.

Activity 4.1

Using a garden hose with a nozzle on it, squirt a stream of water nearly straight up and as high as possible. (You can do it straight up, if you want a bath.) By doing this alongside a familiar object, such as a house, you ought to be able to estimate the maximum height the stream attains. Now squirt the water into a bucket of known size and, with a stopwatch, measure how long it takes to fill. (Calculate the number of kilograms of water in the bucket

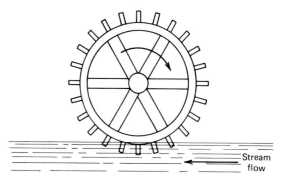

Figure 4.4 An undershot water wheel converts some of the kinetic energy of the stream to useful work.

by multiplying gallons by 3.8.) From this information, you should be able to calculate how many joules per second are available at the hose nozzle. Just pretend that a full bucket of water is at the top of the stream all at the same time, and find its potential energy. Dividing by the number of seconds it took the bucket to fill will give the number of joules per second, or the power, in watts. Doing a similar experiment, I got an answer of about 15 W.

In an undershot wheel, shown in Figures 4.3 and 4.4, the *kinetic* energy of the water is converted to useful work by the wheel. As the water flows by the wheel, it hits the paddles and exerts a force which causes the wheel to turn. For this wheel, one wants a fast-flowing stream. In theory, the paddles could move as fast as the stream but, of course, this will not happen in practice because of the resisting forces.

Figure 4.3 This is a reproduction of a tenth-century copper engraving by Agostino Ramelli of an undershot water wheel providing power to a smelting furnace. (The Bettmann Archive.)

Questions

Energy is never just magically produced out of thin air. It must generally be converted from one form to another. The kinetic energy of the water flowing in a stream must come from somewhere. Where does it come from? Why doesn't the water in a stream flowing down a mountain side simply flow faster and faster as it goes farther down the mountain?

EXAMPLE

Water flows by an undershot water wheel at a rate of about 1000 kg/s and a speed of about 2 m/s. Assuming an overall efficiency of conversion of kinetic energy to work of about 10 percent, what is the power produced by this wheel? Does this help to explain why overshot wheels were more popular?

Solution

The rate of flow given in this example corresponds roughly to a wheel whose paddles are 1 m wide and dip into the water to a depth of about 0.5 m. That is, the cross-sectional area of water flowing past is 0.5 m². Then, if the rate it flows through that area is 2 m/s, in one second one cubic meter of water will flow past the blades of the wheel. As you already know, the mass of 1 m³ of water is 1000 kg.

The kinetic energy of 1000 kg of water flowing at a rate of 2 m/s is

$$KE = 1/2 \ mv^2$$
$$= 1/2 \times 1000 \ kg \times (2 \ m/s)^2$$
$$= 2000 \ J$$

Since 10 percent of the kinetic energy carried past the wheel each second is converted into useful power, the power produced will be

$$P = 200 \ J/s = 200 \ W$$

This certainly does point out the advantage of the overshot wheel. With the same efficiency (10 percent), the overshot wheel was capable of producing about 750 W with a flow of only 150 kg of water per second. The un-

dershot wheel, on the other hand, requires a flow of 1000 kg/s to produce only 200 W of usable power.

Questions

To get further insight into this last example, compare the velocity of the water in the stream bed to the velocity the water used to power the overshot wheel would have had if it had simply fallen the 5 m, rather than running down the wheel. How does the energy available per kilogram of the water compare in the two cases? (Note that the overshot wheel uses the potential energy of the water, and the undershot wheel uses kinetic energy. But that makes no difference: It is only the *amount* of energy stored in the water that is important.)

DAMS

Activity 4.2

Get a plastic jug of 1-gal capacity and make a hole, perhaps ¼-in. in diameter, in the side, near the bottom. Now fill the jug about a quarter full with water while plugging the hole with your finger; place the jug on the edge of a support such as a table; and take your finger off the hole. (Perhaps it would be wise to do this outside, or on the edge of the bathtub?) About how far from a spot directly beneath the jug does the water hit? Now fill the jug halfway and try the same thing. Try it for a three-quarters-full jug and a full one. Figure 4.5 shows a similar experiment. What can you conclude about the comparative velocities of the water as it leaves the hole in these four trials?

As a "thought experiment," imagine filling a 1-gal paint pail with water and weighing it. See Figure 4.6. You would find that the weight is roughly 37 N, ignoring the weight

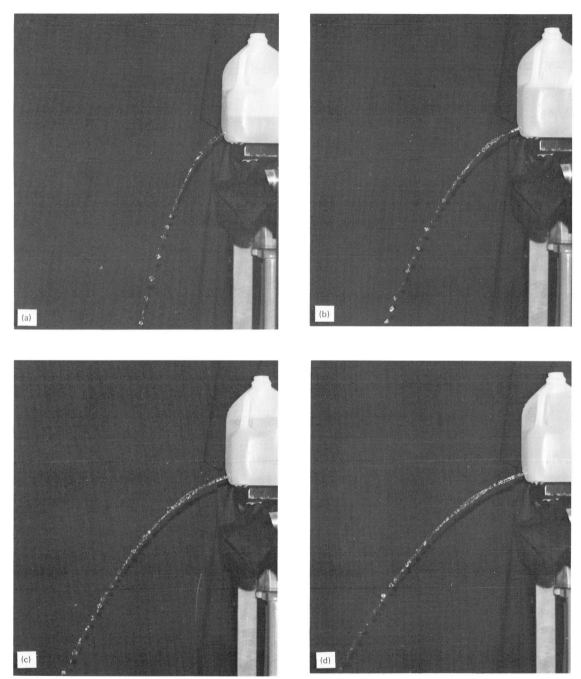

Figure 4.5 The velocity of the water streaming from a hole in the jug depends on the level of the water in the jug.

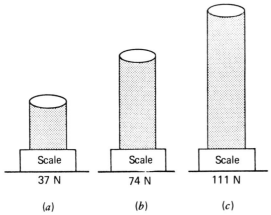

Figure 4.6 The weights of pails of water of different heights.

of the pail itself. If you measured the bottom, you would find it had an area of about 0.02 m². Let us define a quantity called *pressure*:

Pressure = force/area

$Pr = F/A$

The unit here is newtons per square meter, also called a *pascal* (Pa). For the paint can, the pressure is 37 N divided by 0.02 m², or about 1850 Pa. (We are more accustomed to dealing with pounds per square inch (lb/in²). One pascal ≈ 1.45 × 10⁻⁴ lb/in², so 1850 Pa is a pressure of about 0.27 lb/in²—not very great.)

Continuing with our thought experiment, suppose we can get a 2-gal pail that has the same diameter but is twice as high. When this is filled with water and placed on the scale it weighs 74 N, but the area of the bottom is the same, so the pressure must be doubled to 3700 Pa. Put a third gallon on and it goes to 5550 Pa, and so on.

Now think of a lake 10 m deep. You can mentally draw 10 stacked boxes in the lake, each measuring a meter on a side, as indicated in Figure 4.7. The box at the top contains 1000 kg of water, whose weight is 9800 N

($Wt = mg$). Since the area of the bottom of the box is 1 m², the pressure is 9800 Pa. The next box down also contains water weighing 9800 N. Thus the total weight to be supported at its bottom is 19,600 N, and the pressure there is 19,600 Pa. At the fifth box down, the pressure is 5 × 9800 = 49,000 Pa, and at the bottom of the lake it is 98,000 Pa. The weight of the water contained in two columns of boxes, side by side, would be twice as great. However, the area at the bottom of the columns of boxes would also be twice as great, and the pressure, weight divided by area, would be the same. Three columns of boxes, or four, or however many you like, would all give the same result: The pressure at the bottom is 98,000 Pa. This is about 14 lb/in², and is added to the normal atmospheric pressure of about 15 lb/in². If you are a scuba diver, you have experienced this increase of pressure as you dive deeper into the water.

This thought experiment and the activity that precedes it should help to give you some

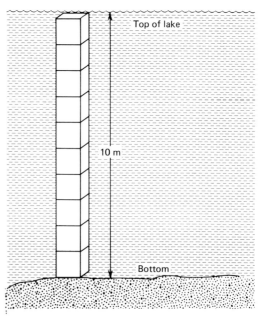

Figure 4.7 Water pressure in a lake.

Figure 4.8 The Grand Coulee Dam. The base of
the dam is much wider than the top. (U.S. Bureau
of Reclamation photo, from the U.S. Department
of Energy, by H.S. Holmes.)

insight into the construction of dams. First,
the pressure under water increases as the depth
of the water increases, getting to be quite large
rather quickly. Although 14 lb/in² might not
sound large, consider the total *force* that would
exert on your body if you were to dive that
deep. The area of your body might be 16 square
ft—over 2300 square in.—or more, depending on how large you are. Multiplying this by
the pressure, the extra force on your body due
to the water would be over 32,000 lb! (Here,

extra means in addition to the usual force exerted by the atmosphere on your body.) Second, the pressure is exerted not only in the
downward direction, but to the sides, as well.
This is what pushes the water out of the hole
in the side of the jug.

Because of these two facts, dams are generally built relatively narrow at the top and
very wide at the bottom. As is apparent in
the photograph of Grand Coulee Dam (Figure
4.8). This is also illustrated in cross-section in

figure 4.9. Dams could be built very wide all the way up, but that would waste concrete, because the pressure is lower near the top. It would also give the dam more of a tendency to topple over. A dam must be designed so that, at every depth, it is strong enough to withstand the tremendous forces of the water, so that it will not topple, and so that it will not slide. Figure 4.10 shows the location of the power house at the base of the dam, and the channel, called a *penstock*, through which the water comes rushing to the power house. The *head* is the depth of the water (above the level of the turbine in the power house).

HYDROELECTRIC POWER (HYDROPOWER)

Although modern power plants are much larger and differ in appearance and detail from the old water wheels, they use exactly the same principles to convert energy stored in water to useful work—the generation of electric power.

Every electric power plant works on the same principle. A "motor" of one kind or another provides work to turn a large generator which generates electric power, as indicated in Figure 4.11. In a conventional power plant the energy to make the motor run comes from a fuel, such as coal. In a hydroelectric plant,

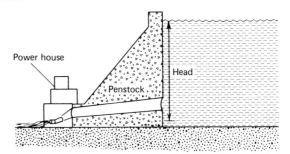

Figure 4.10 Power house and penstock of the Grand Coulee Dam.

it comes from the energy stored in water because of a *head*. The head is just the height of the water behind the dam, measured from the level of the power plant, as indicated in Figure 4.10. The "motor" in this case is a turbine, which is a collection of blades or buckets on a rotating axis. In a steam-driven power plant, the steam turbine and generator appear as shown in Figure 4.12. One type of water-driven turbine, which looks like a ship's propeller, is shown in Figure 4.13. In this turbine, the potential energy of the water causes it to "fall" past the blades, turning them in the process. This is much like the overshot water wheel. An interesting feature is the "draft tube," which may extend as much as 30 ft below the turbine. The weight of water falling in this tube "pulls" on the propeller blades and adds to the power imparted to the turbine by the

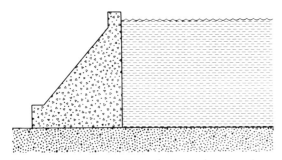

Figure 4.9   Cross section of a typical power dam.

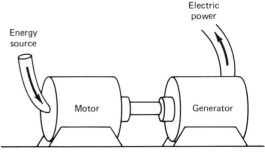

Figure 4.11   In a power plant a "motor" drives a generator to produce electric power.

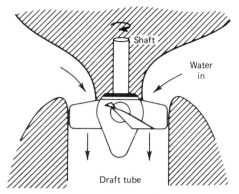

Figure 4.13   Cross section of a propeller-type water turbine.

Activity 4.3

It is very easy to see a hydraulic turbine of the impulse type in action. Just turn a bicycle upside down, so that the front wheel is free to turn. With the garden hose nozzle set for the fastest possible jet of water, direct a stream of water at the front wheel, as indicated in Figure 4.15. You may be surprised at how fast the wheel gets up to speed and how high that speed is. A little practice will show you what conditions will produce the highest speed.

Figure 4.12   Partially assembled giant steam turbine (the "motor") to be connected to a large generator in a power plant. (Courtesy of the General Electric Company, Schenectady, N.Y.)

water "pushing" from above. The advantage of the draft tube is that it keeps the power plant above the flood level without giving up the extra energy of the further drop to the usual level of the stream. An "impulse type" turbine is shown in Figure 4.14. In this type of turbine, the potential energy of the water is first converted to kinetic energy by squirting it out of nozzles, much like the nozzle on your garden hose. The high-speed water jets then strike a series of buckets or scoops on the rim of a wheel and impart power to the wheel. With properly designed nozzles and scoops, the energy of the water jets can be converted to energy of the rotating wheel with almost 100 percent efficiency.

Hydroelectric power plants operate on the same principles of power generation as do other types, but their source of power gives

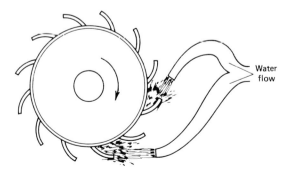

Figure 4.14   An impulse-type water turbine.

Figure 4.15 A bicycle-wheel "turbine."

them some distinct advantages. First, the source of energy is *renewable*: The water vapor produced by evaporation of water from the oceans is pushed over the land by winds, falls as rain, and eventually collects in the reservoir behind the dam. Even better, the water continues on downstream and may be used for other purposes, including powering another power plant.

Further, the plant is pollution free. Its only effluent is a stream of clean water; it does not send tons of particles and noxious chemicals into the air or water. Although the initial cost of building a dam is great, a hydropower plant is relatively cheap to run and maintain and it should last a long time, thus making it less expensive in the long run than a conventional plant. The overall efficiency of such a plant in converting the potential energy stored in the water to electric energy is 85 to 90 percent, two to three times greater than for coal-powered or nuclear plants.

One final advantage of hydropower plants is their ability to cope with varying power demands. Figure 4.16 shows a typical summertime power load on an urban power plant for the period from midnight to midnight on a weekday. Generally, the highest demand comes from about 10 o'clock in the morning until about 6 o'clock in the afternoon, with a peak around 3 o'clock. This is when industries are operating and people have air conditioners turned up to maximum. The lowest power demand comes at about 4 or 5 A.M., corresponding to the coolest part of the day, and is only about 40 percent of the peak demand. The power companies must have enough equipment to meet the peak demand and yet be able to reduce production at other times. In conventional plants, the powering up of equipment is difficult and costly; in the hydropower plant it is much simpler and faster because, in essence, it amounts to allowing the water to flow through the turbines or stopping it, as appropriate. We shall see more later about the difficulties of meeting peak demands.

If hydroelectric power plants are so great, why don't we just produce all of our electricity that way? In fact, in 1978 the United States produced only about 9 percent of its electric power in hydroelectric plants, a figure that

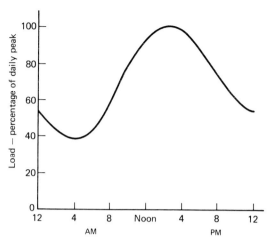

Figure 4.16 The varying demands on a typical power plant on a summer day.

has decreased from about 30 percent in 1959. This power was produced largely in a few major plants, starting with the granddaddy of them, the Hoover Dam on the Colorado River at the Arizona–Nevada border, which was completed in 1935. Table 4.1 lists the major dams in the United States in order of eventual planned capacity. Those capacities are in *megawatts* (1 MW = 10^6 W). It should be pointed out that many of those dams, including the Hoover Dam, have as their primary purpose water storage and management, and

electric power is produced when and as it is compatible with that primary purpose. For this reason, some of the dams listed usually produce much less power than their rated capacities.

These large plants can produce roughly as much power as typical large coal-powered or nuclear plants, which are about 1000 MW. Interestingly enough, on a worldwide basis, the Grand Coulee station, which was for a long time the largest hydroelectric plant in the world, is the only one in the United States to make it into the top ten. It is now second to the Itaipu plant in Brazil. The eventual planned capacity of the Brazilian plant is 12,600 MW. The Chief Joseph station is number 14 in the world, and the John Day is number 22. There is a clue here about the future growth of hydroelectric power, with the greatest potential appearing to be in many of the less industrialized countries, such as Brazil, Mozambique, Zaire, Venezuela, Argentina, and Paraguay. Perhaps the greatest potential of all is in the People's Republic of China.

Yet, there remains the question of why in the United States the percentage of electricity produced by hydropower has consistently declined over the past 30 years, or so. It is not due to a lack of willingness to use hydropower, nor even to the great cost of constructing a dam. Most of the remaining potential dam sites are in the northwestern part of the country and Alaska, where there are mountains to produce sufficiently high heads and enough rain to keep the reservoirs adequately full most of the time. However, the population centers are far removed from those locations and, as we shall see later, there are problems associated with transmitting electric power over long distances. Besides, there are some other drawbacks not yet mentioned.

What? This paragon of a power source has more drawbacks? Even where dams are well situated, a drought can have serious consequences. A few year ago, when rains were

TABLE 4.1 *Some Large Power Dams in the United States*

Name	Location	1978 capacity (MW)	Planned capacity (MW)
Grand Coulee	Washington	3460	9780
Chief Joseph	Washington	1500	3670
John Day	Oregon-Washington	2160	2700
Robert Moses	New York	1950	1950
The Dalles	Oregon-Washington	1810	1810
Hoover Dam	Arizona-Nevada	1345	1345
Rocky Reach	Washington	1210	1210

not great enough, the Pacific Northwest experienced extensive "brownouts" because power production had to be cut down. A dependence on the elements for something that has become so essential to life can be chancy. Another danger of dams is their possibility of breaking, with disastrous effects downstream if the collapse comes suddenly. This is unlikely, and dams are stressed for all eventualities, including earthquake, but it does happen occasionally, as we see in the papers. The resulting flash floods usually kill people and destroy property.

Finally, although hydropower plants do not pollute, they do affect the environment. Obviously, large areas of land have to be flooded. This has both positive and negative effects: usually valuable land is put underwater, but a recreational lake comes into being. Many dams are also used, in addition to power production, to control flooding, for irrigation, and to remove silt from the water.

The Aswan Dam in Egypt has had harmful effects, some of which were not anticipated. For example, many works of ancient peoples had to be moved from their original locations before the dam filled; some which were not moved are suffering unexpected water damage. Even more serious, the annual flooding of the Nile valley, along with its deposits of silt, which had much to do with the upspringing of early civilization and which was most important to farming in that area, has been "controlled" by the dam. As a result, the farming and the fishing in the Mediterranean have been harmed.

Another example is provided by the two dams being built in the Kafue River in Zambia. The resulting changes in the flood plains may lead to the extinction of a species of antelope not found elsewhere. In each case, the possible harmful effects must be balanced against the benefits, and thoughtful decision making is in order. Our immediate needs are important, but to meet them at too great a cost may eventually detract from rather than add to the quality of life. Of course, it is possible to go to extremes. For example, lawsuits halted construction for two years on the Tellico Dam in Tennessee. This was done to prevent the possible extinction of the snail darter, a tiny species of perch which was then thought to live only in the Tennessee River, but whose existence was not even known of before that time. (Snail darters have since been found in at least three other streams in Tennessee.) Once again, balanced, thoughtful, nonemotional decisions are in order, as they are in every facet of planning the energy future.

An interesting sidelight is that at one time there were well over 10,000 small hydroelectric plants on streams in upper New England. As electricity produced in oil-fired plants became very cheap, hydropower plants were abandoned. Now that conventional energy is getting more expensive, there is interest in restoring some of the old plants and building new ones. No one of them is likely to be very large but, together, they could make a substantial and low-cost contribution toward meeting energy needs. This is just one example of a lesson to be gained from the past, and we Americans need to start reevaluating resources that we have been rich enough to ignore in our recent history.

EXAMPLE

If the Hoover Dam power plant were to produce an average of 1000 MW of electric power, how much energy would it produce in a year? Assuming about 85 percent efficiency in converting the energy stored in the water to electric energy, how much energy would have to be stored in the water to produce a year's worth of electric energy? With a head of about 200 m, how much water then would have to pass through the turbines in a year?

Solution

First, note that 1000 MW is the same as 10^6 kW. Then, the energy produced is power times time:

$$E = Pt$$
$$= 10^6 \text{ kW} \times 365 \text{ days} \times 24 \text{ hr/day}$$
$$\approx 8.8 \times 10^9 \text{ kWh}$$

Since 1 kWh = 3.6×10^6 J, this can also be expressed as about 3.2×10^{16} J. If the power is produced with 85 percent efficiency, then

$$Eff = \frac{\text{useful energy}}{\text{energy stored in water}}$$

or

$$E(\text{stored}) = \frac{E(\text{produced})}{Eff}$$
$$= \frac{3.2 \times 10^{16} \text{ J}}{0.85}$$
$$\approx 3.8 \times 10^{16} \text{ J}$$

But the energy stored is just the potential energy of the water:

$$PE = mgh$$

or

$$m = \frac{PE}{gh}$$
$$= \frac{3.8 \times 10^{16} \text{ J}}{9.8 \text{ m/s}^2 \times 200 \text{ m}}$$
$$\approx 1.9 \times 10^{13} \text{ kg}$$

Since there are about 3.8 kg/gal, this is roughly 5×10^{12} gal of water.

Questions

Say that, on a given day, the Grand Coulee station generates electricity at the average rate of 2000 MW. This is about 60 percent of its capacity, and a reasonable average. Most power plants will run at an average of 50 to 70 percent of their capacity. How much energy is produced in the course of a day? Do this both in kilowatt-hours and joules. How much energy is this in a year? If the plant is 85 percent efficient, what is the daily energy change in the water as it goes from the upper to the lower level? With an average head of about 150 m, what mass of water in kilograms must flow through the generators daily? Annually? Convert your answer to cubic meters and gallons to get some idea of the tremendous volume of flow (1 m^3 of water has a mass of 1000 kg; 1 gal of water has a mass of about 3.8 kg.)

PUMPED STORAGE

When you start your automobile, large amounts of electric energy are consumed, especially in the wintertime if you have to crank the engine for a long time. This depletes the energy stored in the battery. When you are driving along the highway the car's generator ("alternator") produces electric energy, and it is capable of doing so at a rate greater than the electric power consumed by the auto. If the battery is "low," the extra energy goes into the battery, where it is stored until needed. This is a good example of changing load conditions, with the peak load coming at the time of starting the car and much lower loads coming at other times. Of course, the lowest load of all is when the car is not running, but let's ignore that.

The nice thing about this source of electric energy is that by generating just a little extra during times of average load we are able to store energy for later use when the demand is large. It would be very nice if we could do the same thing with power plants. But there are no batteries that can handle the storage of electric energy in the huge amounts required by the changing demands on power plants, nor are there likely to be any such batteries in the foreseeable future. This causes serious problems for power companies and adds complexity to the economics of power production. In the past, power plants have used their regular generators, adjusting their rates of generation as well as possible, along with *peak-load generators.* Often these peak-load generators have been powered by turbojet engines, like those used in some aircraft. That is a very expensive way to generate electricity. Because of our extensive nationwide power network, it is sometimes possible to "borrow" power from other companies. However, if the other company is far enough away to have a different peak-load time, then losses in transmission become important. The lack of ability to meet the greatest peak demands leads to brownouts and occasionally, because of a severe overload on the system, to widespread blackouts. Unless steps are taken to reduce power consumption or to increase the maximum available supply, those blackouts may become more frequent. Pumped storage can provide one partial answer.

Questions

On a cold day, it might take a minute to cranking to start an automobile engine, and the starter requires power of 1 kW to operate. What is the total energy consumed? Driving along the highway at 55 mph, the generator may produce about 300 W, and the ignition and accessories require about 200 W to operate. Assuming that all of the excess goes into the battery, how long will it take to get it up to the same level of charge that it had before starting the car?

Pumped storage is a form of hydroelectric power, and it operates on exactly the same principles as does any hydropower plant. The difference is that the head is created artificially rather than by nature. In periods of low use, the extra energy available from the conventional (steam-powered) generators is used to pump water uphill, where it is stored for later use. When the peak load comes, the water is allowed to come back downhill to run a generator. The added power produced gives assistance to the other generators in the plant. Figure 4.17 shows a schematic picture of the general layout of a pumped-storage plant, and Figure 4.18 is a photo of the plant at Ludington, Michigan. There are two reservoirs, a lower and an upper. During the periods of pumping, water is pumped from the lower reservoir to the upper, thus gaining potential energy. In the high-use periods, the water flows down through the power plant and into the lower reservoir. No water is used up, except by evaporation, and the cycle can be repeated the next day.

Figures 4.19 and 4.20 indicate the way in which pumped storage might be used in a 1000 MW plant. Looking again at Figure 4.16, you will see that this fairly typical plant has a load that is less than 80 percent of capacity for much of the day and it peaks for only a short period. The usual plant will need gen-

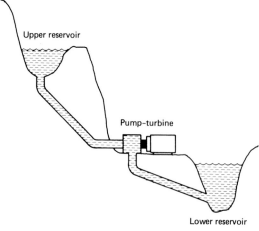

Figure 4.17   Pumped storage.

Figure 4.18 The pumped storage plant at Ludington, Michigan. The upper reservoir is a pond over 2 miles long and nearly 1 mile wide. The lower reservoir, in the foreground, is Lake Michigan. (Photo courtesy Detroit Edison Company.)

erating capacity at least equal to the peak load, and will have one or more generators running at reduced capacities for most of the day.

In the pumped-storage plant, the hydroelectric generator, which is much easier than a steam turbine to shut down or to run at reduced rates, can be used to provide the

portion of the load over 80 percent of capacity. It derives its power from water in the upper reservoir running through the turbine and causing the generator to turn.

When the load drops below peak, the hydroelectric generator can be run at a reduced rate, allowing the steam-powered generators

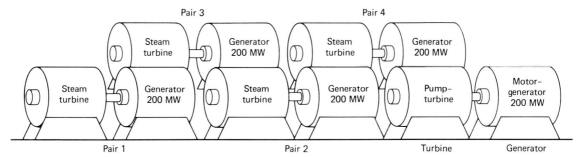

Figure 4.19 When the load is above 80 percent of capacity, the hydroelectric pair is run as a turbine–generator, using the water in the higher reservoir.

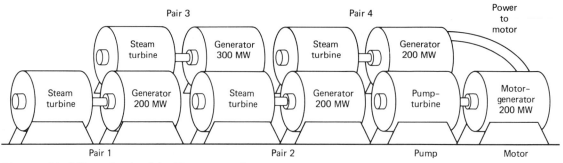

Figure 4.20 When the load is 60 percent of capacity of less, one of the generators supplies power to the motor–pump pair, which pumps water to the higher reservoir from the lower.

to continue to run at full capacity, their most efficient mode of operation. When the load drops to below 80 percent, the hydro generator can be shut down and converted to a motor–pump pair. As the load goes below 80 percent, part of the power produced by steam generator number 4 can be used to run the hydro unit as a motor–pump pair, starting to pump water to the upper reservoir. When the load is 60 percent or less, all of the power from number 4 can go into pumping water to the upper reservoir. Thus, in the range of 60 to 100 percent of capacity of the plant, all four steam powered generators can run at full capacity, and the reductions required of them because of lower loads are much less drastic. The excess energy produced at times of low demand is saved in the potential energy of the water pumped to the upper reservoir for use in times of high demand.

As always, there is a price to pay, and this analysis is oversimplified. For example, the pumping process is not perfectly efficient. Present pumped storage plants operate at an overall efficiency of about 70 percent, and 75 percent may be realized in the near future. That is, only about 70 percent of the electric energy produced by generator number 4 for the purpose of pumping is returned as electric energy by the hydroelectric generator; the rest is lost to various frictional forces. Nonethe-less, the increased efficiency that results from running the stream-powered generators nearer to capacity can more than make up the difference. Analyses based on current economics indicate that it would be cost-effective to have about 10 to 20 percent of our generating capacity in pumped-storage generators. At present, there are a few pumped-storage plants scattered about the country, and construction of more is planned.

EXAMPLE

Suppose a hydroelectric generator running on pumped storage is required to provide its full power of 200 MW (2×10^8 W) for a period of 8 hours. What is the total energy, in joules, of the output? Assuming that the generating part of the cycle is 85 percent efficient, how much energy must be stored in the water in the upper reservoir? If the upper reservoir is 100 m (about 330 ft) above the power plant, what is the mass of the water needed? Recalling that 1 m^3 of water has a mass of 1000 kg, what is the volume of the stored water? If the water level in the reservoir is allowed to drop 2 m in the 8 hours, what is the area of the reservoir and, if the reservoir is square, how long are the sides of the square?

Solution

The energy output is equal to power times time:

$$E = Pt = 2 \times 10^5 \text{ kW} \times 8 \text{ hr}$$
$$= 1.6 \times 10^6 \text{ kWh}$$
$$= 1.6 \times 10^6 \text{ kWh} \times 3.6 \times 10^6 \text{ J/kWh}$$
$$= 5.8 \times 10^{12} \text{ J}$$

With 85 percent efficiency in the generating phase, the energy stored in the water must be

$$PE = \frac{5.8 \times 10^{12} \text{ J}}{0.85} = 6.8 \times 10^{12} \text{ J}$$

Since $PE = mgh$, the mass of water needed to provide this much stored energy is

$$m = \frac{PE}{gh}$$
$$= \frac{6.8 \times 10^{12} \text{ J}}{9.8 \text{ m/s}^2 \times 100 \text{ m}}$$
$$= 6.9 \times 10^9 \text{ kg}$$

The volume of the stored water is just the mass divided by 1000 kg per m³, or 6.9×10^6 m³. The amount by which the reservoir drops is given by:

$$V = AH$$

Where A is the area of the lake surface and H is the amount it falls when emptied. Solving this equation for A and substituting the numbers yields

$$A = \frac{V}{H}$$
$$= \frac{6.9 \times 10^6 \text{ m}^3}{2 \text{ m}}$$
$$\approx 3.5 \times 10^6 \text{ m}^2$$

If the reservoir is square, $A = L^2$, where L is length, and

$$L^2 \approx 3.5 \times 10^6 \text{ m}^2$$
$$L \approx 1{,}900 \text{ m}$$

This example gives an indication of one other problem with pumped-storage plants. The upper reservoir in this case turns out to be a square with sides a little more than 1 mile long. That is not too large and it would be a nice little lake, except for one thing. Who wants a lake that is full for half of every day and empty for the other half, with the water level changing by more than six ft twice a day? In fact, unless the lower reservoir is a river or large lake, there will be two lakes doing this, with one full when the other is empty. One can see why some environmental groups have already raised objections. Once again, the cost must be balanced against the gains.

A further limitation of pumped storage is the lack of suitable sites in appropriate locations. One way around this might be to put the pumped-storage plant deep underground, as indicated in Figure 4.21. This can be done in far more locations than are suitable for pumped storage on the surface. Further, since the head could be made quite large, say 1 kilometer (1 km) or more, a smaller volume of water would be needed, thus making the requirements on the storage reservoirs somewhat easier to meet.

Questions

Suppose that the turbine for the pumped-storage generator of the previous example is 1 km (1000 m) below the surface, which is the location of the upper reservoir. If the upper reservoir is a square lake with sides 2000 m long, how far will the water level fall in going from full to the lowest level of the day? This is a lot better than the preceding example, isn't it? The lower reservoir will probably have a shape very different from the upper one, one that will be easy to excavate. Supposing it to be roughly cubical, how big will it be?

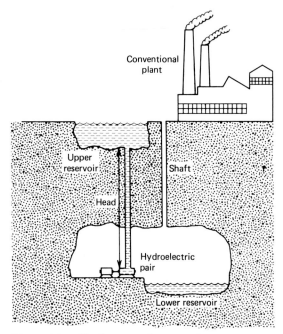

Figure 4.21 Underground pumped storage.

One other form of pumped storage that should be mentioned is compressed air. By pumping air into a natural or artificially created cavern, it can be compressed to a pressure up to 80 times the normal atmospheric pressure. Potential energy is stored in the "springiness" of compressed air, and it can be released later to run an air turbine. There is a small experimental plant of this type in West Germany and a working plant is to be built in Decatur, Illinois. Pumped-air storage is attractive because more geological formations are appropriate for it than for pumped-water storage. However, highly compressed air is dangerous to handle. Another difficulty is that in the pumping process heat energy is lost to the surroundings and has to be replaced, thus reducing efficiency.

Finally, I would like to mention once again the possibility of storing large amounts of energy in flywheels. These are just very large and massive wheels mounted on axles so that they can spin at a very high speed. When they are rotating very rapidly, such wheels are capable of storing large amounts of kinetic energy, with little loss to friction. This stored energy can be converted to electric power when it is needed. But it is very difficult to build huge wheels that are strong enough and well enough balanced not to tear themselves to pieces when rotated at high speeds. Perhaps technology will advance to the point that such wheels will be practical in the near future.

TIDAL POWER

Anyone who lives near an ocean is familiar with the *tides*, a daily variation in the level of the ocean near the shore. This variation is caused by the gravitational forces of the moon and sun acting on the water, with the moon's effect being the greater one. Figure 4.22 shows how the force of attraction of the moon varies at different points on the Earth; the closer to the moon a given mass is, the greater will be the force exerted on it. The Earth, being a relatively rigid body, is attracted as though all of its mass were concentrated at it's center. A body on the side closest to the moon will feel a greater force per kilogram, and a body on the opposite side will feel a lesser force than the average force on the Earth. Of course, at any location on the surface of the Earth, the force exerted on a body by the Earth is much greater than the force of the moon; if

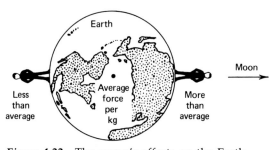

Figure 4.22 The moon's effects on the Earth.

the moon were to exert the greater force, people would go flying off the surface of the Earth whenever the moon rises high in the sky!

To see how the moon's gravitational force affects the oceans, imagine an earth that is a smooth sphere covered with a thick layer of water. A property of water, a liquid, is that it is able to flow freely in response to forces. Then, as shown in greatly exaggerated form in Figure 4.23, the water close to the moon gets pulled with a greater than average force, and tends to pile up somewhat. On the opposite side, the water gets pulled with less force than the average, and it too piles up a bit. (One way to visualize this is to think of the Earth as being pulled away from under the water by the force of the moon.) Then, if there were no friction between the water and the Earth, as the Earth turned the point marked A would slide under this strangely shaped water, causing two high and two low tides each day, with a maximum or minimum level coming every six hours.

Of course, there is friction between the Earth and the water, and the water does not cover all of the Earth. But, in most places there are two high tides and two low tides daily. Because of the friction, the tides generally lag behind the passage of the moon by several hours. The sun's force also complicates the

matter. When the sun, Earth, and moon are in a straight line the difference between high tide and low tide is the greatest; this is called *spring tide*. When the sun and the moon are in *quadrature* (when the lines connecting sun–earth–moon form an angle of 90°), the difference is the least, producing a *neap tide*. Also, each high tide or low tide runs about 50 minutes later than the high or low for the previous day because the motion of the moon about the Earth causes it to appear about 50 minutes later each successive day. As a final complication, there are some places, such as areas in the Gulf of Mexico, where the effects of the land masses cause considerable variations in the tides, sometimes causing a single high tide in a day, sometimes causing there to be no observable high tide.

Nevertheless, in most places there are two high tides separated by two low tides, with the entire sequence repeating approximately every 25 hours in most places on the coast, as indicated in Figure 4.24. In some areas the effect is magnified by the shape of a coastline that tends to funnel the tides into a narrow area. In a place like the Bay of Fundy, an inlet on the North Atlantic coast, the tides may be

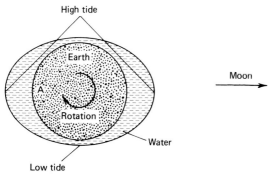

Figure 4.23 The "piling up" of water on the Earth's surface due to the moon's pull (greatly exaggerated).

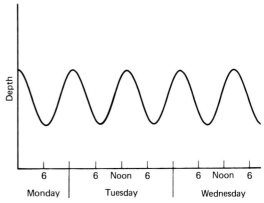

Figure 4.24 A graph of the tides in a given location. The height of the graph corresponds to the average depth of water. The average is used to eliminate the effects of the waves.

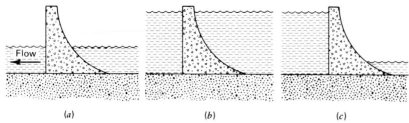

Figure 4.25 Operation of a tidal power plant. (*a*) Tide incoming; open dam. (*b*) High tide; dam closed. (*c*) Low tide; trapped water is released, generating power.)

as great as 70 feet, and they average 50 feet between high and low tides. If a dam were built across the mouth of the inlet, it could be opened as the tide comes in. When high tide was reached, the dam could be closed, trapping the water behind it. Then, when the tide went down, the trapped water would be released by passing it through a hydroelectric generating plant. This is indicated in Figure 4.25. Like hydroelectric power generated on a river, the source of energy is "free"; once the dam and the plant are built, the cost of operation is relatively small.

Questions

The tidal power hydroelectric plant shown in Figure 4.26 is located at Rance, on the coast of France. It has been in operation since 1967, and it has a peak capacity of about 240 MW. What is this in kW? Multiplying by the number of hours in a year, what is the total energy, in kWh, that that plant could produce in a year, if it were able to operate at full capacity at all times? The actual output of the plant is about 500×10^6 kWh per year. What percentage is this of full capacity, and how does this compare to the figure of 50 to 70 percent for conventional power plants?

The main reason that a relatively small percentage of capacity is produced in a tidal power plant is that the plant is not generating at all times. Further, because the daily times of the tides vary, the power may be available at times when it is not needed and unavailable when it is needed. A partial answer to this problem might be to build the dams so that they work in both directions; then power could be generated on both the incoming and outgoing tides. Also, in some locations, a two-pool arrangement would allow there to be storage when peak power happened to be produced during low-demand times.

The total worldwide production possible at known tidal power sites is estimated to be about 10^{12} kWh per year, not insignificant, but not large enough to solve the world's energy problems. Tidal power has been used to power mills since the middle ages in Europe and since colonial times in America. However, at the present time, aside from the plant at Rance, there is only one small experimental plant in the Soviet Union. There has been much talk for years about a project to be undertaken jointly by Canada and the United States in the Bay of Fundy region, but it is not clear that this will ever come to pass.

WIND POWER

Energy provided by the wind has been used for centuries to lift water for irrigation and other uses. A few decades ago, hundreds of thousands of windmills dotted the landscape of rural areas in the United States, pumping

Figure 4.26 The tidal power station at Rance, on the northern coast of Britanny. (Michel Brigaud, French Embassy.)

water and providing electricity to isolated farmhouses. However, when cheap electricity became available to rural areas, largely in the 1930s and 1940s, it spelled the end of windmills, and for a time a windmill in working order was a real rarity. The last manufacturer of wind-powered electric generating systems went out of business in 1953. In Europe, however, windmills have continued to be an important source of power.

Obviously, the present energy situation and the uncertain future have revived America's interest in the use of wind machines. Figures 4.27 and 4.28 show two of the many experimental designs. Today a number of companies sell complete wind power sets on a small scale. They include generators and storage batteries and can produce from 500 to 1000 kWh per month, enough to supply the entire needs of some homes. The storage batteries ensure that energy is available when it is needed and not just when the wind is blowing. Naturally, the degree of usefulness of such a system depends on how much wind

Figure 4.27 Wind power. This 200-kW wind turbine provides about 20 percent of the electric power for the people of Culebra Island, Puerto Rico. The rotor blades span 125 feet. (Courtesy of the U.S. Department of Energy.)

blows. Figure 4.29 indicates the estimated annual average wind power available for various locations in the United States. The lines on the map connect regions with the same available power. The power is measured in watts per square meter of surface perpendicular to the wind. For people living in southern Wyoming and other high-wind areas, wind power is now available at a cost that is competitive with the cost of other commercially produced electricity. In other parts of the country, where the winds are not as great nor as consistent, the cost is not yet competitive, but it may be in the not-too-distant future. Since, in general, the strongest winds blow at the times of the coldest weather, wind power may supplement conventional home heating systems, thereby saving expensive fuel oil and making the wind machine relatively economical.

To find out just how dependent upon wind velocity wind power is, take a look at Figure 4.30. Here a "tube" of air moves past a windmill with a velocity v. The circular area at the end of the tube is the circle that the tips of the blades of the windmill produce as they

turn. Call that area A. The tube has a length L. If the tube's length is such that the whole thing moves past the windmill in t seconds, then $L = vt$. Therefore, the volume of the tube—the area of the circle times the length—is

$$V = A \times L = Avt$$

Now we need the density of the air. Remember that the density of water is 1000 kg/m³. Air, of course, is much less dense—about 1.3 kg/m³ at sea level. Using k for density, with $k = 1.3$ kg/m³, the mass of the air in the tube is

$$m = kV = kAvt$$

Figure 4.28 A Darrieus wind turbine. This is one of the types of experimental "vertical axis" wind machines. (Courtesy of the U.S. Department of Energy.)

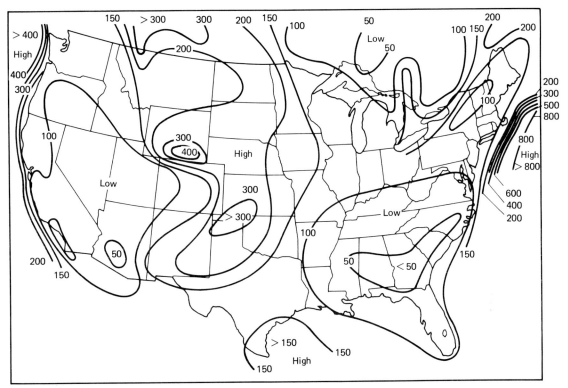

Figure 4.29 Average wind power in the United States. The lines connect points where the average annual power available is about the same. The figures represent watts per square meter of a surface that is perpendicular to the wind. (From the *McGraw-* *Hill Encyclopedia of Energy*, Second Edition, Sybil P. Parker, Editor in Chief, copyright © 1981. Used with the permission of the McGraw-Hill Book Company.)

Now look at the kinetic energy of the air in the tube. It is

$$KE = \tfrac{1}{2}mv^2$$

Substituting *kAvt* for mass, this becomes

$$KE = \tfrac{1}{2}\,kAv^3t$$

Finally, we want the *power* available. It equals the kinetic energy per unit time going past the windmill:

$$P = \frac{KE}{t}$$
$$= \frac{\tfrac{1}{2}\,kAv^3t}{t}$$

or,

$$P = \tfrac{1}{2}kAv^3$$

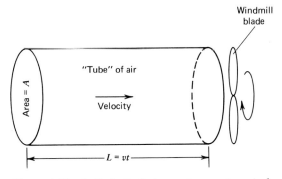

Figure 4.30 A "tube" of air moving past a windmill propeller with a speed *v*.

Remember A is the area in square meters through which the wind passes, v is the velocity in meters per second, and k is the *density* of the air. The density is the number of kilograms of air in a cubic meter, about 1.3 kg/m^3 at sea level. The power, P, is in watts. It can be thought of as the amount of kinetic energy in a chunk of air that moves past an area A in 1 s.

Questions

Suppose one has a propeller-type windmill with blades 1 m long (radius of the circle). What is the area swept out by the blades? (It is not 1 m^2.) It can be shown from theoretical considerations that even a perfect, frictionless windmill can capture no more than 59 percent of the energy in the air passing it. (If the windmill were to capture 100 percent of the kinetic energy of the air, the air would come to a stop. Then, where would the next batch of air go?) If the overall efficiency of this windmill is 50 percent, what is the power produced when the wind speed is 1 m/s (about 2.2 mph)? At 2 m/s? At 10 m/s? Make a graph of the power produced versus wind speed.

The power produced by a wind-powered generator very strongly depends on wind speed; 1000 times more power is delivered at 10 m/s than at 1 m/s. That is why the range of average power available at different locations is so great. Figure 4.29 reveals some areas where large-scale power production from the winds may be feasible. One such region is offshore along the north Atlantic coast; the winds are consistently large and the power needs of the Northeast are great. Large wind-powered generating machines could also be placed in various land regions. One estimate—admittedly an optimistic one—is that the United States could eventually produce from wind power up to 2×10^{12} kWh per year, an amount about equal to the U.S. consumption of electricity in 1975.

This additional source of "free" energy seems to hold great promise but there are unsolved problems associated with large-scale wind-powered machines. First, the state of the art in the design of wind-powered machines is not highly advanced, although some new designs are now coming forth. Present-day propeller-type machines are expensive to build in large sizes, and they are difficult to maintain, tending to shake themselves to pieces. It may be some time before large machines that are affordable and dependable can be produced in great numbers. Then, too, although the annual energy available at a given site varies only about 15 percent up or down in a given year from the average, there is a great variation in winds at a given site from hour to hour or day to day. In the production of energy for a home, batteries can be used to store energy for use in low-wind periods; in the large-scale production of electricity there is no simple way to store the energy, and we run into the same problems we have already discussed. In addition, one can imagine that a landscape dotted with hundreds of windmills might eventually become as hideous as the oil fields of the 1930s.

One other caution is in order. In recent years, quite a few home owners have bought wind-powered generators, expecting to save a lot of money on electricity. Such a system will usually cost $2000 to $5000 or more. The map of Figure 4.31 is the same as that of Figure 4.29, except that the numbers have been converted to dollars. Each number represents the approximate cost of the electricity produced per year by a windmill running 24 hours a day. (This assumes that the windmill has an efficiency of 50 percent and that it has blades 1.4 m in radius. Also, a cost of $0.06 per kilowatt hour is assumed.)

Questions

Suppose you can buy a wind generating system and have it installed for just $2500. Also suppose that it will run for many years with

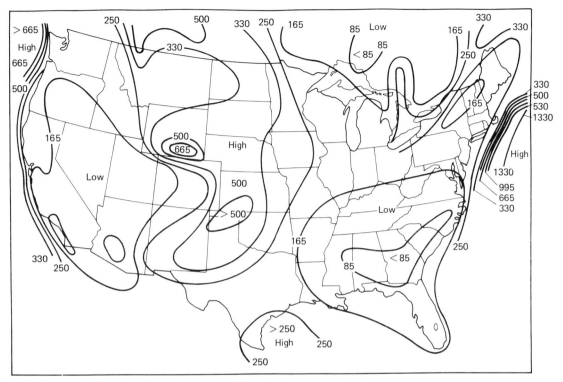

Figure 4.31 Dollars saved per year by a full-time home windmill in various areas of the country. (From the *McGraw-Hill Encyclopedia of Energy*, Second Edition, Sybil P. Parker, Editor in Chief, copyright © 1981. Used with the permission of the McGraw-Hill Book Company.)

no maintenance and that it runs 24 hours a day. Using Figure 4.31, find the payback time for such a machine in your part of the country. (The payback time is the time it takes the machine to generate enough electricity to save its own cost.) Is it longer than you expected? Do you see why a careful evaluation of the local winds is essential before buying such a machine?

SUN POWER

You may wonder what a section on the sun is doing in this chapter. Well, think about it for a minute. If there is a moral to this book,

it is that you can't get something for nothing. All of the energy derived from water and wind must come from somewhere, and the sun is the only available source. Through heating and evaporation the sun pours enormous amounts of energy into the atmosphere.

For example, near the seashore, the sun shines equally on land and water. However, the land absorbs more of the sun's energy than does the water, and therefore the land and the air above it become warmer than the ocean and the air above it. The warmer air above the land expands and rises and cool air then comes in from the ocean, as indicated in Figure 4.32. This is the origin of the "sea breeze" that people who live near the shore

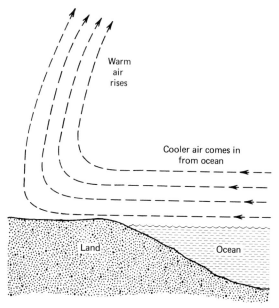

Figure 4.32 A sea breeze.

experience on sunny days. At night, when the sun is not shining, the land area cools faster than the sea. Then the warmer air over the ocean rises and, as shown in Figure 4.33, a "shore breeze" is produced. All winds re-

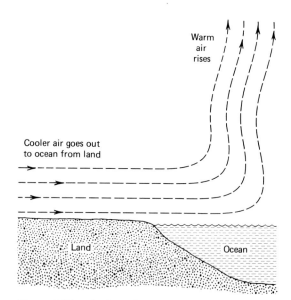

Figure 4.33 A shore breeze.

ceive their energy from the sun in a similar manner, with "highs" (high pressure areas) and "lows" (low pressure areas) being formed on a large scale and with air being pushed from one area to another.

Water power also derives its energy from the sun. It begins with the sun shining on the Earth's oceans, lakes, and streams, and evaporating water. Now, water vapor in the air is invisible and it seems unsubstantial; it may be that you have difficulty in imagining that very much potential energy is stored this way. However, if 1 kg of water is raised to a height of 1000 m it gains 9800 J of potential energy, whatever form it happens to be in. The energy needed to lift it comes from the sun. Figure 4.34 is a simplified sketch of the rain cycle that brings energy to the power dams.

SUMMARY

This chapter has presented some of the alternatives to oil- and coal-fired and nuclear power plants. None of the alternatives is likely to solve the problems of itself, but each, intelligently applied, can help to reduce the problems.

You have learned that electric power can be produced by using energy available in water and winds. Water power (hydropower) is available because of the potential energy stored in water having a head. The head may accumulate behind a dam on a stream, it may be produced by tides, or it may be created by pumping water up to a higher location. The potential energy stored in the water is converted first into the mechanical energy of a turbine, then into electric energy by a generator. The amount of hydropower available is limited because the number of suitable sites is limited. "Free" energy may also be derived from the winds using both large-scale and small-scale wind machines. An important point is that, except for tidal power, all of the energy available for water or wind power comes orig-

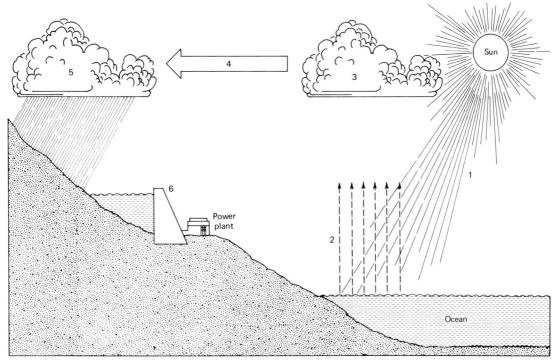

Figure 4.34 The rain cycle. The sun's energy (1) evaporates water (2) from oceans, lakes, and streams and raises it to great heights to form clouds (3). Then the wind, which also receives its energy from the sun, blows the moisture inland (4). Rain falls in the mountains (5) and forms streams which eventually flow to the ocean. If a dam (6) is built on a stream, it may convert some of the potential energy of the water to electric energy.

inally from the sun. Finally, of utmost importance, energy *never* is created out of nothing; it is always converted from one form to another.

ADDITIONAL QUESTIONS AND PROBLEMS

1. In the discussion of the overshot water wheel, it was found that a flow of about 2400 gal/min (9000 kg/min) of water was needed to run the mill. Suppose this water is running down a long flume that is 1 m wide, and that the depth of the water in the flume is 0.2 m. What is the volume of the water, in cubic meters, that must flow over the wheel in 1 min? In 1 s? How long a section of flume would be needed to hold that much water? Calling that last answer x, think of a "chunk" of water x meters long in the flume, just about to fall over the wheel. In 1 s, the back end of the "chunk" moves a distance x down the flume, to its edge. If that back end goes a distance x in 1 s, what is the speed of the current in the flume? Since 1 m/s $\approx$ 2.2 mph, what is the speed of the current in miles per hour?

2. In the activity in which you squirted water nearly straight up with a hose, suppose that the water reached a maximum height of 12 m. Then, when aiming the same stream into a bucket, the water reached a mark at 2.6 gal

(10 kg) in 40 s. In the 40-s period, how much energy was carried from the hose nozzle by the water? (*Hint:* Imagine that all 10 kg are 12 m above the starting point at the same time. Then, what is the potential energy of the water?) What is the power (rate of production of energy) generated at the hose nozzle? What is the velocity of the falling water just as it reaches the level of the hose nozzle? (*Hint:* Use conservation of kinetic plus potential energy.) Then, what is the velocity of the

water leaving the nozzle? (Again, use conservation of energy.)

3. Sketch a graph of the water pressure acting on the side of a dam as a function of depth. That is, as the water gets deeper and deeper, what happens to the pressure?

4. In a power plant, as everywhere else, energy is *never* created; it is converted from one form to another. What are the conversions of energy that occur in a hydroelectric

Figure 4.35 Hoover Dam at a time when water is being released without going through the penstocks and turbines. (U.S. Bureau of Reclamation photo, from U.S. Department of Energy, by E.E. Herzog.)

power plant, starting with the energy stored in the water behind the dam and ending with electric energy leaving the plant?

5.   Hoover Dam is shown in Figure 4.35 at a time when water is bypassing the penstocks and turbines and being released through valves. The distance across the canyon is nearly 500 feet. What visible indication(s) of energy or power do you see in the picture?

6.   What is the difference between a renewable and a nonrenewable source of energy? Give some examples of each.

7.   List some of the advantages and disadvantages of hydroelectric power production.

8.   Using Figure 4.29, find locations in the United States where wind power would be least feasible. What would be one of the best locations on the land?

9.   A home-sized wind-power generator is found to produce electric power at the rate of 2000 W when the wind is blowing at 10 mph. Assuming that the efficiency stays the same, how much power will it produce at 15 mph?

10.   In several parts of the country, the wind averages 100 W/m^2 of power available. Suppose that a windmill with blades 2.8 m in diameter and with an efficiency of 50 percent is used in one of those locations. On average, how much power does it produce? What is this in kilowatts? If it runs for an hour, how many kilowatt-hours of energy are produced? How many in 24 hours? How many in a year? If the cost of electricity is 6¢ per kWh, how much money is saved in a year? (The answer should be about the same as the dollar figure on Figure 4.31 corresponding to the lines labeled 100 on Figure 4.29.)

11.   During a heavy thunderstorm 1 in. of rain might fall in a relatively short period of time. If this were to happen, what volume of water would fall on a square mile of land? (1 mile ≈ 1600 m and 1 in. ≈ 0.025m) What is the mass of that water? If the water came from a cloud at 3000 m (almost 10,000 ft), how much potential energy was stored in it? Did it have that much kinetic energy when it reached the ground? Why? Why don't people build machines to convert "rain power" to electric power? (Or do they?)

5 *Heat*

Except for the mechanical energy sources discussed in Chapter 4, heat plays a major role in virtually every form of energy production and use. This is fairly obvious in such applications as heating a home with a wood stove. Less obvious but equally important applications include the production of electricity in coal-fired and nuclear power plants, the consumption of electricity, the operation of gasoline and diesel engines, refrigeration, solar energy, and many others. Understanding the basic ideas of heat will help in the understanding of nearly every aspect of energy. This chapter will help you to gain that basic understanding.

TEMPERATURE

Temperature is like sex appeal: We all know what it is, but it is difficult to define. Our instinctive notions about temperature are very helpful, but they are quite subjective and can sometimes be misleading. For example, if the air around us is 100°F, we say that it is "hot." However, water at that same temperature would be considered to be "lukewarm," and an element on an electric range would be downright "cold." One way in which we can compare the temperatures in different locations without having to make subjective guesses is illustrated by the following activity.

Activity 5.1

Fill a narrow-necked bottle with water, so that the surface of the water is about at the middle of the neck. This is illustrated in Figure 5.1. Use water that is near room temperature. Then heat the bottle; you can either put it into a warm oven (set the oven at 200°, let it warm up, and turn it off before putting the bottle in) or, if it is a warm day, put it into the sunlight. What happens to the level of the water? After that, put the bottle into the refrigerator and let it cool for a couple of hours. Now where is the water level? How do you explain this?

As you saw, the water level rises as the water gets warmer, and it falls as it gets cooler. The water-filled bottle is a crude *thermometer*, a device to measure temperature. It works because as the water gets warmer it expands, filling more volume, and as it gets cooler it

Figure 5.1 A water-bottle "thermometer." (*a*) Warm. (*b*) Cool.

the working fluid, and it was quite like thermometers used today, although now some inexpensive thermometers contain colored alcohol. Most mercury or alcohol thermometers are made along the lines of the one shown in Figure 5.2, and they work in the same way

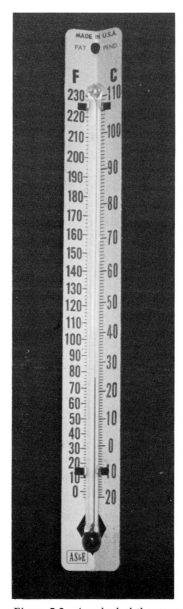

Figure 5.2 An alcohol thermometer.

contracts. In order to be useful, the thermometer should have points of comparison that do not change. In fact, it should have at least two such points, so that the interval between them can be divided up into segments, or *degrees.* We might mark the high spot just found and call it the temperature in the sunlight, but that would change from day to day . The low spot, corresponding to the temperature of the refrigerator, might also change, depending on conditions. Obviously, we need two points that are more dependable in order to "calibrate" our thermometer.

It happens that water provides a convenient means for providing the calibration points. At sea level and in "standard" conditions, pure water freezes at the same temperature every time, and it boils at a temperature that is the same every time. This provides the two calibration points that we need. However, we cannot use a thermometer with water as the working fluid because it would freeze or boil as it's temperature reached the two extremes. The first practical thermometer, made by Gabriel Fahrenheit, used mercury as as the bottle you used for a thermometer. At the bottom is a reservoir, called the *bulb,* which contains most of the working fluid. Above is a *capillary,* a tube with a very fine hole along its length. As the fluid in the bulb is heated, it expands and is forced higher in the capillary, which has points of known temperatures marked off on it. The glass of which the thermometer is made also expands as it is heated, but not as much as the working fluid.

In Fahrenheit's scale, 32 degrees (32°F) is taken as the temperature of freezing water, or an ice–water mixture. The temperature of boiling water at sea level is taken to be 212°F. Thus there is a spread of 180° between the temperatures of freezing and boiling water on the Fahrenheit scale. For temperatures above boiling or below freezing, one simply *extrapolates* the scale. That is to say, the scale is just extended in equal steps above and below those points. There is no definite maximum value of temperature for real objects, but there is a theoretical minimum, below which temperature has no meaning. Because it is absolutely as low as temperature can go, that minimum is called "absolute zero." In the Fahrenheit scale the value of absolute zero is 459.67 degrees below zero (−459.67°F). That temperature has never been reached, but it has been approached rather closely in the laboratory. Obviously, a mercury thermometer is as useless as a water thermometer if the temperature goes below the freezing point of mercury, about −38°F, or above its boiling point of nearly 674°F. For extreme temperatures, different kinds of thermometers are required; some of them are quite sophisticated and difficult to calibrate accurately.

The temperature scale used in the scientific community and in most of the world other than the United States, is the *Celsius* scale, previously called *centigrade.* On this scale, the freezing point of water is arbitrarily taken to be 0° and the boiling point is 100°, so there are 100°C between the two calibration points.

Absolute zero on the Celsius scale is −273.15°C.

A third scale of interest is the *Kelvin* temperature scale. In this scale, the degrees are the same size as the Celsius degrees, but they are called *kelvins* (K), and the zero of the scale is at absolute zero. For this reason, the Kelvin scale is called an *absolute* temperature scale. This is the scale used in the SI system of units, but we shall often use the Celsius or Fahrenheit scales, depending on which is more convenient. Figure 5.3 shows the relationship of the three scales. From it you can see that every 100° interval on the Celsius scale corresponds to 180° on the Fahrenheit scale. Thus a Celsius degree is 9/5 of a Fahrenheit degree. To convert from one scale to the other, one can use the following relationships:

$$5(T_F - 32°F)/9 = T_C$$
$$9T_C/5 + 32°F = T_F$$

where T_F is the temperature in Fahrenheit de-

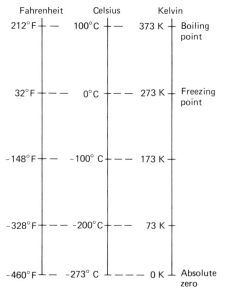

Figure 5.3 Comparison of three commonly used temperature scales.

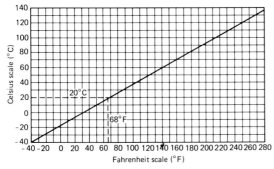

Figure 5.4 A graph for converting between Fahrenheit and Celsius scales.

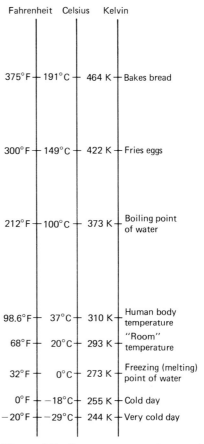

Figure 5.5 Some common temperatures in the three scales.

grees and T_C is the same temperature in Celsius degrees. A little algebra will show that the two relationships are really just a single relationship, somewhat rearranged for the two conversions.

Another way to convert from one temperature scale to the other is to use a graph (Figure 5.4). Just pick a temperature on one scale, move straight over or straight up to the diagonal line on the graph, then move straight down or straight over to the other temperature scale.

Questions

"Room temperature" is often taken to be 68°F. From the relationship discussed, what is this on the Celsius scale? Does this answer agree with the one sketched in Figure 5.4?

Figure 5.5 shows several temperatures from everyday life expressed on all three scales, and Figure 5.6 indicates the range of temperatures found in the universe.

You may have noticed that there has not yet appeared a zingy, one-line definition of temperature. Undoubtedly, however, you now have a good understanding of what temperature is. That is, you know how to change the temperature of an object and how to mea-

sure temperature. This is an "operational definition," one that defines a property by describing its effects. An example of a somewhat oversimplified operational definition of temperature is that temperature is whatever you measure with a thermometer. As you will soon see, temperature is related to the internal energy state of the object being measured.

THERMAL ENERGY

Now that you know something about temperature and how to measure it, you can do the following thought activity. (It could easily

T (kelvins)

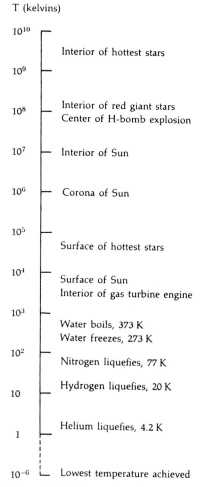

Figure 5.6 The approximate range of temperatures found in the universe. (From J. B. Marion, Physics and the Physical Universe, 3rd ed., John Wiley & Sons, 1980.)

be done as an actual activity.) For this activity, you would need a thermometer that measures in the range from the freezing to the boiling points of water, and that can be put into water. You would also need a styrofoam cup and an immersion heater of the type used to heat water in a coffee mug (or a pan of water and a kitchen range.)

Measure the temperature of the water. Then, with constant stirring, heat the water and measure its temperature periodically, say every minute or so, until the water boils. Then let the water cool and continue to measure the temperature periodically for a few minutes.

As you know, if you actually do this little experiment, the temperature of the water will rise as it is heated. When the water is not being heated it will cool. You also know that energy is involved. The power company expects to be paid for the energy used whenever an electrical appliance is used. (If you used a gas range for this, then the gas company will expect to be paid, also for energy provided.) The electric energy or energy of the gas is certainly not transferred to the water. Yet, the hot water must have more energy than the cold. It does not appear to be any of the types of energy we have discussed previously, such as potential or kinetic.

That is to say, the water does not go flying through the air as a result of being heated, nor does it get moved from one level to a higher one. The energy it gains is *in* the water itself. Thus, it is called *internal energy*. Another name for it is *thermal energy*, the energy associated with the temperature of an object. If thermal energy is added to the object, then its temperature rises; if it loses thermal energy, then its temperature falls.

To better understand internal, or thermal, energy, we must imagine that we can see the atoms and molecules of the substance we are examining. Every object has thermal energy; if we could examine the molecules, we would see that they are continuously in motion. This is true for solids, liquids, or gases. In a solid, the molecules remain in particular locations and their motion is essentially vibrational around an average position. In a liquid, the molecules can slip past one another, and their vibrational motions are added to motions within the liquid. In a gas, the distance between molecules is much greater than the diameter of a molecule, and molecules will bounce around all over the container. This is indicated in Figure 5.7. If the thermal energy of a substance is increased, the vibrational motion of the molecules also increases. That is to say, the kinetic energy of the molecules increases. Thermal energy, the internal energy of the object, is just the total kinetic energy of all of the molecules.

It should be understood that, in a given sample of material at a given temperature, a wide range of molecular velocities are present. It is the *average* molecular kinetic energy that is directly related to the temperature of the object. To see this, imagine putting a cold thermometer into some hot water. The molecules of the cold mercury in the thermometer have less kinetic energy, on the average, than the molecules of hot water. When the thermometer is immersed, the water molecules bang into it, transferring to it some of their energy. As the thermometer gains energy, its molecules move faster (it gets warmer) and, as a result of their increased motion, they tend to spread farther apart. That is, the material of the thermometer expands. As explained earlier, this means that a higher reading for temperature is observed. The hotter the water is, the greater the average kinetic energy of its molecules, and the more energy will be transferred to the thermometer, raising its reading. Thus, temperature is a measure of the average kinetic energy of the molecules.

Since we cannot see molecules, coming to this understanding of internal energy a century or so ago was difficult and required a different way of thinking about material objects. Because of this past difficulty, thermal energy was long thought to be different from mechanical energy and it has traditionally been measured in units called *calories* (cal), rather than in joules. Realizing that the two kinds of energy were simply different forms of the same thing and finding the exact relationship between the two units required years of painstaking effort. Because of our modern knowledge, we can cut that effort short, as in the following activity in which you will learn the definition of a calorie and will find its rela-

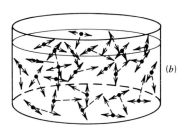

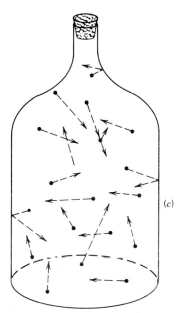

Figure 5.7 (*a*). In a solid, the molecules are tightly bound in place and vibrate about fixed positions. (*b*) In a liquid, the molecules are free to change position and to vibrate. (*c*) In a gas, the molecules are widely separated and move about in the container.

tionship to the joule. Although very similar to the preceding thought experiment, it is done quantitatively, (that is, with measurement of the relevant numbers).

Activity 5.2

For this activity you will need a thermometer, a styrofoam cup, an immersion heater, and a timer, such as a wristwatch.

Measure and pour an amount of water at about room temperature into a styrofoam cup (150 or 200 grams of water should fill the usual cup.) Suspend an immersion heater in the water so that it does not touch the styrofoam. Insert the thermometer in the water and note the temperature. Have the timer ready, and, keeping the thermometer in the water, plug in the immersion heater. (*CAUTION: Never* plug in the immersion heater unless it is immersed in water. Putting current through it while it is in the air will cause it to "burn up" in a most spectacular manner.) Stir the water continuously and vigorously and take readings of the temperature of the water at 30-s intervals until the temperature reaches 90°C, or so. Make a graph of the results.

Is your graph a linear one? Pick a time interval where the graph is linear, say 30 s somewhere in the middle of the graph. How much did the temperature of the water rise in that 30-s interval? A *calorie* is defined as the change in thermal energy required to raise the temperature of 1 g of water by 1°C. In calories, how much did the thermal energy of each gram of water increase in the 30-s interval? What was the increase in thermal energy of the entire 150 or 200 g?

Find the power output of the immersion heater; it should be stamped somewhere on the heater, and it will be in watts. How many joules does the heater supply in 1 s? In 30 seconds? Assuming that all of the energy produced in the heater goes into the water (not exactly correct, but a reasonable approximation), the energy that was produced by the heater must be equal to the increase in thermal energy of the water for the same time period. Using this fact, about how many joules do you find are equal to 1 cal?

In the preceding activity, you should, indeed, have discovered that the graph of temperature versus time is a straight line, like the one of Figure 5.8. The heater provides equal amounts of energy in each 30-s interval, so the thermal energy of the water increases by the same amount in each interval. Thus, following from Figure 5.8, the graph of temperature change (ΔT) versus the change of thermal energy (ΔTE) is a straight line, shown in Figure 5.9. In equation form, this is

$$\Delta TE = Cm\Delta T \qquad (5.1)$$

Here m is the mass of the material: twice as much water will require twice as much added thermal energy to change the temperature by the same amount. (If you are not persuaded completely that this is correct, try the experiment with twice as much water.) C is a constant, depending on the material and it is called the *specific heat* of the substance. Using the definition of the calorie given in Activity 5.2, the specific heat of water is

$$C_{water} = 1.000 \frac{cal}{g\ °C}$$

That is to say that water requires the addition of exactly 1 cal of thermal energy per gram to

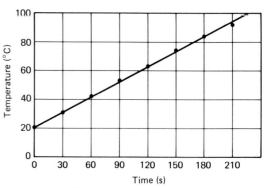

Figure 5.8 Warming water with an immersion heater. (mass = 200 g; heater rating = 300 W.)

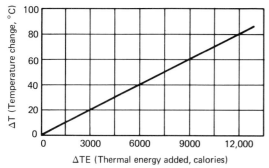

Figure 5.9 The plot of change of temperature of an object against the change of its thermal energy (internal energy) is a straight-line graph.

TABLE 5.1 *Specific Heats of Common Substances*

Material	Specific heat (cal/g°C)	(J/kg°C)
Air	0.17	711
Alcohol, methyl	0.600	2,510
Aluminum	0.214	895
Carbon dioxide	0.153	640
Concrete	0.16	669
Copper	0.0921	385
Gold	0.0316	132
Iron	0.107	448
Lead	0.031	130
Mercury	0.0332	139
Helium	0.749	3,134
Hydrogen	2.404	10,060
Oxygen	0.156	653
Rocks	0.2	840
Water	1.000	4,184
Wood	0.4	1,670

raise its temperature by 1°C. A different substance will require a different amount of thermal energy to raise the temperature of one gram a like amount. In other words, different substances have different specific heats. Table 5.1 gives the approximate specific heats of several common substances.

Note that, except for hydrogen, the specific heat of water is considerably greater than that for any other substance listed. As we shall see later, this is an important characteristic of water which can be exploited in some solar heating systems.

EXAMPLE

When you calculated the number of joules per calorie, did you get about four? What result do you get if you do this calculation from Figure 5.8, which is a graph of some student data?

Solution

Arbitrarily pick two points on the graph, say at 150 s and 30 s. The difference in these two is 120 s, and in that period of time, the heater produces an amount of energy given by $E =$

Pt. Since the rating of the heater used was 300 W, this becomes

$$E = Pt = 300 \text{ W} \times 120 \text{ s} = 36,000 \text{ J}$$

In the same period of time, the temperature increased from about 31°C to 74°C, so $\Delta T = 43°C$. Thus, the amount of thermal energy going into the water was

$$\Delta TE = Cm\Delta T = 1 \text{ cal/g°C} \times 200 \text{ g} \times 43°C$$
$$= 8600 \text{ cal}$$

Assuming that all of the energy of the heater goes into the water, this means that

$$8600 \text{ cal} \approx 36,000 \text{ J}$$

or

$$1 \text{ cal} \approx 4.2 \text{ J}$$

(Table 5.1 uses the accepted value of 4.184 J/cal.)

Very often, it is convenient to use *kilo-calories* (kcal) instead of calories. A kilocalorie is the amount of thermal energy that must be added to a *kilogram* of water to raise its temperature by 1°C. It is thus 1000 cal, and is sometimes referred to as a *Calorie* or a *large calorie*. This Calorie is the unit used by dieticians when describing the energy content of food, and is widely used elsewhere. To avoid confusion, we shall stick with *kilocalorie*.

EXAMPLE

A plumber heats 10 kg of lead in a pot. How much thermal energy must be added to the lead to raise its temperature 20°C? What is this in joules? What would the answer be if copper were being heated, rather than lead?

Solution

From Table 5.1, the specific heat of lead is 0.031 cal/g°C. Therefore

$$\Delta TE = Cm\Delta T$$
$$= 0.031 \text{ cal/g°C} \times 10^4 \text{ g} \times 20°C$$
$$= 6.2 \times 10^3 \text{ cal}$$
$$= 2.6 \times 10^4 \text{ J}$$

For copper, the calculation is the same, but with a different specific heat,

$$\Delta TE = 0.0921 \text{ cal/g°C} \times 10^4 \text{ g} \times 20°C$$
$$= 1.8 \times 10^4 \text{ cal}$$
$$= 7.7 \times 10^4 \text{ J}$$

Questions

In calories, how much thermal energy must be added to 2 kg of aluminum in order to raise its temperature 10°C? What is this in joules?

Many solar heating systems require that thermal energy be "stored" for use when the sun is not shining. Both water and rocks are frequently used as the medium for such storage. Suppose that a particular installation can accomodate one metric ton (1000 kg) of either rocks or water, and that the temperature of the medium can be raised from 20 to 50°C. How much thermal energy would be stored in the water? In the rocks? If the amount of thermal energy stored were the only consideration (it is not), which would be the better storage medium? How much better?

HEAT

So far we seem to be pussyfooting around the subject of "heat," getting close, but never quite there. In fact, what we define as thermal energy is just what is often called heat in everyday terminology. Strictly speaking, the two are not identical, but they are closely related. To get a handle on the concept of heat, imagine pouring boiling water into a cold coffee mug. The water cools as the mug warms, and it is reasonable to believe that they soon come to the same temperature. Careful measurements would bear this assumption out. In fact, when any two objects with different temperatures are placed in contact, internal energy is transferred from one to the other until they are at the same temperature, as indicated in Figure 5.10. It is this *flow* or *transfer* of energy that is called heat.

It one sticks to this definition, everyday use of the word *heat* is often not quite correct. For example, the term *the flow of heat* is redundant, because it means *the flow of the flow of thermal energy*. Also, when we talk about "adding heat" to an object, we really mean increasing its thermal energy. "Heating up a room" also means adding thermal energy. However, since these terms are in such common use, and there is little chance that their use will cause a serious misunderstanding, I will use them in this book when it seems appropriate. Besides, if you talk to your friends about adding thermal energy to a bowl of soup

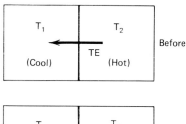

Figure 5.10 Heat is the transfer of thermal energy. T_3 is somewhere between T_1 and T_2.

to improve its taste, they will probably think you are a little strange.

THE FIRST LAW OF THERMODYNAMICS

That is a rather formidable title, but don't let it scare you, it came about as the result of people in the past not knowing the true nature of heat and thermal energy. It is really just another name for the law of conservation of energy. Now that we know about thermal (internal) energy, the law of conservation of energy has to be extended to include it. We can now say that, for an *isolated* system,

$$\Delta PE + \Delta KE + \Delta TE = 0$$

This equation simply says what we saw before in Chapter 3: In an isolated system the total energy remains constant. The only difference is that now we are including thermal energy as a form of energy. The energy may change from any one of the three forms into the others, but the total does not change. To make the law perfectly general, one might say

$$\Delta PE + \Delta KE + \Delta TE + \Delta A.O.K.E. = 0$$

Where *A.O.K.E.* stands for "any other kinds

of energy," which may include electric energy, nuclear energy, and even undiscovered kinds of energy. This may seem like cheating, for if we ever run into a situation in which total energy is not conserved, then it would seem possible just to invent another kind of energy to make up the difference. But that would not be good because, after all, a "law" of this type is a law only because we have never found an exception to it. If an exception ever occurred, the law would be disproved once and for all, and we should not make up new kinds of energy just to prevent this from happening. However, by having faith that the law works for all kinds of energy, physicists have been able to discover new particles and processes in situations where the law *seemed* to be violated. Their faith in the law made them look harder for reasonable explanations. I will not carry *A.O.K.E.* along any further, but I will continue to have faith that the law is valid for all kinds of energy.

You have seen that the thermal energy of an object can be increased by providing heat to it. To get some further insights into other processes that produce changes in thermal energy, try the following activities.

Activity 5.3

Rub your palms briskly back and forth together, pushing them tightly against one another. Describe the sensation you feel.

Hammer a nail into a two-by-four. Pull it out. Feel the nail. Is it hot?

Saw a piece of wood, the faster the better. Feel the saw. Is it hot?

Feel the hose leading from a bicycle pump and then pump up a flat tire with it as rapidly as you can. Is the hose warmer now than it was before pumping?

Stretch a heavy rubber band as far as you comfortably can and hold it stretched for a time, perhaps 30 seconds. Then, release the tension suddenly and hold the rubber band across your upper lip, a fairly sensitive thermometer. What can you say about the temperature of the rubber band?

Earlier you saw that heat flowing into or out of an object changes its thermal energy, and thus its temperature. In the activities, you can see that by *doing work* on an object you can increase its thermal energy. Further, in one case, the system under study did work on the outside environment: By letting the rubber band contract suddenly, you allowed it to do work on your fingers. Since this happened too suddenly for much heat to flow into the rubber band, it lost thermal energy and felt cooler on your lip. That is, thermal energy was converted to work.

These activities indicate that a complete description of the First Law of Thermodynamics must include work and heat. In other words, many systems are not isolated systems. Work may be done on them by something outside the system, or the system may do work on the outside. Heat may flow either into or out of the system. To take this into account the First Law of Thermodynamics can be written as

$$W + H = \Delta KE + \Delta PE + \Delta TE \qquad (5.2)$$

That is, the work (W) *done on* the system plus the heat (H) *put into* the system equals the change in total energy of the system. If the system does work on the outside, W is *negative*; if heat flows to the outside, H is *negative* in this equation. This is the complete version of the First Law of Thermodynamics (except for *A.O.K.E.*), a very powerful and far-reaching law.

Questions

When you pumped up the tire, the effect was to compress the air trapped in the pump. Since a force acted through a distance in pushing the pump handle down, work was done on the air. Did the kinetic energy of the system change? (For example, did the pump or the tire go flying through the air?) Did the potential energy of the system change? (What about the compressed air now in the tire? Can it do work?) What indication was there that some of the work put into the pump was converted to thermal energy?

THE SECOND LAW OF THERMODYNAMICS

Think again about pouring hot water into a cold mug. The water cools and the mug warms until they are the same temperature. Now imagine that you continue to watch the mug. At first, the temperature of the water continues to go down, more slowly than when it was first poured. Then, as you watch, the water temperature begins to rise. It gets hotter and hotter, soon reaching the boiling point, while the mug gets cooler and is quite cold to the touch, with frost forming on the outside.

That's absurd, isn't it? You can imagine the sequence of events just described, but can you imagine that it would actually occur? Your common sense and experience tell you that it could never happen. It is no more possible than your going to bed tonight and waking up yesterday morning, or that an egg might unscramble itself. Such events make good science fiction, but they *never* happen in the real world. That is why a movie run backwards looks so weird; it may portray events that we know from experience never happen.

There is the essence of the Second Law of Thermodynamics. One way of stating it is that, in an isolated system, heat will never flow from a cold region to a hotter one. The First Law of Thermodynamics does not cover this situation. For example, heat may flow from the cold mug into the hot water without violating the first law; the first law simply says that the total thermal energy does not change. So this new "law" is, as are the other laws we have discussed, a statement of what we see in everyday life. It is valid as long as no exception is found in nature. Thus far, none has been found, nor is such a discovery likely.

Note that the second law does not state that heat cannot *be made* to flow from cooler to hotter regions. An air conditioner is an obvious example of a machine that moves heat from a cool room to a hot outdoors. But to do so, work must be done by the air conditioner, using electric energy in the process.

Questions

Processes that can occur spontaneously in one direction only are called *irreversible* processes. The flowing of heat in an isolated system is such a process; it always goes from hot to cold, never in the opposite direction. Which of the following are irreversible processes and which are reversible?

You pull on a spring, stretching it. In the process you do work and potential energy is stored in the spring.

A stick of dynamite explodes.

You push a cart up a long hill.

You let the air out of your friend's tires.

You toss water out of a bucket onto the ground.

Entropy

Suppose you place two blocks of brass at different temperatures in contact, as shown in Figure 5.11. Then, as you already know, heat flows from the warmer to the cooler block. If we look at the situation for a short time—before any significant change takes place in the temperatures of the two blocks—a small amount of heat H flows from warmer to cooler. Now, let's define a new quantity, called *entropy*:

$$S = \frac{H}{T} \tag{5.3}$$

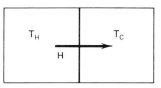

Figure 5.11 Heat flows from the hotter block, at temperature T_H, to the cooler at T_C.

The entropy, S, is heat divided by temperature. The temperature used must be expressed in an *absolute* temperature scale. In the SI system of units, the Kelvin scale is used.

An interesting thing happens to entropy in the process of heat transfer. For the blocks of Figure 5.11, an amount of heat H leaves the block on the left, taking with it an amount of entropy:

$$S_H = \frac{H}{T_H}$$

The block on the right receives the same amount of heat H but a different amount of entropy:

$$S_C = \frac{H}{T_C}$$

What is important, is the *change* in entropy of the system:

$$\Delta S = S_C - S_H$$
$$= \frac{H}{T_C} - \frac{H}{T_H}$$
$$= H\left(\frac{1}{T_C} - \frac{1}{T_H}\right)$$

Remember that T_H is greater than T_C; therefore $1/T_H$ is *less* than $1/T_C$. This means that, in the process of heat transfer, the entropy has increased, with more arriving at the cool block than left the hot block. That is

$$\Delta S > 0$$

We then have an isolated system: two blocks at different temperatures in contact with one another. The thermal energy (and the total energy) does not change, but the entropy does. Entropy is not conserved in this isolated system. Further, the change is such that the entropy of the system increases.

This happens to be an extremely important result; when all sorts of systems are analyzed carefully, the general law is discovered that entropy *never decreases* in an isolated system. It increases for all irreversible processes, and it stays the same for all reversible processes. There is no reason this result should be obvious to you, and it is quite difficult to prove. But it is important because it is a very general statement of The Second Law of Thermodynamics. In the example just given, the connection between the two different ways of stating the second law seems clear: Heat flows from hotter to cooler and the entropy increases. In the examples to come, the connection may not be as obvious, but it is there, nonetheless.

One way to think about entropy is to consider it to be a measure of how well ordered a system is. That is to say, it tells us something about how well the parts of the system are neatly arranged. When the two brass blocks are at different temperatures, there is a fair amount of orderliness; the faster (on the average) molecules are in the hotter block, and the slower ones are in the cooler block. After the blocks have been in contact for a while,

they are at the same temperature, and the molecules have the same average energies. So, in a sense, there has been a mixing process, and the faster molecules are no longer confined to one region. We say that the entropy of the "neat" system is lower and the entropy of the "messed up" system is higher, so in the process the entropy has increased.

Let's take a look at some other systems that undergo change. For example, imagine a room from which all of the air has been pumped into an attached box, as in Figure 5.12a. This is a well-ordered system, with all of the air in one small region, and it has a relatively low entropy. Then the cover holding the air in is popped off by the pressure, and the air rushes into the room, as indicated in Figure 5.12b. The entropy is larger than before because the system now has less order. This process is an irreversible one: We have to pump the air into the box; it would never go there of its own accord.

A good system to look at is the Earth. Someone once called it Spaceship Earth—a good reminder that many of the resources on Earth can run out and cannot be replaced. Millions of processes continually occurring on Earth increase entropy. They include the burning of fuels, the running of factories, and even the eating of food. The bad news is that isolated systems whose entropy continually increases eventually reach a state of maximum entropy, where no further change can occur. This is called heat death, for then everything enclosed within the system is at the same temperature, and transfers of thermal energy are no longer possible. The good news is that the Earth is not really an isolated system. It loses energy into space, and it gains energy from the sun. On average, the energy gained is equal to the energy lost, and the Earth remains stable at about the same average temperature. The same is not true for entropy. The energy coming from the sun—sometimes called high-grade energy—is at a relatively high temperature, and it therefore carries relatively little entropy with it

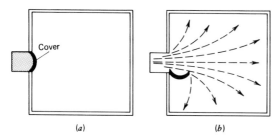

Figure 5.12 (*a*). All of the air that was in the room has been pumped into the small box. (*b*). The cover of the box blows open and air rushes into the room.

($S = H/T$). The energy leaving the Earth—called low-grade energy—is at a lower temperature and has greater entropy associated with it. Thus, the Earth loses entropy to space faster than it gains it from the sun. In fact, the rate of net entropy loss is just great enough to get rid of the excess entropy generated on the Earth. The energy balance of the Earth is indicated in Figure 5.13.

The energy/entropy balance of the Earth is important for our continued functioning. We are approaching a time when society's activities may have noticeable effects on this balance, with possible serious consequences.

HEAT ENGINES

An *engine* is defined as a machine that converts energy into mechanical work. A *heat engine* is a machine that converts thermal energy into work.

Thought Activity

Suppose you plug up the hose of a bicycle pump, with the handle about halfway up, so that no air can flow in or out. Now, if you put the bottom part of the pump into very hot water, as in Figure 5.14, you will heat the trapped air. The air will expand, pushing the handle upward and doing work on it. Then, if you put the pump into cold water, the trapped air will contract and the pump handle will move back down. This could be repeated over and over, first in the hot water, then the cold, with work being done on the pump handle each time the trapped air is heated.

Try this as a real activity, if you wish. However, I found that, even with a well-oiled and sealed pump, the effect is small, and the experiment has to be done with care. Nonetheless, if you measure carefully, the effects described are definitely there.

This thought activity describes a very crude heat engine. It goes through repeated cycles, doing some work in part of each cycle. You might even imagine hooking it up to some levers and wheels to do useful work. Many

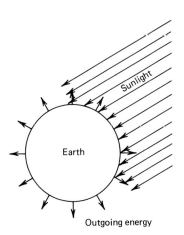

Figure 5.13 The energy balance of the Earth. Outgoing energy is equal to the energy coming in from the sun, but outgoing entropy is greater than incoming entropy.

Figure 5.14 When you immerse the lower part of a bicycle pump alternately into hot and cold water, the handle moves up and down, respectively.

heat engines, including automobile engines and the giant turbines that run generators to produce our electricity, operate on the same general principles. One important principle is that some heat must always be thrown away when a heat engine does work. With the bicycle pump, the handle would stay in its uppermost position and no further work could be done if you did not cool it off. You cool it by placing it in the cold water, losing heat into the cold water in the process. The same thing is true for every heat engine, no matter how sophisticated: Heat must be discarded in the process of going through a complete cycle. Since energy is unavoidably lost in each cycle, no heat engine could be 100 percent efficient, even if there were no losses to friction.

Figure 5.15 shows the energy flow in a perfect (ideal) heat engine, that is, an engine that has no frictional losses and that has the highest possible efficiency. This is called a *Carnot engine*, and we can only approximate it in the real world. Heat comes from a *reservoir* at a high temperature T_H. The reservoir ensures that a continuous supply of heat is available. In the real world, the reservoir might be a furnace in which coal is burned to supply energy. In a given time, an amount of (high-grade) heat H_H flows into the engine, carrying entropy $S_H = H_H/T_H$. In that same time, waste (low-grade) heat H_C flows into a cold reservoir, along with entropy $S_C = H_C/T_C$. The engine does an amount of work W.

There are several things to note about this process. First, the engine may go through one or several cycles in the given time, arriving back at its *original state*. This means its kinetic energy, potential energy, and thermal energy are all the same as they were at the beginning. The First Law of Thermodynamics tells us that the work done is equal to the heat lost in the process. That is:

$$W = H_H - H_C$$

Next, the Carnot engine cycle is a reversible

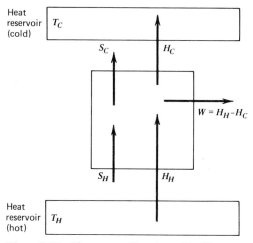

Figure 5.15   The energy flow in an ideal heat engine.

process; it could be run "backwards," with work being supplied to the engine and heat being transferred from the cold reservoir to the hot one (like a refrigerator). Since the process is reversible, there will be no change in entropy during a cycle. This means that the entropy decrease in the hot reservoir is equal to the entropy increase in the cold reservoir:

$$S_H = S_C$$

or

$$\frac{H_H}{T_H} = \frac{H_C}{T_C}$$

Finally, let us go back to the definition of efficiency and use it for this engine:

$$Eff = \frac{\text{useful work out}}{\text{energy in}} \times 100\%$$

$$= \frac{W}{H_H} \times 100\%$$

Putting all of this together, we can find the efficiency of the engine in terms of the temperatures of the two reservoirs. First, substitute for work from above:

$$Eff = \frac{H_H - H_C}{H_H} \times 100\%$$

$$= \left(\frac{H_H}{H_H} - \frac{H_C}{H_H}\right) \times 100\%$$

$$= \left(1 - \frac{H_C}{H_H}\right) \times 100\%$$

Then, rearranging the entropy equation, we get $H_C/H_H = T_C/T_H$, and substituting this gives

$$Eff = \left(1 - \frac{T_C}{T_H}\right) \times 100\%$$

or, equivalently,

$$Eff = \frac{T_H - T_C}{T_H} \times 100\% \qquad (5.4)$$

This is called the *Carnot efficiency*, and is the efficiency of an ideal, perfect heat engine. Remember that because of the way entropy was defined, the temperatures used here *must* be expressed in an absolute temperature scale. (Either Kelvin or Rankine scales may be used. See Problem 2 at the end of this chapter.)

There are three extremely important points to note. First, even the ideal, perfect Carnot engine will never run at 100 percent efficiency. Second, real heat engines are not reversible because they lose energy to friction. Therefore, the efficiency of a real engine will always be less than the efficiency of the ideal Carnot engine, given similar operating temperatures. Finally, at a given high temperature, the efficiency of a heat engine increases when the difference in the temperatures of the hot and cold reservoirs increases.

EXAMPLE

A Carnot engine is operated with a hot reservoir at 1000°F (538°C) and a cold reservoir at 0°F (-18°C). What is its efficiency in converting heat to work? If 10^6 J of useful work are produced, how much heat is used? How much waste heat is discarded?

Solution

First remember that the temperatures needed to calculate efficiency are absolute temperatures. Thus

$$T_H = 538°C + 273°C = 811 \text{ K}$$
$$T_C = -18°C + 273°C = 255 \text{ K}$$

Then

$$Eff = \frac{T_H - T_C}{T_H} \times 100\%$$

$$= \frac{811 - 255}{811} \times 100\% = 69\%$$

To find the required heat input, use

$$Eff = \frac{W}{H_H} \times 100\%$$

Then

$$H_H = \frac{W}{Eff} \times 100\% = \frac{10^6 \text{ J}}{69\%} \times 100\%$$
$$= 1.4 \times 10^6 \text{ J}$$
$$\approx 3.5 \times 10^5 \text{ cal}$$

and the waste heat discarded is

$$H_C = 0.31 H_H = 0.31 \times 1.4 \times 10^6 \text{ J}$$
$$= 4.3 \times 10^5 \text{ J}$$
$$\approx 1.0 \times 10^5 \text{ cal}$$

Questions

Suppose that a heat engine were operating between two heat reservoirs, one at the boiling point of water and the other at the freezing point. What are T_H and T_C? (Remember that these must be Kelvin temperatures.) Show that the maximum theoretical efficiency of the engine, assuming no frictional losses, is only about 27%. What could be done to increase the efficiency?

In a modern, coal-fired power plant, the hot "reservoir" is a supply of high-pressure steam at temperatures as high as 500°C. The cold reservoir will typically be a river at a temperature of 5 to 25°C. Assuming that 5°C is the wintertime temperature and 25°C the summertime temperature of the river, what is the difference in efficiency in the two seasons? (Incidentally, real power plants are considerably less efficient, with the most modern running with an efficiency of a little better than 40 percent.)

In a 1000 MW plant that is 30 percent efficient, how much "waste" heat is thrown away in the course of a day? Can you think of any ways to use this low-grade energy rather than just dumping it?

SUMMARY

Although difficult to define, temperature is a property that is easy to understand in an operational sense. In this chapter, you learned that the temperature of an object is a measure of the average kinetic energy of its molecules. Internal energy—sometimes called thermal energy—is the sum of the kinetic energies of all of the molecules. When the thermal energy of an object is changed, its temperature also changes according to the equation $\Delta TE = Cm\Delta T$. Heat is the transfer of thermal, or internal, energy from one object or region to another.

The law of conservation of energy, extended to include all kinds of energy, is called the First Law of Thermodynamics. The Second Law of Thermodynamics tells us that, in an isolated system, irreversible processes can proceed spontaneously in only one direction. That is, heat flows from hot to cold or, equiv-

alently, the entropy of an isolated system never decreases. These concepts lead to the formulation of an equation for the efficiency of an ideal (Carnot) heat engine which requires knowledge of only the temperatures of the hot and cold heat reservoirs. A real heat engine will always have an efficiency that is less than that of a Carnot engine operating under the same conditions of temperature.

ADDITIONAL QUESTIONS AND PROBLEMS

1. Define temperature. (Hard, isn't it? Try it in the form of an "operational" definition: What effects does a change of temperature produce?)

2. The Rankine temperature scale is an absolute scale, like the Kelvin scale, but it uses units the same size as Fahrenheit degrees. That is, its zero is absolute zero, but with degrees only 5/9 the size of kelvins. The freezing point of water is at about 492°R and the boiling point is 672°R. Since it is an absolute scale, Rankine temperatures can be used for calculating the efficiencies of heat engines. (Just add 460° to the Fahrenheit temperature to get the Rankine temperature.) Make a graph, like that of Figure 5.4, for converting Kelvin temperatures to Rankine and Rankine to Kelvin. What is room temperature (about 68°F) in the Rankine scale? In kelvins?

3. Describe the flow of thermal energy (heat) for each of the following:

    a. A warm house in the cold wintertime air.

    b. The same house, air-conditioned, on a hot summer day.

    c. A refrigerator (not running) in a warm kitchen.

    d. The same refrigerator while running.

    e. A soft drink that has just been poured into a glass full of ice cubes.

f. A water-cooled automobile engine that is running.

g. A pan of soup that is being heated on the stove.

4. Running a mile "burns up" roughly 100 dietary Calories (100 kcal). What is this in joules? If you run the mile in 6 min, about how much power do you generate? Since 1 horsepower is about 750 W, what is this in horsepower? What happens to the energy you produce? (*Hint:* Besides being tired, what other changes do you feel after running the mile?)

5. Suppose you swim a mile in 30 min and, in the process, "burn up" about 500 kcal provided by the food you have eaten. The pool is 10 m wide, 20 m long, and 2 m deep. Assuming that all of the energy you have used is turned to heat by your body, how much is the temperature of the pool raised by your exertions? (Remember that 1 m^3 of water has a mass of 1000 kg.)

6. Physics students sometimes do strange things. For example, the other day a student dropped a chunk of hot aluminum with a mass of 100 g into a container containing 200 g of water. The water was originally at a temperature of 20°C and, after the aluminum was dropped in and the water stirred, it was at 41°C. How much heat was added to the water? Where did it come from? What is the final temperature of the aluminum in the water? Thus, what was the initial temperature of the aluminum?

7. Give some examples of reversible and irreversible processes.

8.a. People continue to try to invent "perpetual motion machines," which will produce useful work with no input of work or energy. Explain why the First Law of Thermodynamics prohibits the possibility of such a machine existing.

b. Another kind of machine, sometimes called a "perpetual motion machine of the second kind," uses the same reservoir for both the hot and cold sides. For example, a power plant might scoop up water from a river, extract the heat from the water and convert it to electric power, and dump the water back into the river. According to the First Law of Thermodynamics, this could be possible, for no energy is created; energy from the river is just converted to electric energy. Show that the Second Law of Thermodynamics prohibits the possibility of this machine. (*Hint:* The easiest way is to use the expression for the efficiency of a heat engine, which is a result of the second law. In this case, T_H and T_C are the same.)

9. A nuclear power plant uses steam at a temperature of 500°F to run the turbines and gives off waste heat to a river whose water is at 50°F. What efficiency would the plant have if it were a perfect heat engine? How does this compare to the actual efficiency of about 30 percent for such plants? What are some factors that could account for the difference in the ideal and the actual efficiencies?

10. In a nuclear power plant that is 33 percent efficient and that produces 1000 MW of electric power, what is the rate at which "waste" heat must be disposed?

6 *Home Energy Conservation*

In the 1970s more than one-fourth of all energy use in the United States went for space heating and cooling and refrigeration. Better than one half of that was for home heating and cooling. Much of the energy that was used could have been saved had there been a great enough incentive to do so. Now we, as a nation, are becoming increasingly aware that energy is a valuable commodity which should not be squandered mindlessly. Furthermore, this point is being brought home to us by the fact that the price of energy is already much higher than it was a decade ago and it is still rising (despite occasional temporary decreases). Home heating and cooling will become even more expensive than they are now, unless we learn of ways to keep comfortable with less energy. Those ways are available to us and, intelligently applied, they could have a tremendous impact on the energy future of the United States. In this chapter we will concentrate on ways to reduce the amount of home heating and cooling needed by making homes more energy efficient. The principles learned here can be applied to almost any kind of space heating and cooling problems.

Earlier in this book, we discussed conservation of energy. The law of conservation of energy states that, in an isolated system, the total energy remains the same. The popular use of the term *conservation of energy* has a very different meaning. It refers to efforts to "use up" as little energy as possible to do needed tasks. For example, home energy conservation can mean using a more efficient furnace or air conditioner, insulating walls, ceilings, and floors, and sealing up air leaks so that less fuel is needed to heat a home comfortably. When we "lose" energy, that energy still exists, and the total energy is still the same. It is just no longer available to us in usable form. For example, the thermal energy that escapes from a house warms up the outdoors a negligible amount; we cannot benefit from it, but it is there. Or, if an electric motor runs without doing any useful work, the energy it uses is turned to heat. That energy can no longer be used to do work, but it does heat up the room slightly.

Before we discuss ways of conserving energy we need to have some idea of what we are up against. The first sections of this chapter will help you to understand the energy needs of a home and ways to keep those needs as low as possible.

DEGREE-DAYS

One commonly used measure of how much heat will be needed in any given locality is the *degree-day*. A *heating degree-day* is based on the assumption that, when the outdoor temperature is 65°F, it will be comfortable indoors. For temperatures lower than 65°F heat must be supplied for comfort. Of course, 65°F is not a comfortable temperature. It's use is based on the knowledge that human bodies, electric lights, appliances, dogs, cats, and so forth, all give off heat, thus raising the temperature indoors a few degrees higher than that outdoors, even without heating.

If the average outdoor temperature for a day is 65°F (or higher), there are *zero* degree-days on that day. If the average is 64°F for a day, there is 1 degree-day. If the average is 55°F for a day, there are 10 degree-days, and so on. The total number of degree-days for several days is found by adding up the number of degree-days for each day. For example, the number of heating degree-days for a week may be found as shown in the following table

Day	Average temperature (°F)	Degree-days
Sunday	60	5
Monday	65	0
Tuesday	68	0
Wednesday	64	1
Thursday	59	6
Friday	52	13
Saturday	46	19
Total		44

Figure 6.1 shows the average number of degree-days for the month of January in the United States. The figures come from measurements made during the period from 1931 to 1960. Each line on the map connects regions with about the same number of degree-days. For example, Albuquerque, New Mexico; Dodge City, Kansas; St. Louis, Missouri; and Pittsburgh, Pennsylvania, are all near the 1000 degree-day line. Fort Worth, Texas, and Birmingham, Alabama, are considerably warmer, with 600 heating degree-days for January, and Fargo, North Dakota, is very cold with 1800. That means that, all other things being equal, it will take about three times as much fuel to stay warm in Fargo as it will in Fort Worth. Figure 6.2 is a map of the number of heating degree-days in the United States for a full year.

Cooling degree-days are defined in a similar manner. The base is often taken to be 75°F, and temperatures above this are counted. That is, it is assumed that when the outdoor temperature is 75°F or lower, it is reasonably comfortable indoors with no cooling. Then, a day with an average temperature of 80°F will generate 5 cooling degree-days. Figure 6.3 is a map showing the annual number of cooling degree-days for the United States. As we might expect, this map shows a reverse situation from that shown on the heating degree-day map, with the larger numbers being farther south. (Strangely enough, some government agencies take 65°F as the base for cooling degree-days. This does not seem reasonable, since 65°F is far below the temperature needed for comfort cooling; we shall stick with 75°F.)

One caution is in order: The degree-day maps are only approximate and they represent past experience. They cannot show local variations, and they cannot predict what will happen in any given year. In some cases, two localities that are quite close together will have significantly different weather. Also, as you know, some years have strange weather, and no-one knows when we will have a year like that. Thus, these maps are just a guide, and more detailed local information would be needed for the design and installation of heating or cooling systems.

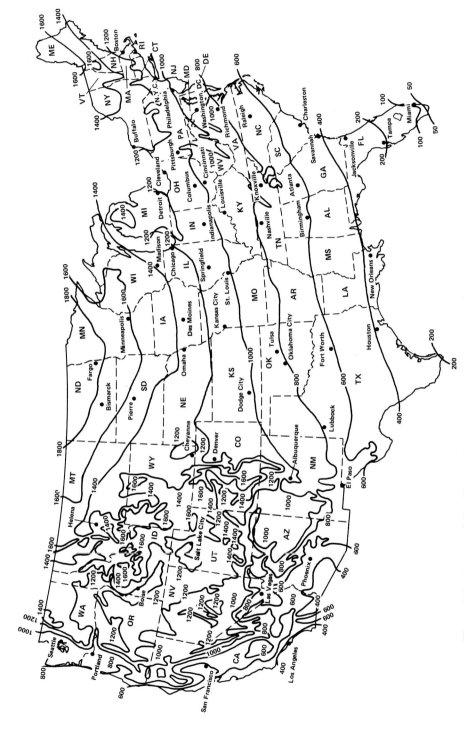

Figure 6.1 A map of the United States, showing the approximate number of heating degree-days in January. Locations on the same line have the same number of degree-days. (Adapted from *Climatic Atlas of the United States*, U.S. Department of Commerce.)

94

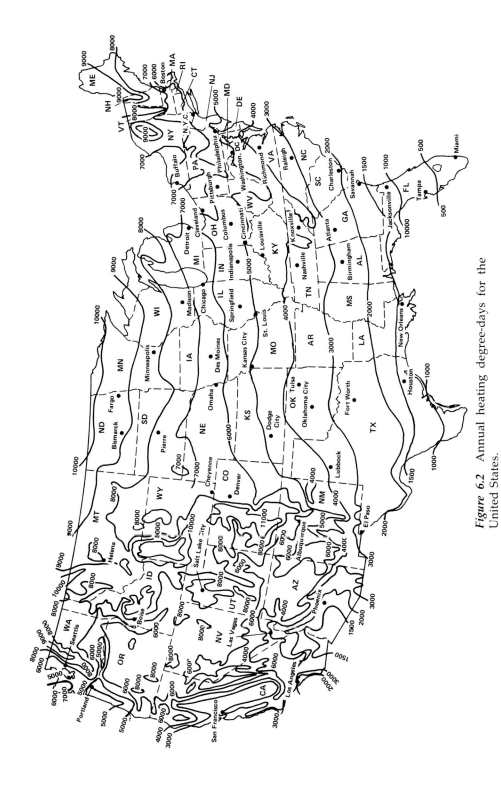

Figure 6.2 Annual heating degree-days for the United States.

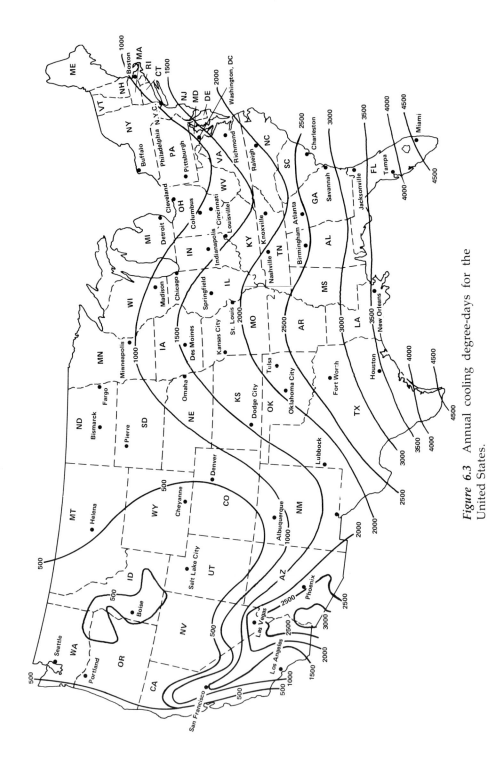

Figure 6.3 Annual cooling degree-days for the United States.

Questions

What is the approximate number of heating degree-days for a year at Tulsa, Oklahoma? At Cheyenne, Wyoming? At San Francisco, California? At Seattle, Washington? At Boston, Massachusetts?

What are some cities that have about 2000 cooling degree-days annually?

With all other factors being equal (and they never are), how much more would it cost to heat a home in Boston than one in Tampa? How would the costs of cooling the two homes compare?

Assuming that fuel or electricity costs to keep a home comfortable are the same for a heating degree-day as for a cooling degree-day (they generally are not) and that they are the same all over the country, what sections of the country would have the lowest annual energy costs for heating and cooling combined? The highest?

HEAT TRANSFER

Before now, we have discussed heat, the transfer of thermal energy, without saying anything about *how* that transfer occurs. In what follows, we discuss ways in which heat flows. Quite frequently, the discussion is about heat losses from homes. However, you should keep in mind that when it is hot outdoors and homes are being cooled, *heat gains* occur. Everything said about heat losses also applies to heat gains. Both refer to a transfer of thermal energy and only the direction—between outdoors and indoors—is reversed.

There are only three ways in which heat can flow; they are called *conduction, convection,* and *radiation.* Conduction and convection are the main ways in which heat is lost from buildings. Radiation is the way energy reaches us from the sun, and it may be more important to our comfort indoors than we have tended to realize in the past.

Activity 6.1

Turn an electric range heating element on at its highest setting. When it gets red hot, rest the bowl of an old spoon on it, as illustrated on the left in Figure 6.4, while holding the handle. After a short time, you will feel the handle of the spoon getting hot. The heat reaches your hand by the process of *conduction* through the spoon.

Now place one hand about 3 feet above the heating element. Move it from side to side at that same height. Directly above the heating element the air is much warmer than it is to the side. The heat is reaching your hand by *convection*; it is being carried by air rising from the region just above the heating element.

Finally, place one hand near the range top 8 or 10 in. to the side of the element, and you will feel intense heat. This heat is not being carried by moving air; that air is going upward. This heat transfer is by *radiation*, directly through the space between the element

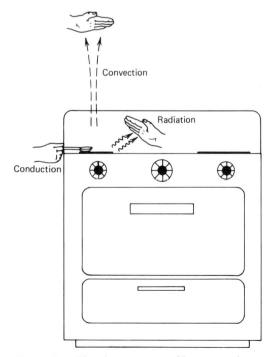

Figure 6.4 The three means of heat transfer.

and your hand. There will be more about radiation in the next chapter.

Convection

Convection is the transfer of thermal energy from one location to another by means of a moving fluid. The fluid may be a liquid, like water, or a gas, like air. A volume of the fluid moves from one location to another, carrying its internal energy along with it. For example, a cold room will be "heated" if warm air is introduced to replace the cold air. The following activity shows you how natural convection currents occur.

Activity 6.2

Place a pan of cold water on a hot range element so that just one side is on the heating element, as indicated in Figure 6.5. Wait a couple of minutes for the sloshing motion of the water to stop completely. Then drop several tiny, moistened balls of facial tissue in on the side of the pan over the heating element. Watch until the water on that side of the pan starts to get hot, and you will see the tiny pieces of tissue move. Observe them until you see a definite, repeating pattern to the movement.

In this last activity, *convection currents*, were shown rather crudely by the pieces of facial tissue floating around in the water. As the water over the hot element warms up, it expands slightly. This warmer water thus has a somewhat lower density than the cold water, and it tends to rise to the top of the pan. When it reaches the surface, it gets pushed to the side by more warm water rising, and it is also cooled by the surrounding cooler water. It then sinks to the bottom and is pulled into the heated area to replace the newly warmed water that is rising, and the whole cycle starts again. The result is the circulating current you observed by means of the floating pieces of tissue. This current carries (thermal) energy from the water near the bottom of the pan to the surface.

Convection currents also exist in air, and they are one of the main reasons warm air circulates around in a room. Figure 6.6 shows the circulation caused by convection currents in a room with two heat sources but with no fans or blowers. The warm air rises near the walls, drifts to the center of the ceiling, and falls near the center of the room as it cools. Then it goes back to the walls and the process starts all over again. You can follow such currents for short distances by watching the movement of smoke in a room. These air cur-

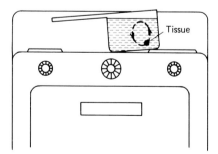

Figure 6.5   Observing convection currents in water.

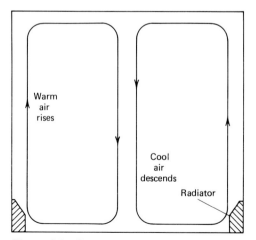

Figure 6.6   Convection currents in a room.

Figure 6.7 When strongly heated, the molecules in the bowl of a spoon vibrate furiously.

rents carry warm air to the center of the room. Convection currents in a room behave very much like the air currents that produce winds outdoors. As you learned earlier, they are caused by air in one region being warmer, and thus less dense, than the air in neighboring regions. The lighter air rises, thus starting a convection current on a large scale.

Conduction

Heat conduction is the transfer of thermal energy through a material. When you placed the bowl of the spoon on the hot range element, it became very hot, and its molecules vibrated furiously, as indicated in Figure 6.7. As they did so, they hit neighboring molecules, transferring energy to them. Those neighboring molecules then became hot, and they hit their neighbors, passing the energy on, and so it goes to the end of the handle. It is something like a long line of dominoes, as shown in Figure 6.8. By giving a little push to the domino on the end, all of them can be made to fall. Each domino transfers energy to the one next

to it. Thus energy is transferred down the line, but no domino changes its place in line. In the spoon, the process of conduction transfers internal energy from molecule to molecule along the spoon, even though the molecules themselves do not move the length of the spoon.

To simplify the discussion, let's switch from talking about heat conduction in a spoon to heat conduction through a slab of material, such as that in Figure 6.9. The piece of material might be part of a wall. The Second Law of Thermodynamics tells us that heat will naturally flow from the warmer side to the cooler. The rate of heat flow depends on the properties of the material and the difference in temperature of the two sides. Notice that I said *rate*. If the process goes on for an hour, a certain amount of thermal energy will go through the wall. In two hours, twice as much energy will pass through, and in three hours, three times as much energy will pass through. Thus we are interested in *power*, the energy transfer per unit time and we can write an equation for the power through the wall:

$$P = \frac{H}{t} = K\Delta T \tag{6.1}$$

Figure 6.8 Energy is transferred along the line of dominoes, but no domino leaves its position in line. (Leif Skoogfors/Woodfin Comp.)

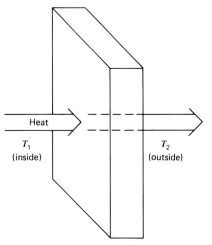

Figure 6.9 The flow of heat through a slab of material.

Where K is called the *thermal conductance* of the slab, and ΔT is the difference in temperature between the two sides. Note that we are using the symbol H for heat, the thermal energy that flows through the wall. The conductance, K, is a constant that includes features of the slab, such as its area and its thickness. It also includes a property of the material called *thermal conductivity*. The conductivity of a material tells how well it conducts heat. Brass has a much higher conductivity than does wood, so a piece of brass will conduct much more heat than a piece of wood under the same conditions. The following activity illustrates this point.

Activity 6.3

Get several cups or glasses of about the same size, but made of different materials. For example, one might be glass, another aluminum, a third styrofoam, and perhaps two cups can be put together to make a double thickness of styrofoam. Put some water on to boil. Make a cover for each cup from heavy cardboard, or some such material, with a hole for insertion of a thermometer. Fill the smallest cup with cold water to within $\frac{1}{2}$ in. of the top and mark the level of the water inside the cup. Measure the amount of water in the cup in milliliters. (One ml of water has a mass of 1 g.) Use the same amount of water in each of the cups, marking the level in each. The following may be done one cup at a time or all at once, depending on how many thermometers you have available. Use thermometers that will measure up to 100°C.

Fill a cup to your mark with boiling water and let it stand for a minute to heat the cup. Then, pour out the hot water and immediately fill the cup again with boiling water. Take the temperature of the water at intervals: At first the intervals will have to be close together, say every 10 s, then as the rate of cooling slows, they may be farther apart. Stir the water continously with the thermometer to keep it the same temperature throughout.

When you have done this for all of the cups, make a cooling curve for each cup, with all of the curves on the same piece of graph paper. A cooling curve is just a graph of temperature versus time as the water cools. How do the curves compare? Which cup has the highest thermal conductance? Which has the lowest?

Do the following just for the cups with the highest and lowest conductances. Pick a small section of the curve, somewhere near the high end. What period of time does that section of the curve represent? What is the initial temperature of the water in that time period? What is the final temperature? Approximately what is the average temperature for this period? (Halfway between the two extremes.) What is the difference between that average water temperature for that period and the room temperature outside the cup? How many grams of water were in the cup? Using $\Delta TE = Cm\Delta T$, find the thermal energy in calories lost by the water in the time interval corresponding to that section of the cooling curve. This lost thermal energy is equal to H, the heat through the cup. Now, divide H by the time to find the average power loss, P, in cal/s. Finally, using this value for P and the equation $P = K\Delta T$, find the thermal conductance, K, of the cup. What units do you get for conductance? How do the conductances of the two cups compare?

(Note that you have to use *two different* ΔTs. In the expression $\Delta TE = Cm\Delta T$, ΔT refers to the *change* in temperature of the water over a time span. In $P = K\Delta T$, ΔT refers to the *difference* in temperatures between the inside and outside of the cup. This can lead to confusion, but I do not think you will run into such a situation again.)

As you saw from this activity, different materials have different conductances, and the thicker a particular material is, the lower will be its conductance. A house wall will usually be made of several different materials, each with its own conductance, as shown in Figure 6.10. (Figure 6.11 shows a worker installing insulation in a new house to form one of the wall layers, one with a low conductance.) When the wall is composed of several different layers, one needs to know the overall conductance of the wall. If the individual conductances are known, that overall conductance turns out to be

$$\frac{1}{K} = \frac{1}{K_1} + \frac{1}{K_2} + \frac{1}{K_3} + \frac{1}{K_4} + \ldots$$

This is a somewhat clumsy expression to work with, and it is helpful to define a *thermal resistance*, $1/K$. Whereas the thermal conductance is a measure of how well heat is conducted by an object, the thermal resistance measures how well the object resists the passage of heat. A material with a high conductance is a good conductor of heat; a material with a high resistance is a good *insulator*. When several layers are put together in a wall, the overall resistance of the wall is just the sum of the individual resistances of the var-

Figure 6.11 A workman installing insulation into the walls of a new home. The insulating material comes in rolls, like that shown in the foreground. (DOE Photo.)

ious layers. That is, the total thermal resistance is

$$R = R_1 + R_2 + R_3 + R_4 + \ldots$$

R-VALUE

What is this *R-value* business, anyway? It seems, these days, that everyone is talking about it but few understand it. It is not really all that mysterious, and it is closely related to the thermal resistance already discussed. However, because of current practices in American engineering, we shall have to define a different energy unit before we can define *R*-value.

Back in the days when the sun never set on the British Empire, most of the civilized world used British Engineering Units, or some version of them. Now, the United States is virtually alone in using them regularly; even the British have gone metric. It may be some years before we Americans convert completely to a metric system, so you shall still have to learn some of the old units. You have obviously already done so, and in fact are probably more comfortable with some of

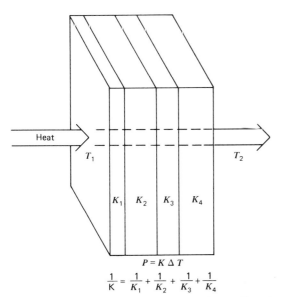

Figure 6.10 The flow of heat through a wall with several layers.

them—pounds for force, Fahrenheit degrees for temperature, and feet for length. Undoubtedly, you are not quite as familiar with slugs for mass or foot-pounds for mechanical energy.

The only new unit you need to know just now is the unit for thermal energy, the *British thermal unit*. You may not have heard that term before, but you certainly have seen its abbreviation—*Btu*. A Btu is the amount of thermal energy that must be added to a *pound* of water to raise its temperature one degree, *Fahrenheit*. This definition is similar to the definition of the calorie. As a result, the specific heat of water is 1.000 in either system of units, which means that the specific heat of any material will have the same numerical value in the two systems of units. For example, the specific heat of copper is 0.092 cal/g°C or 0.092 Btu/lb°F.

This unit of energy can be compared to the units you are now more familiar with:

1 Btu = 1054 J

1 Btu = 252 cal = 0.252 kcal

1 Btu = 2.93 × $^{-4}$ kWh

A related unit of power which is used in discussing heating and cooling is the Btu per hour (Btu/hr). It is sometimes abbreviated as Btuh, but we shall stick with Btu/hr to avoid confusion. When someone says that an air conditioner is "a 6000-Btu unit," that means that the unit is rated to be able to remove 6000 *Btus per hour* of heat from a room and dump it to the outdoors. One Btu/hr is about 0.29 W. Another unit of power, often used when dealing with large-scale energy consumption, is 1 million Btus per year (MBtu/yr).

EXAMPLE

In the olden days, when I was a lad, irons for pressing clothes were made of iron and they were placed on the stove to be heated. If a 5-lb iron were heated from 75 to 375°F, how much thermal energy would it gain? If the heating process takes 5 min, what is the average power input to the iron? How does this compare to a modern electric iron rated at 1100 W?

Solution

The gain in thermal energy is given, as before, by

$$\Delta TE = Cm\Delta T$$

Only, in this case, the specific heat is in Btu/lb°F; m stands for the weight in pounds; and ΔT is in °F. Thus, the thermal energy added is

$$\Delta TE = 0.107 \; \frac{Btu}{lb \; °F} \times 5 \; lb \times 300°F$$
$$= 161 \; Btu$$

If the heating takes place in 5 min ($\frac{1}{12}$ hr), then the rate of heating (power input) is

$$P = \frac{\Delta TE}{t} = \frac{161 \; Btu}{\frac{1}{12} \; hr}$$
$$= 1930 \; Btu/hr$$

To put this into watts use 1 Btu = 1054 J:

$$P = 1930 \; Btu/hr = \frac{1930 \times 1054 \; J}{3600 \; s}$$
$$= 565 \; W$$

So heat is added to this iron at about one-half the rate that heat is added to the modern iron.

With this background, we are finally ready to define *R-value*. It is simply the thermal resistance of 1 ft^2 of material, as indicated in Figure 6.12. When using *R-value*, heat losses

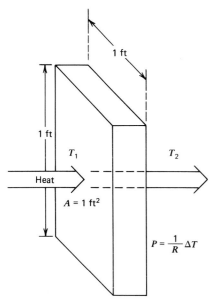

Figure 6.12 *R*-value is the thermal resistance of 1 square foot of material.

are expressed in Btus per hour and temperatures in degrees, Fahrenheit. Thus, the rather strange units for *R*-value are ft² °F/Btu/hr. (Read this as "square feet times degrees, Fahrenheit, per Btu per hour." The reciprocal of *R*-value has units that are easier to understand. They are Btu per hour, per square foot, and per degree, Fahrenheit. You will see more of this in the next section.) Note that, the greater the *R*-value of a material, the greater its resistance to the flow of thermal energy. In other words, a material with a high *R*-value is a good insulator. Table 6.1 shows the *R*-values of some common building materials.

Note that these values are approximate and they will vary considerably, depending upon the exact material, methods of fabrication, and the like. Also, since the conductivity of any given material is slightly different at different temperatures, an average had to be used for the figures. However, the *R*-values listed do represent some reasonable estimate in each case, and they can be used for guidance in calculating heat losses.

TABLE 6.1 *Typical R-Values of Some Building Materials*

Material	Thickness (in.)	R-value ft² °F Btu/h
Asbestos–cement board	$\frac{1}{4}$	0.06
Gypsum board	$\frac{1}{2}$	0.45
Plaster board	$\frac{1}{2}$	0.45
Plywood (Douglas fir)	$\frac{1}{2}$	0.62
Plywood (Douglas fir)	$\frac{3}{4}$	0.93
Asphalt roll siding	$\frac{1}{16}$	0.15
Acoustic tile	$\frac{1}{2}$	1.25
Wood subfloor	$\frac{3}{4}$	0.94
Carpet and rubber pad	$\frac{3}{4}$	1.25
Hardwood floor	$\frac{3}{4}$	0.68
Fiberglass insulation	3	12.0
Rock wool insulation	3	12.0
Expanded polystyrene	3	15.0
Cellulosic insulation	3	10.2
Concrete (average)	6	3.60
Brick	3	0.33
Concrete block (average)	8	1.11
Glass (single)	$\frac{1}{8}$	0.9
Glass (double pane)	$\frac{3}{4}$	2.0
Glass (triple pane)	1	3.2
Window plus storm window	2	2.0
Gypsum plaster (average)	$\frac{1}{2}$	0.32
Asphalt shingles	$\frac{1}{8}$	0.44
Wood shingles	$\frac{3}{8}$	0.87
Wood clapboard siding	$\frac{3}{8}$	0.90
Wood two-by-four	$1\frac{5}{8}$	2.03
	$3\frac{1}{2}$	4.38
Aluminum siding	$\frac{3}{8}$	0.61
Aluminum (insulated)	$\frac{3}{8}$	1.82

EXAMPLE

What is the *R*-value of the wall shown in cross section in Figure 6.13?

Solution

First, we need to list the *R*-values of each section, as in Table 6.2, then add them up. Note that a calculation had to be done to find the

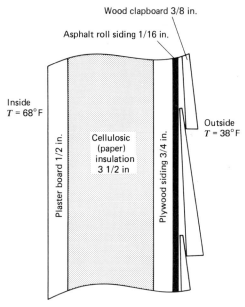

Wood clapboard 3/8 in.

Asphalt roll siding 1/16 in.

Inside
$T = 68°F$

Plaster board 1/2 in.

Cellulosic
(paper)
insulation
3 1/2 in

Plywood siding 3/4 in.

Outside
$T = 38°F$

Figure 6.13 Cross section of an exterior wall.

TABLE 6.2 *Adding the R-Values of Wall Components*

1.	Plaster board,	$\frac{1}{2}$ in.	0.45
2.	Cellulosic insulation,	$3\frac{1}{2}$ in.	11.9
3.	Plywood siding,	$\frac{3}{4}$ in.	0.93
4.	Asphalt roll siding,	$\frac{1}{16}$ in.	0.15
5.	Wood clapboards,	$\frac{3}{8}$ in.	0.90
	Total R-value		14.33

R-value for the cellulosic insulation. (This is just paper that has been ground up and treated to resist fire, and it can be "blown" into existing walls.) The table gives a value of 10.2 for 3 in. of the material. Dividing by 3 gives the value per inch, and multiplying that result by 3.5 gives the value of 11.9 listed in the table. Figure 6.14 shows insulation being "pumped" into a home to produce an insulated wall.

Questions

The *R*-value just calculated is for an area between wall studs, where the wall is filled with insulation. At a stud, the same space is filled with a two-by-four. What is the *R*-value at that location? Does the presence of the studs tend to increase or decrease the heat flow through the wall?

In a typical wall, the studs are nearly 2 in. wide with spaces 14 in. wide between them. (They are said to be "on 16-inch centers.") To

find the average *R*-value of the wall, one can imagine dividing the wall into vertical strips 2 in. wide. Then for every seven strips of insulation, there will be one strip of wood. The average *R*-value will then be 14.33 added in seven times plus the value at the stud, with the entire sum being divided by 8. (This is called a "weighted average.") I get a value of about 13.4; do you agree?

U-VALUE

Another quantity that is often used by heating contractors in the *U-value* of a wall. The *U*-value is simply the reciprocal of the total *R*-value, and it represents the conductance of a square foot of the wall. That is, $U = 1/R$ and the rate of heat loss through a wall is

$$P = UA\Delta T$$

where *A* is the area of the wall and ΔT is the difference in inside and outside temperatures. The units of *U* are Btu/hr/(ft^2 °F). (Btus per hour per square foot and per degree, Fahrenheit.)

EXAMPLE

Given that $\Delta T = 30°F$, as indicated in Figure 6.13, and that the wall is 12 ft long and 8 ft high, what is the heat loss through the wall?

Figure 6.14 Cellulosic insulation being "pumped" into an older home. (Courtesy Wadsworth Heating and Cooling.)

Solution

As you calculated in the previous questions, the average *R*-value of this wall is about 13.4. Then

$$U = 1/R = \frac{1}{13.4} \frac{Btu/hr}{ft^2 \, °F}$$
$$= 0.075 \frac{Btu/hr}{ft^2 \, °F}$$

If the wall is 12 by 8 ft, and the temperature indoors is 68°F and outdoors it is 38°F, then the rate of heat loss for the wall is

$$P = UA\Delta T$$
$$= 0.075 \frac{Btu/hr}{ft^2 \, °F} \times 96 \, ft^2 \times 30°F$$
$$\approx 215 \, Btu/hr$$

Questions

A room has two exterior walls, constructed as in Figure 6.15. Each wall is 15 ft long and 8 ft high, and each has two windows measuring 3 ft by 5 ft, glazed with double-pane glass.

What is the total *R*-value of each wall? What is the *U*-value? Not counting the window area, what is the area of each wall? If the temperature outdoors is 28°F and indoors it is 68°F, how much heat per hour is lost through the two walls? How much is lost through the windows?

If the ceiling consists of 0.5 in. of gypsum board and 6 in. of fiberglass insulation, how much heat is lost through the ceiling?

If the floor consists of 1 in. of plywood and 6 in. of fiberglass insulation, how much heat is lost through the floor? (In both ceiling and floor, neglect the extra loss through the wooden joists.)

The two interior walls are next to rooms that are also heated to 68°F. What is the heat loss through those two walls?

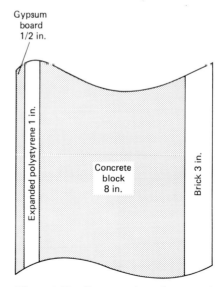

Figure 6.15 Cross section of an exterior wall. The interior of the house is to the left.

AIR FILMS

The calculations for heat losses through walls have left out one additional factor: The wall is not generally at the same temperature as the air a short distance from it. This is true for both the indoor and outdoor surfaces of the wall. The reason is that a thin "skin" or "film" of air clings to the wall and provides a certain amount of additional insulating value, depending on how much wind is blowing. Wind tends to blow away the still air near the wall, thus reducing its insulating value. Rather surprisingly, even a fairly stiff wind does not completely eliminate this insulating effect. Table 6.3 gives approximate R-values at different wind velocities for the thin film of air. It is common practice to use an average value of 15 mph for the outdoor wind velocity when making calculations of R-values for building design.

TABLE 6.3 *R-Values of Surface Film at Different Wind Velocities*

Wind velocity (mph)	R-value of surface film
0	0.68
5	0.31
10	0.25
15	0.17
20	0.14
25	0.12
30	0.10

Questions

For the wall of Figure 6.13, what is the total R-value when including the effect of a surface film of air? (*Hint*: Remember to include a surface film on both the indoor and outdoor surfaces of the wall. The average R-value you found earlier was 13.4.) Using the new R-value, what is the heat loss through the wall? (The wall is 12 ft long and 8 ft high and ΔT is 30° F.)

INFILTRATION

Infiltration is another of those catch words one often hears these days. It simply refers to leakage of air into a house from outdoors (and from indoors to outdoors). In a sense, infiltration is the same thing as convection: Air moving from one place to another carries thermal energy with it. Being aware of infiltration and its sources is important to us because stopping those air leaks is one of the easiest and least expensive ways to save fuel.

Infiltrating air currents are usually caused by winds from the outdoors, and thermal energy is carried out through any crack or hole in the walls. Because of drafts, we tend to notice cold air coming in. But, if cold air comes in, then warm air must be escaping. The net result is that we lose heat to the outdoors and have to provide heat to warm the cold air coming in.

Activity 6.4

As you look around your home, you may think there is not any significant infiltration (there are no leaks) of air. Unless your home is very unusual, indeed, this is not true; there is plenty of air coming in. One way to detect this, on a cold, windy day, is to carefully feel suspected areas, such as the edges of windows and doors, and near electric outlets. Also, dangle a long (10 in.), thin strip of light tissue paper near some of the same places. If the paper blows about slightly, there is serious infiltration. (This can also be done by observing how smoke from a piece of burning incense is blown about.) Map out one room of your home, finding the larger air leaks, and suggest ways in which they could be reduced.

Some of the sources of infiltration, such as cracks around doors, are obvious; others are less obvious. A recent study indicates the

TABLE 6.4   *Sources of Infiltration in Homes by Percentage*

Soleplate	25
Wall electrical outlets	20
Exterior windows	13
Duct system	13
Range vent (hood)	6
Fireplace	6
Recessed spotlights	5
Exterior doors	4
Clothes dryer vent	3
Sliding glass door	2
Bath vents	1
Other	2

sources of infiltration for a large number of homes surveyed, shown in Table 6.4.

Some parts of these results are surprising, and should be a good lesson for us. We often think that the cracks around doors are the major source of infiltration, but that source represents only 4 percent of the total. A whopping 25 percent comes from a usually unsuspected place: under the soleplates around the exterior walls. (The soleplate is the bottom member of the wall that rests on the subfloor.) Another 20 percent of the infiltration comes from around the electrical outlets in exterior walls. The loss up fireplace chimneys, at 6 percent, is quite substantial. One thing many people do not realize is that burning in an open fireplace often loses more heat up the chimney than the fire puts into the room. The hot air rushing up the chimney carries a great deal of warm room air with it—this is appropriately called the *draft* of the fireplace. The room air must be replaced from somewhere, and the needed air comes from an increased rate of infiltration of outside air while the fire is burning. (That is, air is literally sucked into the house by the fireplace draft.)

A widely used method for calculating the effects of air infiltration is the "air change method." An "air change" represents a complete replacement of the air in a room. Under average conditions in homes, the number of air changes is surprisingly high. For example, in a room with two exterior walls that have doors or windows, there are about 1.5 air changes per hour. To calculate the amount of heat lost during an air change, if the temperature stays the same, we can use the following formula:

$$P = \frac{H}{t} = knV\Delta T$$

where P is the heat loss per hour, k is a constant, n is the number of air changes per hour, V is the volume of the room, and ΔT is the difference between inside and outside temperatures, which is the temperature change needed to warm the incoming cold air. The constant k is the amount of thermal energy needed to raise the temperature of 1 ft³ of air 1°F. It is related to the specific heat of air and it is approximately 0.018 Btu/(ft³ °F).

EXAMPLE

Suppose you return from a winter vacation to find your house at 35°F, the outdoor temperature. If the house has a total volume of 16,000 ft³ and a furnace which can supply 60,000 Btu/hr, how long will it take to heat the air in the house to 68°F? What effect will walls, floors, furnishings, and other objects in the house have on this result?

Solution

This is just like bringing one air change into the house and heating it. Since the time is not specified, a slightly modified version of the equation will give us the total heat provided to heat the air:

$$H = kNV\Delta T$$

Here, N is the total number of air changes, in this case just one. Thus:

$$H = 0.018 \frac{\text{Btu}}{\text{ft}^3 \, ^\circ \text{F}} \times 1 \times 16{,}000 \text{ ft}^3 \times 33^\circ \text{ F}$$
$$= 9500 \text{ Btu}$$

Now, to find the time required, just divide by the power—the rate of providing heat—of the furnace:

$$t = \frac{H}{P} = \frac{9500 \text{ Btu}}{60{,}000 \text{ Btu/hr}}$$
$$\approx 0.16 \text{ hr} \approx 10 \text{ min}$$

If you have ever tried to heat up a thoroughly cold house, you know that it will generally take much longer than 10 minutes. This is because, in addition to heating the air, you must heat the walls, floors, furniture, and other contents of the house. This is hard to calculate, but it will generally require much more time than the heating of the air alone.

Questions

Earlier you calculated heat loss through the walls, windows, ceiling, and floor of a 15-by-15-ft room, 8 ft high. There are two exterior walls, both with windows. Thus there will be about 1.5 air changes per hour. What is the volume of an air change (i.e., the volume of the room)? The temperature outdoors is 28°F, and indoors it is 68°F. How much heat is lost per hour because of having to heat the infiltrated air? How does this heat loss compare to the heat losses by conduction through the walls, windows, ceiling, and floor?

As you can see from this last answer, the losses due to infiltration of air are comparable to those due to conduction through the walls, ceilings, and floors. In a home, many of those air leaks may be quite easy to stop, and doing so can result in large fuel savings.

One final point should be made about infiltration, or the lack of it. In recent years there has been considerable trouble with houses that are sealed *too* tightly in an attempt to prevent infiltration. If a house is so tightly sealed that little or no outside air enters, the air inside will soon become stale, smelly, and even dangerous to breathe. In a tightly sealed house, some means of ventilation must be provided. One way, of course, is to bring in measured amounts of outside air directly. Recently there have been developed *heat exchangers* which partially warm the incoming cold, fresh air by transferring to it some of the heat from the outgoing warm, stale air. Thus, fresh air is provided but not all of the heat in the expelled air is lost. A heat exchanger of this type is shown in Figure 6.16. It is used in the "super insulated" house shown in Figure 6.17. That house is tightly sealed, and it has walls a foot thick filled with insulation.

A LITTLE HOUSE

With this background, we are now ready to take a look at some of the details of home energy conservation. To get an estimate of the net savings that can be effected, let us examine the small house pictured in Figure 6.18. The house is of frame construction, it is 25 years old, and it is in Indianapolis, which has about 5000 degree-days annually. The heating season in that area is almost entirely in November through March, a period of about 150 days. The average temperature difference between inside and outside during that period is the number of degree-days divided by the number of days: $\Delta T = 5000/150 \approx 33^\circ\text{F}$. (This assumes that the house is at 68°F, but would be at 65°F if it were not for the heat produced by human and other bodies, appliances, lights, and so forth. Thus, the extra 3°F is "free," in the sense that we do not have to provide extra furnace capacity to achieve it. Here, since the discussion is about energy conservation, we are interested in the energy we must provide

Figure 6.16 A heat exchanger used to bring fresh air into a well-sealed house. The unit is set to accomplish 0.3 air changes per hour, and it is 80 percent efficient in using the heat of the outgoing air to warm the incoming fresh air. (Courtesy Enercon Consultants Limited, Regina, Saskatchewan, Canada.)

by way of a furnace or other heating device.) Our aim is to insulate and seal the house during a summer. Then we can compare the heat losses for the winter before the improvements and the winter after them. In each of the following calculations, the insulating value of the film of air close to the two surfaces will be included in the R-values and an average wind velocity of 15 mph will be used. The R-values used here are taken from Table 6.1. As must be done in any real application of this type, considerable estimation has gone into

the numbers used. Thus, the answers are only approximate.

The little house on the prairie in Indianapolis has 12 windows, with a total area of 180 ft². Counting the layer of still air on the inside and outside of the windows, the R-value for a single-glazed window is about 1.75. Then, using $P = UA\Delta T$, with $\Delta T = 33°$ F, this means that the heat loss through the uninsulated windows is about 3390 Btu/hr. When storm windows are added the R-value becomes about 2.85 and the heat losses are about 2080 Btu/hr.

There are two outside doors, with a combined area of 40 ft². The R-values for a door is in the neighborhood of 2.85 and the heat loss is 460 Btu/hr. With storm doors added, the R-value is about 4.55, and the heat losses are 290 Btu/hr through the two doors.

The walls are 740 ft². (The full wall area minus the areas of the windows and doors.) The uninsulated walls have an R-value of about 4, causing a heat loss of about 6100 Btu/hr. With blown-in cellulose insulation added, the R-value is increased to about 15, with a corresponding heat loss of about 1630 Btu/hr.

The ceiling has an area of 875 ft². Without insulation, the total R-value through the ceiling and roof is about 3. This allows for some insulating value of the air space in the attic. (This is just an estimate.) The heat loss is about 9620 Btu/hr. With 6 in. of fiberglass insulation added in the attic, the R-value becomes about 26, giving a heat loss of about 1110 Btu/hr.

The floor also has an area of 875 ft², but it is a little more difficult to decide how to make an appropriate calculation for it. We shall be in the right ballpark if we estimate that, due to heat from the furnace and ducts and through the floor, the basement stays at a temperature that is about 10°F lower than the temperature of the house. Then, with an approximate R-value of 2 for the floor, around 4370 Btu/hr are lost through the floor. With 4 in. of fiberglass insulation added under the floor, the R-value becomes about 18, and the heat loss is 490 Btu/hr. (Note that if we in-

Figure 6.17 The Pasqua energy efficient house in Saskatchewan, Canada. The exterior walls are 1 ft thick and filled with insulation, giving them an *R*-value of about 40. The ceilings have an *R*-value of about 52. In a climate with about 11,000 heating degree-days per year, the cost of heating this house with gas and electricity is about $105 per year. From March 7, 1979, to March 7, 1980, the 2660-square-foot house was occupied by a family of four and the cost of heating it with electricity was $171. (Courtesy Enercon Consultants Limited.)

sulated the basement walls, we could save even more fuel. But let's not add that complication.)

Next we must consider infiltration. For a house such as the one we are discussing, the infiltration rate is probably about 1.5 air changes per hour. With a total volume of 25 by 35 by 8 ft. (= 7000 ft³) and ΔT equals 33° F, this gives a heat loss of about 6240 Btu/hr. Our aim is to reduce the infiltration rate to about 0.5 air changes per hour, which will reduce the heat loss to about 2080 Btu/hr. Making that much of a reduction is not as difficult as it might seem. For one thing, the blown-in insulation will help. Further, exterior caulking of cracks, joints, and holes will do a great deal of good. Let's take a look at some of the main sources of air leakage, and see what can be done about them.

The leakage under soleplates, at 25 percent of the total, is the biggest offender. In a new home this source of leaks could be very easily eliminated simply by having the build-

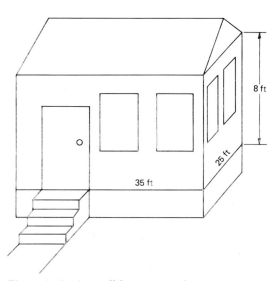

Figure 6.18 A small house in Indianapolis.

ing contractor run a bead of caulk down before putting the soleplates on, but this is rarely, if ever, done. In our existing home, the molding at the bottom of the wall must be removed, the joint between wall and floor caulked, and the moulding replaced. This a little more trouble, but still cheap and effective.

The wall electrical outlets account for 20 percent more. This can easily be stopped by installing in each outlet an insulating insert, available at most hardware stores.

The exterior windows are responsible for another 13 percent. This may seem too large an estimate but, for a poorly fitting window, it is not uncommon for more than 100 ft^3 of cold air to leak in each hour around the edges for *every foot* of edge. Storm windows and weather stripping will reduce this greatly in most cases. But some windows may have to be replaced. The heating ducts in our old house are also 25 years old, and they have cracks at the joints allowing hot air to escape to the outside. This can be reduced by sealing the joints with duct tape in places where they are accessible.

A 6 percent loss can be reduced or eliminated by replacing the drafty range vent with one that seals tightly when not in use. The loss through the fireplace can be greatly reduced, as well. A partial solution is to install a tightly fitting damper and to put tightly fitting glass doors over the fireplace. In a new home, the fireplace should be built so that the air the fire uses comes from outdoors. Then the outdoor air, instead of the room air, goes up the chimney when a fire is burning. To a degree, furnaces cause the same problem as fireplaces: When a furnace gets the air for the burning of the fuel from inside the house, more cold air is pulled in from the outside. So it is a good idea, when installing a furnace, to provide a source of outside air for combustion.

Now that we have most of the heat losses listed, let us put them into a table, so they may be easily summarized. The one heat loss

we have not yet mentioned is the loss from the furnace and heating ducts into the basement and through the outside basement walls. This is not the leakage from the ducts we discussed a couple of paragraphs ago, but rather a loss by conduction through the duct walls. It can be helped by adding insulation around the ducts, but it is difficult to estimate. One fairly common method of estimation is to set it at 20 percent of the total of other heat losses. This is reasonable, since the more heat that is lost by the house, the more heat must pass through the ducts, and thus the greater the loss from the ducts. Better yet, the 20 percent figure is in reasonably good agreement with experience. With this final bit of information, Table 6.5 follows.

This extensive job of home improvement has resulted in a very substantial savings—almost 75% less heat is required than was required before the improvements. This is not a likely result; the whole set of improvements would probably be too expensive to do all at once, and some of them would possibly not give the results expected. But, savings of at least 50 percent are possible with intelligent effort, and the costs of fuel are now getting so high that the effort is worthwhile. Figure 6.19 is a modern example of what can be achieved. In the next chapter, we take a look at the costs of heating.

TABLE 6.5 *Estimated Heat Losses for the House in Indianapolis*

Source of Loss	Loss (Btu/hr)	
	Uninsulated	Insulated
Windows	3,390	2,080
Doors	460	290
Walls	6,100	1,630
Ceiling	9,620	1,110
Floor	4,370	490
Infiltration	6,240	2,080
Subtotals	30,180	7,680
Duct Losses (at 20%)	6,040	1,540
Total	36,220	9,220

Figure 6.19 Hydro Place, in Ontario, Canada, is the first northern office building in the world built without a furnace. The heat from its occupants, lights, computers, and so forth, is enough to keep it comfortable. (Photo courtesy Canada Square Corporation, developer K.R. Cooper, architect, Adamson Associates, consulting architects.)

SUMMARY

The total amount of heating or cooling that a home needs will depend upon the number of degree-days per year in its location. Degree-days are a measure of how far the average outdoor temperature is from a comfortable indoor temperature. Homes lose heat in the wintertime (and gain it in the summertime) mainly because of conduction through the walls and infiltration, a form of convection in which the air leaks through small openings in the house. Heat losses can be reduced by reducing both of these paths. Conduction of heat through the walls, ceilings, and floors, can be reduced by increasing the insulation. The *R*-value of a material is a measure of its insulating properties; the greater the *R*-value, the better the insulation. Poorly insulated houses may have *R*-values of 1 or 2 in some areas, whereas a "superinsulated" house may have *R*-values of over 40 in the walls and 70 or more in the ceilings. The *U*-value is the reciprocal of the *R*-value; thus, the greater the *U*-value, the less the insulation. In the United States, *U*-values are measured in Btu/hr per square foot per degree, Fahrenheit. The air films near the surface of both sides of an exterior wall are also important for their insulating value. Infiltration can be greatly reduced by straightforward measures such as caulking and sealing cracks. It would be best to properly seal the house as it is being built, but caulking is also helpful for older houses. One danger is that a home could be so well sealed that insufficient ventilation results, and this must be avoided. The small, fictitious house used as an example illustrates that the savings of the heating required can be very substantial when a house is properly insulated and sealed against infiltration.

ADDITIONAL QUESTIONS AND PROBLEMS

1. For the section of the country in which you live, find the approximate number of heating degree-days for the year from Figure 6.2. About how long is the heating season, in days, in your area? Thus, what is the average wintertime temperature there? (You might check with your local weather bureau and see if these figures are approximately correct.)

2. Keep a record of high and low temperatures in your area for at least a week. Then, estimate the average temperature for each day (or get it from the local weather bureau), and find the number of heating or cooling degree days for that period. If you can, it would be instructive to repeat this exercise a month or two later.

3. Briefly describe the three mechanisms by which thermal energy is transported from one place to another.

4. On a cold winter day an automobile engine is started and, in a period of 20 min, goes from a temperature of 0 to 180°F. If the engine consists mostly of 300 lb of aluminum, how much heat was added to the engine block in the 20-min period? Where did the heat come from? How much power, in Btu/hr, went into heating the engine during that period? If the engine can provide a total of 100 horsepower, approximately what fraction of its power is used to heat the engine block? (1 hp ≈ 2550 Btu/hr) Explain why this partly accounts for the fact that a cold engine gets poorer gas mileage than a hot one. Describe the flow of heat, after the engine block is thoroughly warmed up, as it continues to run. (Assume that the engine is water cooled, and include both the heat that passes through engine block to the air and the heat removed to the radiator.)

5. What is meant by the "thermal conductance" of a material? How is the *R*-value related to the thermal conductance? (That is, will a material with a greater thermal conductance have a greater or smaller *R*-value?)

6. What is the total *R*-value of the wall sketched in cross-section in Figure 6.20? (This is a simplified example of a "super-insulated" wall.) What is the *R*-value including the air-film effect, assuming a 10 mph wind outdoors? What is the total *U*-value? If the wall is 10 ft wide and 8 ft high and the indoor and

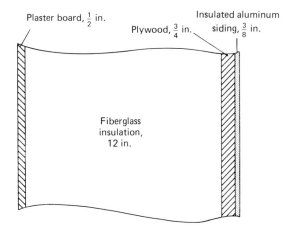

Figure 6.20 A simplified cross section of an insulated wall.

outdoor temperatures are 68 and 0°F, respectively, what is the heat loss per hour through the wall?

7.   Using the materials listed in Table 6.1, design a wall for a home that will have an *R*-value of 40 or more. What will be the *U*-value of that wall? On a cold day, how much heat will pass through the wall. (Make some assumptions about the size of the wall and the indoor and outdoor temperatures.)

8.   Many older two-story homes are roughly box-shaped and roughly 30 ft by 40 ft by 20 ft high. Assuming an infiltration rate of two air changes per hour in such a home, how much heat will be lost to infiltration on a cold winter day?

9.   Carry out an evaluation of your house, or one you are familiar with, using the methods used in the text for evaluating the little house in Indianapolis. Make some reasonable assumptions about *R*-values and infiltration rates presently and what they would be if insulation were added and extensive caulking were done. (For example, if the walls of a frame house are filled with blown-in cellulosic insulation, the *R*-value will be increased by 10 to 12 ft² °F/(Btu/hr). If 6 in. of fiberglass are added to an attic, the *R*-value is increased by about 24 ft² °F/(Btu/hr).) Either from Figure 6.2 or from the local weather bureau, find the number of heating degree-days in the locality of this house. About how many days are there in the heating season? Thus, what is the average outdoor temperature during the heating season? What do you find to be the difference in the amount of heat required to keep the house comfortable before and after insulating and sealing the house?

7 Home Heating

Until recent times, almost all American homes were heated by gas- or oil-fired furnaces or by electric heating. Lately, quite a few homes have switched, at least partially, to wood and coal stoves and furnaces, and heat pumps are becoming popular. A comparison of these various modes of heating is the topic of this chapter. Solar heating is important enough for a separate chapter.

The first thing to consider in choosing a heating system is the size needed. For the little house in Indianapolis discussed in Chapter 6, the average heating requirement was between 9000 and 36,000 Btu/hr, depending upon the amount of insulation and sealing provided. For a typical house in the Indianapolis area with average insulation, the figure might be 20,000 or 25,000 Btu/hr. However, that figure is the *average* for the heating season. A heating system must be sized to handle the *coldest* day likely. For Indianapolis, a design temperature of five below zero ($-5°$ F) is about right. It sometimes gets colder than that, but not for long. On those few very much colder days that do occur, the furnace may not quite keep the entire house warm, and some rooms may have to be closed off. Generally, however, this is better than grossly oversizing the furnace. A furnace that is too large turns on and off frequently in moderate weather and loses efficiency as a result.

In Chapter 6, we used 33°F as the average difference between indoor and outdoor temperatures for the calculations of heat losses. For the same location, the design temperature difference is about 70°F, so the amount of heat to be provided by the furnace is over twice the heat loss calculated previously. To keep the figures rounded off, let us say that the maximum to be provided by the furnace is 50,000 Btu/hr, and the average during the heating season is 24,000 Btu/hr. This is a relatively modest requirement; many homes would need double that, or more.

CONVENTIONAL FURNACES

In general, conventional gas- or oil-fired furnaces heat water or air that is then circulated around the house. Steam heating systems, which once were quite common, have become rather rare now because of their high initial cost and high cost of upkeep. Although many systems have a "flow main," a large pipe through which the hot water leaves the boiler, and a separate "return main" to carry the cooled water back, the simple hot water system shown in Figure 7.1 does not. In a hot-water system, the water is heated to a high temperature and it then circulates through the radiators or convectors. (Strictly speaking, a "radiator" gives off its energy by radiation, and a "convector" heats the air near it, which then rises and produces a convection current. Most modern hot-water heating devices are convectors, but they are very often called radiators.) By giving off heat in the convectors, the water is cooled while the room air is heated. A hot-air heating system circulates warm air through the house, as indicated in Figure 7.2. Note that the air must return to the furnace after cooling during its trip about the house.

Oil versus Gas

Conventional heating systems are between 55 and 70 percent efficient, although some modern high-efficiency furnaces may be 85 percent, or more efficient. If the system is 60 percent efficient, only 60 percent of the heating value of the fuel is usable. The other 40

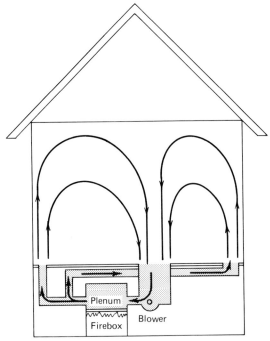

Figure 7.2   A hot-air heating system.

percent is wasted in unburned fuel and losses up the chimney. In what follows, we shall assume an efficiency of 60 percent, which is a reasonable average.

It would be interesting to know about how much it would cost to heat the little house in Indianapolis for a winter, using the various fuels available. The house's average heating requirement is 24,000 Btu/hr. Multiplying this by 24 gives 576,000 Btu/day and, with a heating season of 150 days, the total needed is about 86 million (8.6×10^7) Btu for the season. The *heating value* of a fuel is the amount of heat the fuel can provide if it is burned completely. The efficiency of this heating system is about 60 percent. That is:

$$Eff = \frac{\text{useful heat}}{\text{heating value}} \times 100\% \approx 60\%$$

Thus, the total heating value needed is

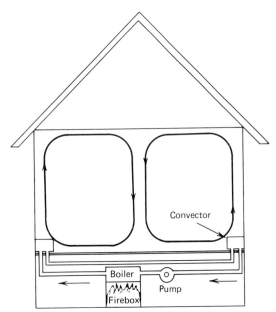

Figure 7.1   A simplified hot-water heating system.

$$H. \ Val. = \frac{\text{useful heat}}{0.6} = \frac{8.6 \times 10^7 \ \text{Btu}}{0.6}$$
$$\approx 1.4 \times 10^8 \ \text{Btu}$$

Fuel oil has a heating value of about 140,000 Btu/gal. Dividing the total number of Btus needed by this figure gives us a total of about 1000 gal of fuel for the season. At $1.25 per gallon, the cost of heating for a season is thus about $1250.

Natural gas has a heating value of about 1100 Btu/ft^3. Dividing this into 1.4×10^8 Btu gives us a total of about 1.3×10^5 ft^3 that must be bought to last the season. At the current rate of about 0.55 ¢ per cubic foot ($0.0055/ft^3), the cost for heating with gas for the season is about $700.

Therefore, with all other factors being equal, heating with natural gas is considerably cheaper than heating with fuel oil. This situation is changing, however, as natural gas becomes deregulated. In January, 1983, the price was about 0.55 ¢/ft^3. A year earlier, natural gas was about 0.37 ¢/ft^3, and the bill was about $470 for the winter. Two years earlier, the same amount of gas cost about $410. In the past, the cost of natural gas has been artificially low because of government regulation. One odd result of this is that in Texas and Louisiana, where over one-half of the country's natural gas is produced, the cost to the consumer has been as much as three times the cost in other parts of the country. This is because the federal regulations apply only when the gas crosses state boundaries and, in those two states, the gas sells for its "real" price. As the regulations are gradually removed, the cost of natural gas can be expected to increase very substantially. Even with the ever-increasing costs of oil, it is likely that the cost of heating with gas will be about the same as with oil. (This is a controversial point. The "experts" seem to disagree about cause and effect. Some of them think that the government should again regulate natural gas—that is, it should place limits on the price at which natural gas can be sold. Other equally qualified persons are sure that reimposing regulation would only make matters worse and that complete deregulation is the only answer. I don't know which group is right—or if, in fact, either one is right.) Heating with gas is somewhat cleaner than heating with oil and it is usually trouble free, so it will still be a preferred method of home heating—as long as the supply of gas lasts. There is more on the topic of supplies of fuels in Chapter 12.

ELECTRIC HEAT

Electric resistance heating has been used widely in this country for the past few decades. In this kind of heating, electricity is simply passed through wires that become hot, as in your toaster. The heat produced may be used to heat water or air in a furnace, but more often the wires are in the walls or in separate units in the various rooms of the house. Major advantages of electric heat are that it is clean, the cost of installation is lower than the cost of installing conventional heating systems, and there is little need for upkeep.

Electric heat is 100 percent efficient in the sense that all of the electric energy used in the house is converted to useful heat. Thus, the total energy needed to heat the little house for the heating season is just 8.6×10^7 Btu. . . . Oops! Electric energy is measured in kWh, so we must convert units. Multiplying 8.6×10^7 Btu by 2.929×10^{-4} kWh/Btu gives about 2.5×10^4 kWh needed. At 6.5 ¢/kWh ($0.065/kWh), the total cost for the season is about $1640.

This price does not seem too bad, but there are some hidden costs to electric heat. A typical power plant may be about 30 to 40 percent efficient in producing and distributing electricity. If the figure is 30 percent and the home furnace burns oil with an efficiency of 60 percent, then an oil-fired power plant must use twice as much oil as the home furnace to produce the same result. The reason the cost to the consumer is not twice as great is that the power company can buy oil for less, since it buys in huge quantities. However, the cost *in oil* is twice as great, and the limited supply of oil will be used up more quickly. Of course, many power plants are not oil fired, and that changes and complicates the considerations. As you know, the cost of electric power is rising rapidly and, in practice, heating by electricity is already quite expensive.

Another factor is pollution. Electric heat is very clean in the home, but pollution is produced at the power plant. Since the power plant can clean its smoke somewhat, the pollution produced to heat one home may not be twice as much as would be produced by a home furnace, but it is likely to be somewhat higher. On the other hand, there may be situations in which it is better to have the pollution concentrated in one lightly populated area than to have it distributed over a densely populated city. This is another one of those give-and-take situations in which there is no clear answer and for which intelligent decisions are needed for each particular case.

WOOD STOVES AND FURNACES

Until the middle of the nineteenth century wood was the world's major source of energy. With the coming of coal, then oil and gas, wood became only a very minor source of fuel. Now that the cost of oil and gas is rising sharply, wood is again gaining popularity for home heating.

In the old days, wood was burned in open fireplaces, a most inefficient means of heating. About the only comfortable place to be was directly in front of the fireplace, where one could be warmed by radiation directly, and a person had to keep turning like a roasting wiener to stay warm all around. Most of

the heating value of the wood went up the chimney. Then, Benjamin Franklin invented the Franklin stove, which although better was still inefficient. Today, we can buy wood stoves, such as the one shown in Figure 7.3, and other types of airtight stoves and furnaces in which the burning is well controlled. In them, most of the wood is burned, and they may have efficiencies of 50 percent or more. (Remember that "efficiency" refers to the fraction of the heating value of the fuel that can be converted to usable heat.)

Suppose that the house we have been discussing is heated with a wood stove that has an efficiency of 50 percent. How much wood would be required for the season? You might guess that the heating value of wood is highly variable, depending on the kind of tree it came from. That is true, if one measures the amount of wood, as it is usually done, in cords. A *cord* is a unit of volume. A standard cord is 128 ft^3 and it is usually thought of as

Figure 7.3 A wood-burning stove. (Buck Stove™ a registered trademark. Reprinted by permission of Smoky Mountain Enterprises, Inc.)

a stack of wood 4 ft wide, 4 ft high, and 8 ft long, as illustrated in Figure 7.4. However, it happens that the heating value of a *pound* of well-cured wood is about the same for different kinds of wood. It is roughly 6000 Btu/lb. A cord of hickory weighs about twice as much as a cord of pine and it provides twice the heating value; the problem is that people usually buy wood by the cord. In general, the hardwoods—hickory, oak, hard maple—have higher densities (weight per cord) than the softwoods, such as spruce, white pine, and aspen. Obviously, if the price per cord is the same, it would be better to buy hardwood rather than softwood. If you are doing the cutting yourself the price may not matter, but you may have to cut and split twice as much softwood as hardwood to get the same heating value.

Getting back to the question, how much wood is needed to heat the house? Remember that the heat to be provided for the season was about 8.6×10^7 Btu. Since the efficiency of the stove is 50 percent, the heating value of the wood will have to be twice this, or 1.72×10^8 Btu. At 6000 Btu/lb, this will require about 29,000 lb of wood. This is 14.5 tons, and if you could buy wood by the ton the question would be answered. To convert this to cords of wood, we have to make some kind of average approximation. For a well-cured hardwood, like oak, a cord would weigh ap-

proximately 3500 lb. Thus, the amount of wood used for the heating season would be about 8 cords. If hardwood costs $100 per cord, as it does in many cities, the cost of heating with wood is then about $800 for the season. If cost were the only consideration, then using this expensive wood as a fuel might be, for the time being, competitive with the other fuels available to us.

Some cautions are in order, however. For one thing, wood is considerably less convenient than other fuels to use. It must be cut and split, and ashes must be removed. Moreover, to be in a decent condition for burning, wood must be dried. When a tree is cut down, the wood has a high content of water, If the tree is burned "green," with the water still in it, a good deal of the energy produced by the burning goes toward boiling away the water. It takes over 1100 Btu to raise the temperature of a pound (a pint) of water from 70°F to boiling and to boil the water away. That heat is carried up the chimney by the steam and is lost for heating purposes. Well-cured (dried) wood can produce up to 20 percent more heat than the same wood burned green. Thus, some arrangments have to be made to cut, stack, and store the wood, preferably under cover, for *at least* a year. In other words, the wood must be purchased or cut at least a year in advance, and two years would be better. Another problem is the risk of a chimney fire. If the wood fire is not hot enough, because of a high water content in the wood or for other reasons, unburned *creosote* is driven out of the wood and it collects on the inside of the chimney. The creosote, a gummy, flammable material, causes a serious fire hazard in the chimney and it must be cleaned out at frequent intervals.

Finally, one must be concerned about future supplies of wood as a fuel. Under good growing conditions, a stand of hardwood may produce as much as two tons of wood per acre per year. That means that the house we have been discussing would need at least 7

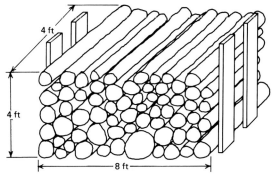

Figure 7.4 A cord of wood.

acres to keep it going, and a 10-acre wood lot would be prefered. Currently, wood prices are relatively low, although we have already seen sharp increases in the past few years as heating with wood has become more popular. In some places one can get wood free for the hauling from the U.S. Forest Service. However, unless you have a 10-acre wood lot, as the competition for wood becomes stiffer, where will you get your wood? For a city dweller it may become very difficult, indeed.

Questions

Along with wood stoves and furnaces, coal stoves and furnaces are making something of a comeback. The attractive thing about this is that coal is likely to be an easily available fuel for a long time in the future. The bad part is that coal is a rather dirty fuel to use.

If coal with a heating value of about 13,500 Btu/lb costs $50 a ton and is burned in a furnace that is 60 percent efficient, what is the cost of heating the home we have been discussing for the entire heating season?

HEAT PUMPS

As you know, heat normally flows from a warmer region to a cooler one. The Second Law of Thermodynamics tells us that, for an isolated system, this must always be the case. However, if work is done on the system, it is possible to reverse this and cause heat to flow from cooler to hotter. A machine to do this is called a *heat pump.* Of course, if you think about it for a minute, a *refrigerator* also does this; heat is removed from the cold interior of the refrigerator and dumped into the room. If you don't believe this, feel the coils behind a running refrigerator sometime, and you will find that they are distinctly warm. That heat came from inside the refrigerator. An *air conditioner* does the same thing, as well.

Actually, all three are the same machine. Whether we call it a heat pump, a refrigerator, or an air conditioner depends only on the use to which we put it. One could say that an air conditioner refrigerates the inside of the house, and a heat pump refrigerates the outdoors. There are also some detailed design differences in the three to make each one as efficient as possible in its function. Each works on the principle that, when a fluid evaporates and the vapor expands, it cools off; when it is compressed and condensed into a liquid, it heats up. This is shown very schematically in Figure 7.5. The two coils, the condenser and the evaporator, are each long lengths of tubing coiled up to take less space and having metal fins to provide more surface to transfer heat. They look a lot like the "radiator" in an automobile and they serve the same function. In a heat pump the condenser coil is the hot one and it is indoors; a fan blows air past it to warm the room air. The evaporator coil is cold and it is outdoors. It also has a fan blowing air over it to speed up the heat transfer from the air.

The working fluid in a heat pump—often Freon—is circulated, around and around a closed loop. Look again at Figure 7.5. The compressor is a pump, operated by an electric motor, and it rapidly pumps the vaporized working fluid to a very high pressure. During this compression process, the fluid becomes

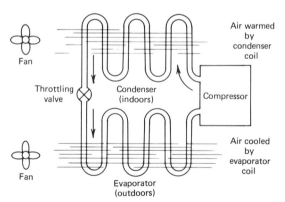

Figure 7.5 Schematic diagram of a heat pump.

a hot, high-pressure vapor-liquid mixture, and it enters the condenser coil in this condition. By blowing room air over the coils, heat is removed from the working fluid and put into the room, and the working fluid condenses— that is, it becomes a liquid, still under high pressure. The fluid then passes through the throttling valve, a small hole. The throttling valve is small enough to keep the pressure in the condenser very high and the pressure in the evaporator low. As the working fluid passes from the high-pressure to the low-pressure region, it suddenly expands and partially evaporates, becoming very cold in the process. In the evaporator coil, the very cold working fluid gains heat from the outdoor air, again with a fan blowing air to aid the process of heat transfer. As it gains heat, the fluid evaporates, and the resulting vapor is pulled into the compressor. Then the cycle repeats itself. The typical heat pump can be run in reverse, so that the cold coil is indoors and the warm one outdoors, thus serving as an air conditioner. The outdoor unit (the evaporator) of a heat pump is shown in Figure 7.6.

Figure 7.7 shows the energy flow in an ideal heat pump. Work is done on the heat pump by an electric motor. An amount of heat H_C is removed from the outdoors, and an amount of heat H_H is pumped into the indoors. From the First Law of Thermody-

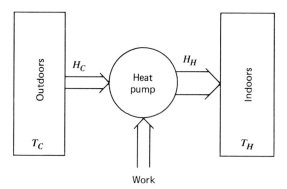

Figure 7.7 Energy flow in a heat pump.

namics, we can write that the total energy into the heat pump is equal to the total energy out:

$$W + H_C = H_H$$

This is a rather startling result: The heat transferred to the indoors is greater than the work done to transfer it! That is, if 1 kWh of electric energy is used to run the heat pump, more than 1 kWh of thermal energy is added to the house. This is the great value of heat pumps, but you may be a bit puzzled by it. When running conventional furnaces or electric resistance heating, the thermal energy added to the house must be produced by burning a fuel or using electric energy to make wires hot. A heat pump, on the other hand, uses heat that is already present in the outdoors, and simply moves some of it indoors. To take the extreme case, if the indoor and outdoor temperatures were the same, it would not take any work at all to move thermal energy indoors. If the outdoors is 1°F cooler, the amount of work needed would be small. With a temperature difference of 10°F the work necessary to move the same amount of thermal energy would be greater, and so on: the greater the temperature difference, the greater the work needed.

If this still seems difficult to grasp, perhaps an analogy will help. Consider, if you will, a freight train moving at 50 mph. It has a great deal of kinetic energy and the job of

Figure 7.6 The outdoor unit of a heat pump looks just like a central air conditioner.

the engine is to maintain that same kinetic energy. If the train is going down a grade, then it will gain kinetic energy without any addition of work from the engine. This is much like the situation when the outdoor temperature is greater than that inside. Then, thermal energy will be gained by the house without any work being done by the heat pump.

If the freight train is moving on a level surface, the engine must only supply enough work to overcome friction; if there were no friction the kinetic energy would remain the same with no work done. Likewise, a heat pump must transfer no heat in order to maintain a constant thermal energy in the house if the indoor and outdoor temperatures are the same.

However, if the freight train is to go up a hill, then the engine must do work in order to maintain the kinetic energy—that is, to keep the speed up. The greater the grade is, the more work will be required of the engine. Likewise, if the outdoor temperature is lower than the indoor temperature, the heat pump will have to move thermal energy inside, thus replacing the heat lost through the walls. The greater the difference in indoor and outdoor temperatures, the more heat transfer will be required of the heat pump.

As you might have guessed, there is an equation to express all of this. The heat pump (refrigerator) is just a heat engine running in reverse. The ratio of heat transferred to the work needed to transfer it (H_H/W) is called the *coefficient of performance* (*CP*). It is similar to efficiency, but is not called that because an efficiency greater than 1 (100 percent) does not make any sense. For an *ideal* heat pump, the coefficient of performance is given by

$$CP = \frac{H_H}{W} = \frac{T_H}{T_H - T_C}$$

This is the *CP* of a perfect heat pump, and the temperatures must be in degrees above absolute zero.

EXAMPLE

What is the coefficient of performance for a perfect (ideal) heat pump that transfers heat in from the outdoors at 40°F to the indoors at 70°F?

Solution

We could use the Kelvin scale and convert the temperatures in degrees Fahrenheit to kelvins. Instead, let us use the knowledge that absolute zero is about -460°F and just add 460° to each temperature. (As pointed out earlier, this is called the *Rankine* temperature scale.) Then, if the temperature indoors is 70°F and outdoors it is 40°F, the coefficient of performance for a perfect heat pump is

$$CP = \frac{460 + 70}{(460 + 70) - (460 + 40)}$$
$$= \frac{530}{30}$$
$$\approx 17.7$$

That's right: For every Btu of work done on the heat pump by its motor, 17.7 Btus of heat are put into the house! Amazing, isn't it?

Questions

What is the coefficient of performance of an ideal heat pump when the difference in outdoor and indoor temperatures is 10°F? 20°F? 50°F?

Naturally, the picture is not quite as good for real heat pumps, but it is not bad. Real heat pumps have a *CP* ranging from about 1.5 to 3.5. This includes the losses to friction and the energy needed to run the compressor, the fans, and the controls. With a range this great, it obviously makes a difference which heat pump you buy. There is an easy way to know what you are getting. Usually, the manufacturer can provide information about the *CP* under normal operating conditions. One should not buy a unit unless the *CP* can be determined in advance. Unfortunately, because of our habit of using different units for thermal energy and electric energy, the information is often somewhat disguised. It is in the form of an *Energy Efficiency Ratio (E.E.R.)*. The *E.E.R.* of a heat pump is the rate of heating in Btu/hr divided by the electric power consumption in watts. That is

$$E.E.R. = \frac{\text{heat power (in Btu/hr)}}{\text{electric power (in W)}}$$

Since 1 Btu/hr = 0.293 W, the *E.E.R.* can be converted to the *CP* by multiplying by 0.293:

$$CP = E.E.R. \times 0.293 \frac{W}{\text{Btu/hr}}$$

Note that the units for the *E.E.R.* are Btu/hr per W, whereas the *CP* has no units at all. The higher the E.E.R. of a heat pump or air conditioner, the better the machine is.

Questions

An air conditioner I looked at recently had an *E.E.R.* of 7.7 listed on its information tag. The tag also stated that it consumed 1000 W of electric power when running. How many Btu/hr will it remove from the house?

Does that air conditioner remove more, or less, thermal energy from the house than it consumes in electric energy? How much more or less? (That is, what is the coefficient of performance?)

If a heat pump or air conditioner had a *CP* = 1 it would be a dreadfully bad one, but the *E.E.R.* does not sound so bad. What is it? (Beware of sales pitches!)

EXAMPLE

Suppose that a heat pump with an E.E.R. of 10.3 were used to heat our house in Indianapolis. What would be the cost for the heating season?

Solution

In this case, the coefficient of performance is

$$CP = E.E.R. \times 0.293 \frac{W}{\text{Btu/hr}}$$
$$= 10.3 \times 0.293 \approx 3.01$$

That is, this heat pump, operating under some average conditions, will provide three times as much (heat) energy to the house as the (electric) energy consumed. Then use

$$W = \frac{H_H}{CP}$$

Since the heating requirement is 8.6×10^7 Btu for the heating season, this is

$$W = \frac{8.6 \times 10^7 \text{ Btu}}{3.01} \approx 2.86 \times 10^7 \text{ Btu}$$
$$\approx 2.86 \times 10^7 \text{ Btu} \times 2.929 \times 10^{-4} \text{ kWh/Btu}$$
$$\sim 8,100 \text{ kWh}$$

At 6.5 cents per kilowatt-hour, this amount of energy would cost about $540 for the season.

That's pretty good! In fact, for the types of heating we have discussed, this is the cheapest heat available. So why doesn't everybody get a heat pump? Well, as usual, there is a catch—or several. For one thing, the cost of a heat pump is much greater than that of a furnace. Also, since the operating conditions for heat pumps put them under much more stress than air conditioners, they tend to break down more frequently, and repairs are expensive.

However, the main problem is that the picture (and cost of heating) I presented is too optimistic. The CP of 3 is an average for moderate outdoor temperatures. A heat pump which has $CP = 3$ at 40°F will have $CP = 2$, or less, at 0°F. Therefore, the heat produced costs more than we calculated if the outdoor temperature goes below 40°F. When operating below the freezing point, the evaporator coil has to be defrosted periodically, using still more electricity. Further, the maximum output of a heat pump also decreases as the temperature falls, whereas the heating needs increase. This is shown in Figure 7.8. The lowest temperature at which the heat pump graphed can produce all of the heat needed is 30°F. Below that temperature some extra heat must be provided. In the past, the extra heat usually has been provided by electric resistance heating, but there is no reason a furnace could not be used. For that matter, when the price of heat pumps becomes competitive, it is conceivable that two heat pumps might be used; in moderate weather they might alternate in providing the heat and in very cold weather they could be used together. This would also provide a backup when one of them is out of order. Alternatively, some heat pumps now being installed have been designed with a very long evaporator pipe being buried 8 or 10 ft

underground, where the temperature never drops much below 50°F, depending on location. A heat pump installed this way runs in a relatively efficient manner. Also, in the summertime, when the heat pump is reversed to run as an air conditioner, it dumps heat into the relatively cool ground and thus runs more efficiently than it would otherwise.

SUMMARY

Homes may be heated in a variety of ways, but all of these ways are likely to become more expensive in the future. In this chapter, you have compared the current costs of various methods of heating, including conventional furnaces, electric resistance heating, wood stoves and furnaces, and heat pumps. You have seen that there are many variables to consider when choosing a form of heating to use, and some of those variables are changing rapidly. The best solution to a heating need may not be obvious. In each case, a thorough analysis of the conditions, along with an intelligent guess about future fuel and electricity costs, would help in the decision making. Experience shows that, very often, simply asking a heating contractor for the answer is unsatisfactory. Certainly, getting advice from more than one contractor is advisable.

Heat pumps are interesting because they do not produce heat by burning a fuel. Rather, they move heat from the outdoors to the indoors. As a result, under moderate temperature conditions, a heat pump can provide two or three times as much energy in the form of heat transferred as the amount of electric energy it consumes. Thus, it is relatively inexpensive to operate. However, heat pumps do not run efficiently in very cold weather, they are quite expensive, and they tend to need a good deal of costly maintenance.

The one source of home heating we have not yet discussed is the sun. I believe that solar heating of homes has great potential in

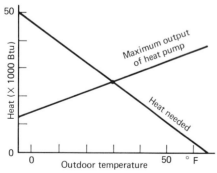

Figure 7.8   Heating needs of a house and heating ability of a heat pump. Below about 30°F the heat pump cannot provide all the heat needed.

many climates, and our society has barely gotten started in that direction. In the next chapters we will take a look at some of the possibilities of using solar energy.

ADDITIONAL QUESTIONS AND PROBLEMS

1. A home in Toronto might need heat at the rate of about 50,000 Btu/hr on a cold day. How much heat would it need for a week of weather this cold? If the furnace in the home is 70 percent efficient, what must be the heating value of the fuel used to provide this heat? If the fuel used is oil, how much must be used, and what is its cost? What if the fuel is natural gas?

2. What is the cost of heating the house of the preceding question using electric resistance heating? Using wood at $100 a cord? (Assume 65 percent efficiency for the wood stove.)

3. Suppose you had to put a new heating system into the house for which we did the heating cost calculations in this chapter. Contact a local heating contractor and find out about how much the installation of a new furnace in a typical home would cost. Get the same information for a heat pump. Assuming that the costs of natural gas and electricity stay the same, what is the "payback time" for the heat pump? That is, how long would it take to make up the extra cost of the heat pump in fuel savings?

4. A "super-insulated" house may well have enough insulation in the walls to give a total R-value of 40, in the ceiling an R-value of 70, and in the floor an R-value of 50. Assume that the area under the floor is 20°F warmer than the outdoors, on the average. The extra insulation may add $1500 to $2000 to the total cost of the house. For a boxlike house measuring 30 by 40 ft, by 10 ft high in your lo-

cality, how long would it take to save $2,000 in fuel costs. That is, what is the "payback time"? (*Hint:* Do this a step at a time. First, how many heating degree-days are there in your area? Thus, what is the average outdoor temperature during the heating season? Then, what is the heat loss through the walls and ceiling and floor? For a "normal" house, with walls and floor at R-15 and ceiling at R-20, what would the heat loss be? Choose a fuel. What is the heating value of that fuel? If the heating system is 70 percent efficient, how much extra fuel is required to heat the "normal" house for the heating season? What is the cost of that much fuel?) Early indications are that, in actual practice, such highly insulated homes being lived in may cost little or even nothing at all to heat in rather cold climates. Why does this practical result differ from the result you calculated?

5. My home used 33,700 ft^3 of natural gas in January, 1982, to stay at an average temperature of about 63°F (68°F during part of the day and 60°F for the rest of the day and at night). There were about 1200 degree-days that month. What was the approximate heating value of the fuel used? Assuming that the old furnace is about 60 percent efficient in delivering heat, how much usable heat was produced during the month? What was the average outdoor temperature during the month? Thus, what was the average ΔT between indoors and outdoors? The house is approximately 30 by 40 ft, by 25 ft high. Assuming a flat roof, what is the total area of the house exposed to the outdoors? Assuming that one-half of the heat provided escapes through the walls and ceiling, what is the heat loss per hour by that route? Thus, what is the overall R-value of the house?

6. Using the same figures as for the preceding question and assuming that the other half of the heat is lost by way of infiltration, about how many air changes per hour occur, on average?

7. Use the answer you got in Question 9 of Chapter 6. Continuing the evaluation of the home you chose, what kind of fuel was used in the house being analyzed? How much fuel was used last winter? (If complete records are not available, make an educated guess based on what records there are.) What was the total heating value of the fuel? If the heating system is 60 percent efficient, what amount of usable heat was produced? How does this compare to the calculation you made of the heat needed in Question 9 of Chapter 6? Would you now modify the estimates you made when you were answering that question?

8. If the range of *CP*s for real heat pumps is about 1.5 to 3.5, what is the corresponding range of E.E.R.s? Would a heat pump with an E.E.R. of 12 be a good one or a bad one? How about one with an E.E.R. of 5?

9. Suppose your present oil-burning furnace can provide 50,000 Btu/hr of usable heat, but it is getting old and you would like to replace it with a heat pump. A heating engineer shows you a heat pump with a (good) E.E.R. of 10 and an electric power consumption rating of 6000 W. (The E.E.R. is the rating for "average" conditions, whatever that means.) Under average conditions, will the heat pump provide as much heat as your old furnace? What will happen when the outside temperature gets very low, say 0°F? How could you provide all of the heat needed in very cold weather?

8 Solar Radiation

Much of the energy we use on Earth comes to us, directly or indirectly, from the sun. That shining orb, 93 million miles away in space, pours its warmth upon us at a nearly constant rate, year after year, century after century. We are using the sun's energy indirectly when we burn fossil fuels, such as coal or oil, which stored the energy millions of years ago. The sun provides the energy for wind and water power more directly. Most directly of all, we can take advantage of the sun's radiation to heat buildings, to provide hot water, to produce electricity, and even to provide summertime cooling—these, and more, are lumped under the term *solar energy*. Knowing how to use solar energy effectively requires a knowledge of the nature and amount of solar radiation reaching the Earth, and the ability to predict the sun's path through the sky. These two topics will be the subjects of this chapter.

WAVES

Activity 8.1

Fill a large round pan or tub with water. Scatter a few tiny pieces of cork or wood across the surface of the water to help you to observe the water's motion. Then let the whole thing settle down for a couple of minutes until the motion stops. Observe the surface of the water by looking at the reflection of light from a window or artificial lights in it. Now dip the tip of your finger once in and out of the water at about the center of the pan while watching the surface of the water. Repeat this several times. Describe what happens on the surface of the water, including the motion of the water and of the pieces of cork.

When the ripple spreads out on the surface of the water, does it seem to do so with the same speed each time? Does the kinetic energy of a piece of cork change as a result of your dipping your finger into the water? Does its potential energy change? (Remember, movement of an object up or down means a change in its gravitational potential energy.) Do the pieces of cork move out to the sides of the pan along with the ripple on the water? Did you notice that the ripple moved out to the sides of the pan, bounced off, then moved back in to the center? (If you observe closely enough, you may see this "bouncing" off the sides happen two or three times after dipping your finger in just once.)

Now put the tip of your finger into the

water at the center of the pan, and move it up and down with a steady rhythm. Try this at different speeds, first quite slowly, then gradually increasing the speed until you are vibrating your fingertip as fast as you possibly can. (Try making your arm rigid until your hand quivers rapidly.)

By looking at the reflected light, do you see a series of rings on the surface of the water? About how far apart are the rings when you are moving your finger up and down slowly? When you are vibrating it rapidly?

What you have been observing is the effect of a *wave* on the surface of the water. There are several different kinds of waves in addition to water waves. Some examples are sound waves, radio waves, and ground waves (as in earthquakes). You can even produce waves on a rope and see some of the same effects you saw on the surface of the water. (Tie one end of the rope to a doorknob, pull it fairly taut from the other end, and send pulses down the rope by giving a fast up–down jerk to the end. Then try shaking the end continuously but at various speeds.)

The different kinds of waves all have several of the same properties:

1. Energy is transmitted from one location to another by the wave, but the material through which the wave is moving (called the *medium*) does not travel with the wave. In the case of the water wave, the medium is water, and it is obvious from the motion of the bits of cork that the water does not travel with the wave. The cork just bobs up and down, with a changing kinetic energy and potential energy, but it does not move to the side of the pan with the wave. This is illustrated for a single ripple in Figure 8.1. You can see the same thing by watching an object floating on the ocean. The ocean waves keep coming in to the shore, but the floating object just bobs up and down, unless

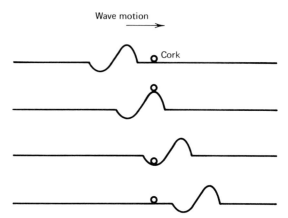

Figure 8.1 The motion of a cork on the water as a ripple goes by.

it is pushed one way or another by the wind. The way in which the wave transmits energy is something like the way energy is transmitted along the long line of dominoes we discussed earlier. Water naturally has a flat surface if left alone. When a pulse deforms the surface of the water, as indicated in cross section in Figure 8.2, the surface tends to fall back to a level shape. As it does so, it pushes water molecules in front of it and behind it. The net result of this complicated interaction is to fill in the trough behind, leaving a new one in a slightly advanced position, and the whole pulse moves along to the right. When it is in its new position, the whole thing happens again and again, giving a smooth motion to the right. (Of course, in reality, a single pulse almost never occurs on water; there is generally a series of them.) Figure 8.3 shows the

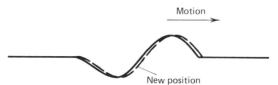

Figure 8.2 The bulge tends to fall down, pushing away water on either side, and thus moves along the surface.

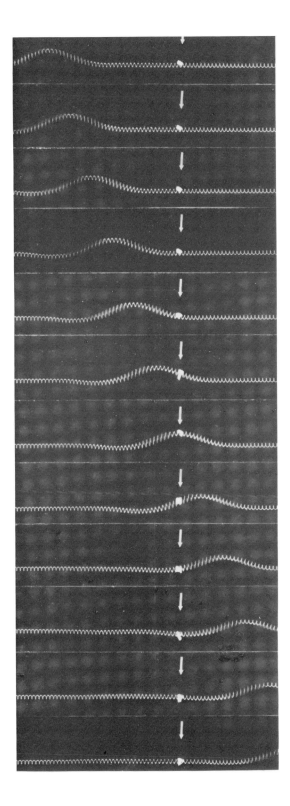

Figure 8.3 Successive frames of a moving picture of a pulse moving along a long, stretched spring. Although the ribbon tied to the string bobs up and down, it does not move to the right with the pulse. (Educational Development Center, Newton, Mass.)

wave pulse motion occurring on a long, stretched spring.

2. When a wave is produced continuously, such as when you dipped your finger in and out of the water rapidly, it has a *frequency*. The frequency is just the rate at which the wave is produced and it is measured in *hertz* (Hz). One hertz is 1 vibration per second, 100 Hz is 100 vibrations per second, and so on. The water waves you produced may have had a frequency of 1 or 2 or 5 Hz. Sound waves in air that most people can hear have frequencies in the range of about 20 to about 20,000 Hz. As the frequency becomes higher, the pitch of the sound becomes higher.

3. A continuous wave also has a *wavelength*. This is written as "λ" (the Greek letter *lambda*) in most textbooks. Figure 8.4 is a cross section of part of a continuous water wave, such as you see on the ocean. The highest points of the wave are called *crests*, and the lowest are *troughs*. A wavelength is the distance from one crest to the next or, similarly, from one trough to the next. For sound waves in air, the wavelength

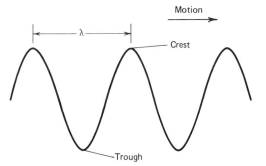

Figure 8.4 The wavelength of a continuous wave is the distance from crest to crest.

is the distance from one point where the air is most compressed to the next point of compression. You may have noticed in the water waves that the greater the frequency of the wave was, the shorter was the wavelength. A slower rate of vibration produced a longer wavelength.

4. The speed at which a wave travels is about the same for all waves of the same kind in the same medium. For example, the speed of a water wave is roughly 0.3 m/s and the speed of sound in the air is about 340 m/s. (This is about 750 mph, but the speed of sound *in water* is 1500 m/s, about 3300 mph.) The speed of a wave is given by the wavelength times the frequency:

$$s = \lambda f \qquad (8.1)$$

5. Waves can be transmitted, reflected, or absorbed. The wave travelling along the surface of the water is *transmitted:* Energy is carried from one point to another. When the wave "bounces" off the sides of the pan it is said to be *reflected:* In a reflection, energy is transmitted oppositely to its original direction. If you were to line the sides of the pan with a fluffy towel, the wave would be *absorbed:* Energy would be caught in the towel and not reflected from the sides.

ELECTROMAGNETIC RADIATION

So, you may ask, what in the world does this have to do with solar heat? The answer, my friend, is everything, just everything. The energy which comes to us from the sun travels in waves, with the same properties as the waves we have just discussed. The difference is that instead of vibrating a medium such as water or air, the sun's energy comes in the form of oscillating electric and magnetic fields.

(You do not yet know what electric and magnetic fields are, but you do not need to know in order to understand the behavior of the waves.) These waves are called *electromagnetic waves.* The remarkable fact is that electromagnetic waves do not need any medium at all to travel through; they can travel through empty space.

One interesting fact about these electromagnetic waves is that they travel at enormously high speeds. In empty space they move at about 3×10^8 m/s (about 186,000 miles per second—That's per second, not per hour!). Another is that they have a very great range of high frequencies, from about 10^3 to about 10^{22} Hz. (1000 Hz is a *kilohertz,* kHz.)

This is such a vast range of frequencies that the electromagnetic waves of widely different frequencies have rather different properties. For example, they all travel through empty space at the same speed, but in air, or glass, or water the speeds of waves of different frequencies are different. In fact, some of the waves cannot travel through air at all, and some of them can easily penetrate such materials as wood and even steel. Some of them are beneficial to human beings and others are deadly in high doses.

As you may have guessed, because of the differing properties of the various frequency ranges, it was not at first recognized that these waves are all basically the same. So they have different names. At the low-frequency end, electromagnetic waves are called *radio waves,* the same ones that bring us rock music and the morning news, as well as television. We cook our food quickly and beam telephone messages at somewhat higher frequencies, by means of *microwaves.* At even higher frequencies is *infrared radiation;* this is also called *radiant heat* and it includes a range of frequencies in which we shall be interested. At increasingly higher frequencies yet, electromagnetic radiation is responsible for *visible light,* with which we see; *ultraviolet light,* which gives us sunburns; *x-rays,* which are used for medical, dental, and industrial purposes; and *gamma rays*, which are associated with radioactive waste materials. Figure 8.5 shows the range of electromagnetic radiation, with both frequencies and wavelengths shown. There are no sharp boundaries when going from one category of radiation to another, and there is considerable overlap between adjacent categories. A chart of this sort, showing fre-

quencies and/or wavelengths, is called a *spectrum.*

One interesting part of the electromagnetic spectrum is the very small fraction of it that is visible light. That is, it is light that is detected by the human eye. Figure 8.6 shows the spectrum of visible light, spread out so that we can see the different colors. Color plate 1 is a photograph of the spectrum of visible light. Note that the color of the light is determined by its frequency (or wavelength).

Activity 8.2

You can do a little experiment to demonstrate the fact that electromagnetic waves come in a range of wavelengths. This involves only radio waves, and you will need an automobile with an AM/FM radio and a bridge with a steel framework over and to the sides of the roadway. Such bridges are common, and you ought to be able to find one fairly near you.

Tune the radio to a loud and clear station near the lower end of the AM broadcast band (at about 550 kHz) and drive across the bridge. What happens? Now try the upper end of the

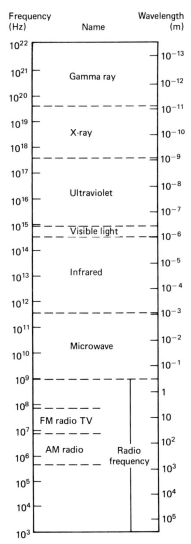

Figure 8.5 Electromagnetic waves.

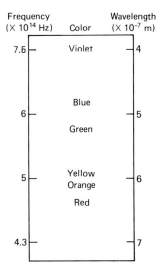

Figure 8.6 The visible part of the electromagnetic spectrum. Each distinct color has a different frequency and wavelength.

AM band (about 1600 kHz) and do the same thing. Are the results the same? Finally, try this with the radio on the two extremes of the FM band (about 88 to 108 Mhz).

Did the radio waves' ability to get through the bridge structure to your radio antenna depend on frequency? The low frequency (long wavelength) waves do not get through at all and the high frequency (short wavelength) waves seem to pass through without being affected at all. The following example will give you some additional insight into the situation.

EXAMPLE

What is the range of wavelengths for the AM and FM radio bands?

Solution

Here, we use Equation 8.1, rearranged to solve for the wavelength:

$$\lambda = \frac{s}{f}$$

Then, for a frequency of 550 kHz (5.5×10^5 Hz),

$$\lambda = \frac{3 \times 10^8 \text{ m/s}}{5.5 \times 10^5 \text{ Hz}} \approx 550 \text{ m}$$

For 1600 kHz, the answer comes out to about 188 m. In the FM band, at the low-frequency end,

$$\lambda = \frac{s}{f} = \frac{3 \times 10^8 \text{ m/s}}{8.8 \times 10^7 \text{ Hz}}$$
$$\approx 3.4 \text{ m}$$

The upper end gives a result of about 2.8 m.

Thus, the AM band has wavelengths of a few hundred meters, much greater than the spaces in the steel framework of the bridge. On the other hand, the FM band has wavelengths of a few meters, comparable to the size of the spaces. A somewhat simplified explanation of the effect is that electromagnetic waves with shorter wavelengths—about equal to the size of the "holes" in the framework—pass through those spaces, whereas, the longer wavelengths are blocked by the steel trusses, which act as huge antennas connected to the ground. That this is not a complete explanation is evidenced by the fact that part of the signal in the upper ranges of the AM band gets through on some bridges, but it is good enough for our present purposes.

Questions

"Ham" radio operators often use the "10-meter band" for short-wave communications. That means that the wavelength of the radio waves used is 10 m. What is the frequency?

Green light has a frequency of about 5.6×10^{14} Hz. What is the wavelength of green light? (Tiny, isn't it?)

BLACKBODY RADIATION

Earlier, you made some observations about the heat coming from a hot element on an electric range. If you hold one hand three or four inches above the element and the other hand the same distance to the side of the element when you first turn it on, you can observe what happens as the element gets hotter. At first, you will feel convected heat above the element, but nothing to the side. Then, as the element warms up, you will feel radiated heat to the side. Finally, as it warms up to its maximum, you will have to move both hands, for the radiated heat will be intense. At the same time, the element will become "red hot," giving off visible light. Most likely, the element will not be the same color all over; there will be some bright red areas, indicating intense heat, and some areas that are more of a "dull" color, indicating that they are somewhat cooler.

As the element changes temperature, the radiation it emits, both as heat and as visible light, seems to depend on its temperature. It is possible to predict the wavelengths of the emitted radiation by imagining what is called a *blackbody* or a *blackbody radiator.* A perfect blackbody is an object that does not reflect any radiation whatever, but absorbs all that falls upon it. We know that the electric stove element is not a perfect blackbody for, if it were, it would not reflect any light and it would appear perfectly black. (That is, we would not be able to see it unless it were placed on a background of something that was reflecting light.) There are no perfect blackbodies on Earth; even objects that look quite black may be good reflectors of radiation other than visible light, such as infrared radiation.

However, the theory for blackbody radiation allows us to make a good approximation of the radiation of a stove element, and an even better approximation of the radiation from the sun. It may seem to be a contradiction that a blackbody gives off light when it is heated, but it is not: Although the blackbody is emitting light, it is still not reflecting any. The theory tells us just what you observed: The color of the light emitted by the blackbody depends on its temperature. If we can heat the object enough, it will go from "red hot" to "white hot," and eventually it will have a brilliant blue–white appearance. A good approximation to a perfect blackbody is found in the furnaces in which steel is smelted. The furnace itself is not the blackbody; the blackbody is a small hole through which one can look into the inside of the furnace, as indicated in Figure 8.7. Very little of the light and other radiation entering the hole would ever be reflected out through it. Steel workers know when the steel is the right temperature for pouring by analyzing the radiant

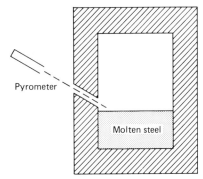

Figure 8.7 The small hole used for looking into a steel-smelting furnace is a good blackbody.

energy coming through the hole with an instrument called a *pyrometer.* Figure 8.8 shows an optical pyrometer in use.

Actually, it is not sufficient to say that the color of the light emitted by a blackbody depends on its temperature: it may not be hot enough to emit visible light. All objects emit radiation unless they are at a temperature of absolute zero, which does not occur on the Earth. The range of wavelengths emitted depends upon the temperature of the object. The total amount of radiation emitted also depends on the temperature, with hotter objects giving off more energy. Figure 8.9 shows the energy emitted by the same object at two different temperatures. The lower temperature, 300 K, is approximately the temperature of the Earth, and the higher temperature, 5800 K, is roughly the temperature of the surface of the sun. Because the peak for the 5800 K curve is over 1 million times higher than the one for the 300 K curve, the two curves have had to be drawn to different scales. However, we are interested in the shape of the curves and that is the same for both. As you can see, the energy is emitted by the object over a range of wavelengths at each temperature. At each temperature, there is a peak wavelength, at which the maximum energy is emitted. This peak is at a much higher frequency (shorter wavelength) for the sun than for the Earth. Much of the energy radiated from the sun is

Figure 8.8 Using an optical pyrometer to determine the temperature of flowing molten iron. (Courtesy of Bethlehem Steel.)

in the visible light range. The Earth gives off no visible light of its own, only the reflected light of the sun.

The formula for the total power (radiant energy emitted per second) from a heated object is

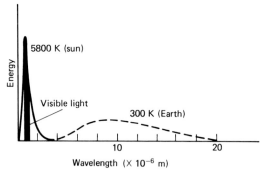

Figure 8.9 The radiant energy emitted by an object at two different temperatures.

$$P = \sigma e A T^4 \qquad (8.2)$$

This is called the *Stefan–Boltzmann law*. The quantity σ (Greek letter *sigma*) is a constant, as follows:

$$\sigma = 5.67 \times 10^{-8} \frac{W}{m^2 \, K^4}$$

The quantity e is the *emissivity* of the object, which tells how well it emits (or absorbs) energy. A perfect blackbody would have an emissivity of $e = 1$; a perfect mirror would have $e = 0$. A is just the surface area of the object, and T is its Kelvin temperature.

EXAMPLE

What is the total power emitted by the sun?

Solution

The sun is very nearly a blackbody, so we'll let $e = 1$. The temperature of the surface of the sun is about 5800 K, so the only thing left to find is the area of the sun's surface. Looking in a reference book, I find that the radius of the sun is about 6.96×10^8 m. Since the area of a sphere is $4\pi r^2$, the area of the sun is about 6.1×10^{18} m². Then, the rate at which the sun emits energy is

$$P = \sigma e A T^4$$
$$= 5.67 \times 10^{-8} \frac{W}{m^2 \ K^4} \times 1$$
$$\times 6.1 \times 10^{18} \ m^2 \times (5800 \ K)^4$$
$$\approx 3.9 \times 10^{26} \ W$$

That's a lot of power! Fortunately, it spreads out in all directions and the Earth is very far away, or we would be fried by it.

SUN POWER ON EARTH

As we have just noted, the sun is, very nearly, a blackbody itself. Figure 8.10 shows the spectrum of the sun's radiation, as measured above the Earth's atmosphere. As you can see, although there are a few "squiggles"—and there would be more if we looked more closely—

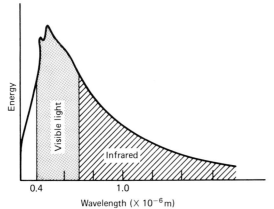

Figure 8.10 The sun's radiation.

the curve for the sun looks very much like the curve for blackbody radiation. When a blackbody is heated in the laboratory, the temperature at which its radiation most closely matches that of the sun is almost 5800 K. Making a match of this sort to objects heated in the laboratory is just the way in which the temperature of the surface of the sun is found. (Although the surface temperature is about 5800 K, or 10,000°F, we believe that the inner core of the sun is much hotter than that.) The spectrum of the sun ranges all the way from x-rays at the high-frequency end to radio waves at the low-frequency end. However, as you can see from the two shaded areas, most of the radiation of the sun is in the visible and the infrared regions.

We are interested more in the *insolation*, the total energy reaching the Earth from the sun, than in the spectrum of the sunlight. Since the energy is coming to us continuously, we need to express the insolation in terms of the energy per unit time—the power. Also, rather than use the power across the entire Earth, the insolation is usually expressed in terms of power per unit area; for example watts per square meter or Btu/hr per square foot. The insolation above the Earth's atmosphere is called the *solar constant*. Problem 6 at the end of this chapter will show you how to find the solar constant. In the two sets of units mentioned earlier:

$$\text{solar constant} = 1395 \ \frac{W}{m^2}$$

or

$$\text{solar constant} = 442 \ \frac{Btu/hr}{ft^2}$$

To rephrase it, the solar constant is the total energy falling per unit time on a square that is perpendicular to the sun's rays, as indicated in Figure 8.11. In the SI system of units, the square is 1 m on a side, and in the English

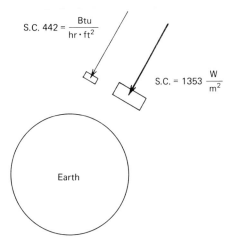

$$S.C. \ 442 = \frac{Btu}{hr \cdot ft^2}$$

$$S.C. = 1353 \ \frac{W}{m^2}$$

Earth

Figure 8.11 The solar constant is the amount of solar radiation above the Earth's atmosphere.

system it is 1 ft on a side. The solar constant is measured for a distance from the sun that is the average distance of the Earth from the sun. (This is called an astronomical unit, *A.U.*, and it is about 149 million kilometers, or 93 million miles.)

You have probably realized by now that the insolation above the atmosphere is not the quantity in which we are most interested. Rather, we would like to know something about the solar radiation reaching the surface of the Earth. Figure 8.12 shows the effect of the atmosphere on the solar radiation. This figure is on the same scale as Figure 8.10, and you can see that only part of the sunlight reaches the surface; the rest is either reflected or absorbed by the atmosphere. In some parts

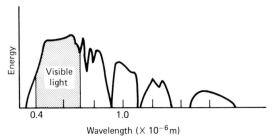

Figure 8.12 The solar radiation reaching the surface of the Earth (typical at sea level).

of the spectrum the sunlight is only partially stopped by the atmosphere, and in others it is totally absorbed or reflected. Almost all of the radiation that gets through is in the visible and infrared parts of the spectrum; virtually all of the x-rays and most of the ultraviolet radiation that leave the sun are stopped by the atmosphere. This is fortunate, for both of those kinds of radiation are harmful to living things in large doses. The absorption and scattering of the solar radiation are due not only to the oxygen and nitrogen molecules which comprise the greatest part of the air, but also to other components of the atmosphere, particularly water vapor and carbon dioxide.

The net effect is that about one-half of the sun's energy reaches the ground directly, as indicated in Figure 8.13. Of the incoming sunlight, about 31 percent is reflected and scattered by the atmosphere and about 3 percent is reflected by the Earth's surface directly into space. The remaining 66 percent is ab-

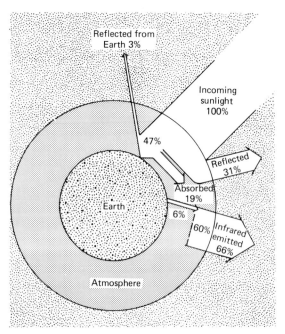

Figure 8.13 The energy balance of the Earth.

sorbed by the atmosphere and the surface of the Earth. But, as you know, if the Earth is to keep its energy balance, it must somehow get rid of that 66 percent. It does so by reradiating it out into space, as indicated on the right side of Figure 8.13. When the sunlight is absorbed, it heats up the Earth and the atmosphere and they both emit infrared radiation. The difference in the incoming and outgoing energy is that the incoming radiation consists of visible light and short-wavelength infrared radiation, whereas the outgoing will be long-wavelength infrared, since it comes from the cool Earth. That is, the incoming energy is "high-grade" energy, having a relatively high temperature associated with it. The outgoing energy is "low grade." Thus, as was pointed out in an earlier chapter, there is net entropy carried away from the Earth. The values given in Figure 8.13 are approximate averages, and they will vary a great deal in different localities, depending upon atmospheric conditions and the surface features of the Earth. Further, there are small but possibly important additions made to the energy balance by geothermal heat (from the Earth's interior) and by the burning of fossil and nuclear fuels. Color plate 3 indicates that different surfaces and materials on the Earth reflect radiation of various wavelengths differently.

The question that most interests us is how much of the sun's power will reach a square foot of a collector of solar energy on the ground. Or, better yet, how much *total* power will reach the ground? The direct rays of the sun are not the only source of energy; the Earth and the atmosphere are continually exchanging energy through infrared radiation. Thus, the collector would get the sum of the *direct* radiation from the sun and the radiation from the atmosphere, called *diffuse* radiation. The total collected per square foot at noon can vary greatly, from over 300 Btu/hr at a point where the sun is directly overhead to as little as 10 Btu/hr near the Arctic Circle. There is

also variation at different times of the year for any given locality and, of course, there are variations during the day, as well. Thousands of measurements have been made all over the country and at different times of the year, and the results are available in tables in handbooks. In this book, we will have need only for approximate values for the total power.

THE PATH OF THE SUN

Activity 8.3

The point of this activity is to trace the path of the sun on a given day. To do so, you can use an oatmeal box cut out as shown in Figure 8.14. The exact center of the bottom of the box is marked, and a sheet of heavy, clear plastic is then taped over the opening of the box.

Figure 8.14 An oatmeal box cut out to admit sunlight.

As the sun hits the plastic, it will shine inside the box. The idea is, at any given time, to mark a spot on the outside of the plastic in such a place that it (the spot) casts a shadow onto the center of the box's bottom. You can do this easily with a felt-tipped marker, as in Figure 8.15. It will take a little trial and error: you may not be able to see the shadow of the spot you mark, but you will see the shadow of the tip of the marker as you make the spot. It is a good idea to mark each spot two or three times, while looking down at the shadow, to be sure you have the right place.

Put the box securely in place, so you are sure it will not move, with the opening facing south. One way to do this is to glue the box to a board and clamp the board to a picnic table. Mark the direction of true south on the box. *True* south is not the same thing as *magnetic* south, and you will have to apply a correction to your compass reading. Figure 8.16 is a map that will help you to do so. It shows the difference between a compass reading of north and true north at any given location in the continental United States. The line marked 0°, passing through the Midwest and the South, represents the locations where the compass reads true north. East of this line the compass points a little to the west of true north, and west of the line the compass points a little to the east of true north. The method of correcting is shown on Figure 8.17. Once you have found true north, you know where true south is, in the opposite direction.

Now you are ready to proceed. Pick a day that is likely to be sunny all day long and on which you can get to the box every hour. You may wish to share the chore with one or more friends. From the time the sun starts to hit the plastic, mark the appropriate spot on it each hour.

If you can spend your lunch break with the box, you might be interested in finding *solar noon,* the time at your locality when the sun is highest in the sky. Figure 8.18 is a map showing the continental U.S. time zones. The time zones are chosen so that solar noon and clock noon are the same at longitudes of 75°, 90°, 105°, and 120°, for the four zones. (The clock time referred to here is *standard* time. If you are in a location with daylight saving time in effect, you will have to subtract one hour

Figure 8.15   Marking the plastic.

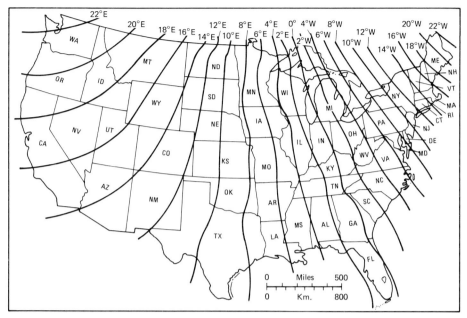

Figure 8.16 A map showing approximate magnetic variation for the United States.

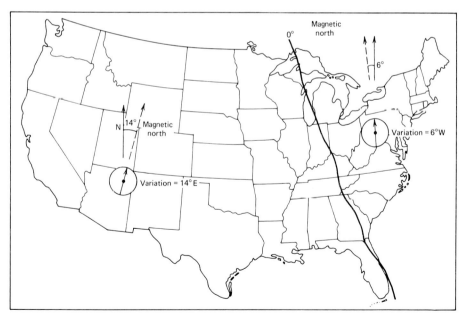

Figure 8.17 In the western part of the country, the compass points to the east of true north. In the eastern part, the compass points to the west of true north.

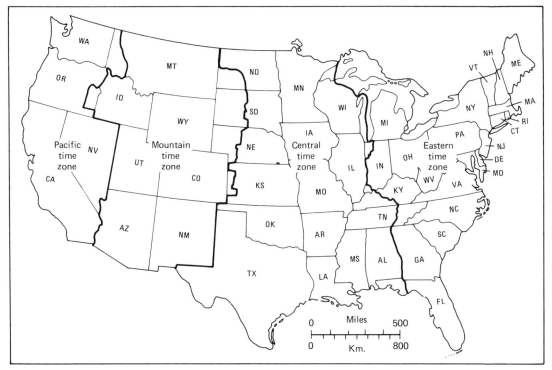

Figure 8.18 The four time zones of the United States.

from the clock time to get standard time.) If you are somewhat to the east of one of these lines, the sun will be highest sometime before noon; if you are to the west, it will be after noon. You can find solar noon for your location approximately from the map. Then, by marking your plastic every 10 min or so during the period you expect solar noon to occur, you ought to be able to find it experimentally. It will be represented by the highest spot on the plastic.

EXAMPLE

Phoenix, Arizona, is at a longitude of 112°. That puts it in the Mountain Standard time zone but 7° west of the 105° line. At what time does solar noon occur in Phoenix?

Solution

Since the sun circles the Earth completely (360°) in 24 hours, it goes 15° (one full time zone) in one hour, or 1° in four minutes. To find solar noon then, use the following:

$$\Delta t = \frac{x}{15°} \times 60 \text{ min}$$

where Δt is the time before or after clock noon and x is the number of degrees in longitude from the place where solar noon and clock noon are the same. In this case, $x = 7°$ and

$$\Delta t = \frac{7°}{15°} \times 60 \text{ min} = 28 \text{ min}$$

Thus, solar noon occurs in Phoenix at 12:28 P.M., Mountain standard time.

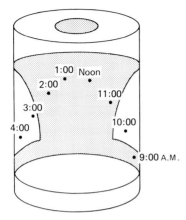

Figure 8.19 The sun's path is traced on a piece of clear plastic mounted on the box.

As another example, Portland, Maine, is at a longitude of about 70°, 5° *east* of the 75° line. Thus, solar noon there is given by

$$\Delta t = \frac{x}{15°} \times 60 \text{ min}$$
$$= \frac{-5°}{15°} \times 60 \text{ min} = -20 \text{ min}$$

Thus, solar noon at Portland occurs at 11:40 A.M., Eastern standard time.

Activity 8.3 will give you some notion of how the sun moves across the sky in the course of a day. The plastic will look something like Figure 8.19. A tracing of an actual sun curve made in this manner is shown in Figure 8.20, reduced to fit the page. This was made in a location where solar noon occurs at about 12:50 P.M.

If you imagine your eye to be at the center of the bottom of the box, you can trace the sun's path through the sky. Better yet, take the piece of plastic off the box and orienting it as it was on the box, look through it toward the south with your eye at the spot where the box bottom was, as shown in Figure 8.21. (Be sure the side representing the morning sun is to your left.) Finally, sketch on the clear plastic the outlines of buildings, trees, and the like that you see to the south through the plastic. This is done in Figure 8.22, and it tells you where the obstructions to sunlight are. If the line representing the sun's path is blocked at any point by obstructions to the south, the sun will not shine on the location of the box during the time it is blocked. If the location is such that the sun is blocked for much of the day in the wintertime, then that location would not be a good place for a solar collector.

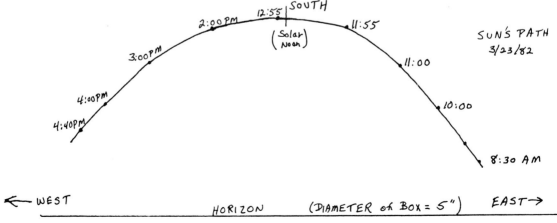

Figure 8.20 A tracing of the path of the sun (reduced).

Figure 8.21 Observing the sun's path through the clear plastic.

If at all possible, it would be instructive to chart the sun's path a week, or, better, a month after your first trial. In general, you will find that the path has moved; the sun is higher or lower in the southern sky at noon than it previously was. (In fact, it is always surprising to me how much the path will move in a few days, particularly in the spring and fall.) A chart showing the path for each month of the year at a latitude of 40° north and a longitude of 75° west (just about the location of Philadelphia) is shown in Figure 8.23. It was made by running a computer program that calculates the points, rather than by using an oatmeal box to mark the points on a piece of plastic. However, the lines correspond very closely to what one would get using a 4-in.-diameter oatmeal box at Phila-

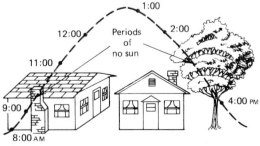

Figure 8.22. In this location, the sun is blocked out at some times of the day by obstructions.

delphia. Note that, except for June 21 and December 21, the sun takes the same path twice in a year, once on its way upward in the sky and once on its way downward. Obviously, the sun is at its lowest in the wintertime, when the need for solar energy would be greatest. For that reason, the obstructions to the south should be carefully considered; a location that gets plenty of sun in the summer may get very little in the winter, and a poorly placed solar collector could be almost totally useless.

Activity 8.4

The reason that the sun has different locations in the sky at different times is related to the way in which the Earth moves around the sun. It is something like a spinning basketball that is also moving along a circular path. To visualize this better, spin a basketball around a vertical axis; that is, so the top and bottom of the ball do not move, as in Figure 8.24. The top of the ball represents the North Pole, and the bottom represents the South Pole. Another person looking at it—playing the part of the sun—will see all the different parts of the ball as it spins, and one complete spin represents one day. Now, if you carry the ball in a circle around the observer, one complete circle will represent one year. If you imagine beams of light coming from the observer (the sun), then the half of the basketball that has light falling on it at any time is in daylight, and the dark half is the nighttime portion.

To complete the picture, you must tilt the axis of the basketball a little, as shown in Figure 8.25, while continuing to let it rotate. When the North Pole is tilted toward the sun, it is summertime in the Northern Hemisphere, and wintertime in the Southern Hemisphere. As the basketball moves in a circle about the "sun," keep it tilted in the same direction. On the opposite side of the circle, the tilt of the "North Pole" will be away from the sun, and the seasons will be reversed.

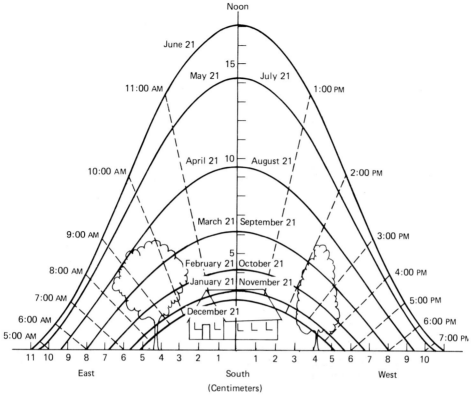

Figure 8.23 A complete chart of the sun's path for Philadelphia. These computer-generated curves correspond to the markings on a sheet of plastic wrapped around a 4-in.-diameter oatmeal box. The units on the horizontal and vertical scales are centimeters.

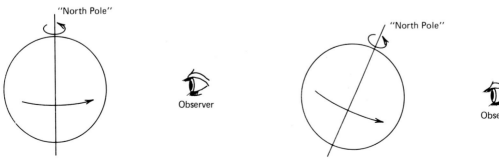

Figure 8.24 Observing a spinning basketball.

Figure 8.25 Observing a tilted, spinning ball.

Figure 8.26 shows the path of the Earth about the sun and its position at different times of the year. As with the basketball, the North Pole is tilted toward the sun in June and away from it in December. Figure 8.27 shows the effect this has on the direction of the sun at noon at a location in the Northern Hemisphere. In the spring and fall, the sun is directly over the equator and, to an observer on the equator, the sun would be directly overhead at noon. To an observer at the spot marked, the sun is at an angle from directly overhead that is equal to the latitude angle. The farther north one goes, the lower the sun is at noon; at the North Pole the sun is on the horizon. (It looks like sunset or sunrise all day long.)

In the summertime, the North Pole is tilted toward the sun, and the sun is more nearly directly overhead for us in the Northern Hemisphere. At this time of the year, the sun's radiation has less atmosphere to get through at the location marked than it does at other times of the year. Also, the sun hits the surface of the Earth more directly, and it is thus more concentrated than it is in other seasons. Thus, it is warmest in the summertime. (By the way, the amount of tilt is about 23.5°.) In the wintertime, the Earth is tilted the other way, and the sun is low in the south at noon. The sun's rays now have a thick layer of atmo-sphere to go through, and not as much heat reaches the ground. The result is cold weather. Having detailed information about the solar radiation at a given location and the path of the sun at all times of the year is very important in the design of solar heating systems.

SUMMARY

All waves transmit energy from one location to another, but the medium through which the wave travels does not move along with it. Continuous waves have a wavelength, the distance between crests, and a frequency, the rate at which the wave repeats. Each kind of wave, moving in a particular medium, has a characteristic speed given by $s = \lambda f$.

Electromagnetic radiation consists of oscillating electric and magnetic fields. It needs no medium, and will travel through empty space. Solar energy reaches the Earth in the form of electromagnetic radiation. The range of frequencies and wavelengths of electromagnetic radiation is very great. A relatively narrow band of those frequencies composes visible light.

A good absorber of radiation is also a good emitter, when hot. A blackbody is a theoretical perfect radiator, with an emissivity of 1. The Stefan–Boltzmann law says that a radia-

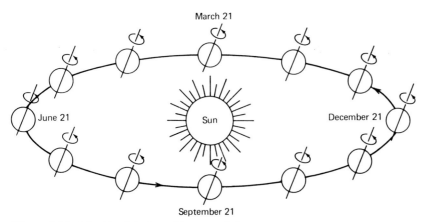

Figure 8.26   The Earth in its orbit around the sun.

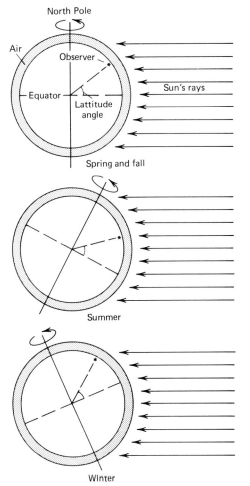

Figure 8.27 The Earth's relationship to the sun's rays for the four seasons.

reaching the surface of the Earth varies greatly, but it may be over 300 Btu/hr per square foot.

To observers on Earth, the sun travels through the sky each day. Knowing its path at different times of the year is important in solar energy applications. The sun is higher in the sky at noon in the summertime than it is in wintertime. In some locations, because of trees, houses, or other obstacles, the sun may strike a solar collector all day long in the summertime. Yet, because of the sun's lower path, it may be blocked for much of the day during the winter at the same location. Such locations would be poor places to put solar collectors.

ADDITIONAL QUESTIONS AND PROBLEMS

1. One day, standing on the shore, I watched waves coming in to the beach. I counted nearly eight complete waves in one minute and estimated that the crests were about 2 m apart. Approximately what was the speed of the water waves?

2. Radar is not explicitly listed in the spectrum shown in Figure 8.5. However, it consists of electromagnetic radiation with wavelengths in the range of about 1 cm (0.01 m) to 1.5 m. What is the corresponding range of frequencies? Where does radar fit in the electromagnetic spectrum?

3. Suppose that you are replacing a roof on your house. Because of the color of the house, and your preferences, the two choices in color for the shingles are black and a very light green. Except for color, the shingles are exactly the same. In the interest of conserving energy, which color would you pick, and why? (*Hints*: In the summertime, which color of shingle will absorb more daytime heat from the sun and pass it on into the attic? In the wintertime, which color will radiate more heat from the house into the night sky?)

tor will emit power at a rate proportional to the fourth power of its Kelvin temperature. Thus, if the temperature of a given object is doubled, the power emitted is 16 times as great.

The sun is nearly a perfect blackbody, emitting energy at a tremendous rate because of its high temperature and huge size. A small fraction of that energy reaches the Earth, which absorbs some of it. The Earth is in an energy balance, emitting the same amount of energy as it absorbs. The total power of solar energy

4. An aluminum sphere, 0.5 m in radius, is highly polished so that its emissivity *e* is 0.2; it is heated to 1000 K. How much radiant energy does it emit? If its temperature is then increased to 2000 K, how much does the emitted energy increase? Now the sphere is covered with lampblack, so that its emissivity becomes 0.8. How much energy does it radiate at 1000 K and 2000 K?

5. A satellite circling Earth in outer space needs an average of 1000 W to continue operation. It is provided with solar cells, which convert about 15 percent of the sun's energy to electric energy, and batteries to store some of the electric energy. If it is in the sun 75 percent of the time, how much area must the array of solar cells cover in order to absorb enough energy to provide the electric energy needed by the satellite?

6. You learned that the sun emits power at the rate of about 3.9×10^{26} W. This power goes out equally in all directions, as indicated in Figure 8.28. Now imagine a huge, hollow sphere with a radius $R = 1.49 \times 10^{11}$ m, the distance of the Earth from the sun, also indicated in Figure 8.28. The Earth will lie on that sphere and on the inner surface of the sphere would fall all of the sun's energy. What is the area of the sphere? What is the total solar power falling on the inside of the sphere? Thus, what is the power per unit area (m²) at that distance from the sun? This is just the solar constant, and it should be close to the measured value given in the text.

7. What is the approximate magnetic variation where you live? Sketch a magnetic com-

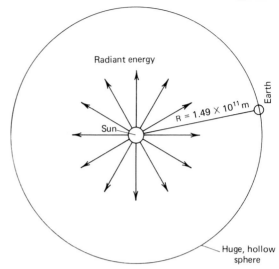

Figure 8.28 Energy radiates from the sun in all directions, reaching the Earth 1.49×10^{11} m away.

pass, with its needle pointing to magnetic north. Then, on your sketch, draw a line pointing to true north in your locality.

8. What is the longitude where you live? At what time (in standard time) does solar noon occur there? At what time in daylight saving time does this occur?

9. Sketch a "beam" of sunlight, say with a cross section of 1 ft². Imagine that a board is positioned so that the "beam" falls directly upon it. It then absorbs a given amount of energy per second, call it P_0. Then, the board is tilted so that the "beam" now covers 2 ft². What is the rate at which energy is now absorbed, per square foot? Use this fact to explain why less solar power is absorbed, on the average, where you live than at the equator.

9
Solar Heating

Few pleasures in life surpass the joy of being outdoors on a beautiful sunny day, summer or winter. In addition to the pleasures it provides, the sun pours energy down upon us, free for the taking to those who know how. Solar energy technology has advanced a lot in the past few years. We are learning to get more of the benefits of the sun's free energy with less effort and expense, and solar systems have been used successfully in a great variety of situations. (See Figure 9.1, for example.) However, there are pitfalls and trying to use solar energy without thought can be a disaster. In many instances large amounts of money and effort have been expended on homes and other buildings to "go solar," but with either the result that only tiny amounts of solar energy are gained in the wintertime, or the result that large but uncontrollable amounts of heat are gained winter *and* summer.

In this chapter we take a look at the two major kinds of solar systems being used for heating. We shall look first at *active* solar systems, using pumps or fans for heat distribution. They are often used on older homes that are having solar added. *Passive* systems use only the sun's energy. They are designed to accept and distribute energy, for the most part,

without electric devices. Some systems may be mostly passive, but include as minor components small fans or other electrical parts. They are sometimes called *hybrid* systems. Using the sun's energy to provide cooling is an undeveloped technology, but the chapter will close with a look at some of the ways in which this is being done.

ACTIVE SOLAR SYSTEMS

An *active* solar system is one that needs pumps or fans and electrical controls for its operation. There are four main parts, indicated in Figure 9.2, to an active system:

1. **The Collectors.** These collect solar energy. They are usually separate from the rest of the system, but connected to it by pipes or ducts.

2. **The Storage.** Since the sun does not shine all of the time, even when the sky is clear, some form of energy storage is needed to save extra heat collected for use at night or in cloudy weather. Some forms of storage involve large tanks of water or bins of rocks.

Figure 9.1 This McDonald's fast food restaurant, in Kingston, Ontario, has a solar system to heat water for the kitchen and rest rooms. (Grumman.)

3. **The Electrical System.**   This includes one or more pumps for liquid systems, one or more fans for air systems, and a control system. In general, the control system senses the temperatures of the collectors, the storage, and the house, and turns on the pumps or fans at the appropriate times.

4. **The Distribution System.**   In an air system, heat is distributed by means of air ducts. In a liquid system, the distribution may be by means of piping to radiators or convectors. Some liquid systems use a heat exchanger to warm air that is then distributed to the house by means of ducts.

A possible fifth part of an active solar system is a backup heating system, to carry the load when the solar system is not able to handle it. This may be needed in very cold weather, or during periods of no sunshine. The backup system may be conventional gas, oil, or electric heat, but more and more persons are turning to wood heat to supplement a solar system. I will not include it as part of

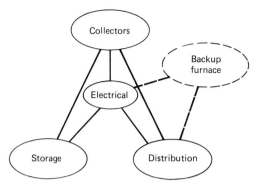

Figure 9.2 The main parts of an active solar system.

Figure 9.3 A simple solar collector made from a cardboard box and a sheet of transparent plastic wrap.

the solar system, but you should remember that, except in unusual circumstances, some sort of backup is needed in cold climates. Perhaps the art of designing solar systems and very well insulated buildings will soon advance to the point that a backup is not needed. Indeed, there are now some solar houses that seem to get along without one.

Here, we shall examine only the collectors and the storage. The control systems and heat distribution systems are interesting enough, but will best be left to a more technical book.

Collectors

Activity 9.1

Cut down a small cardboard box so that its depth is two or three inches. Line the bottom with white paper and attach a thermometer to the inside bottom of the box. Cover the top with transparent plastic wrap, pulled tight and taped to the box so that air cannot enter. Figure 9.3 shows a box like this. Now let the box sit in the shade for a few minutes, so that its temperature becomes the *ambient temperature* (the temperature of the air in the shade). Take a reading of that temperature, then put the box into the sunlight, directly facing the sun.

Take temperature readings every 30 seconds for 5 or 10 minutes. Make a graph of the temperature versus time, a *heating curve*. Now run a trial and graph the heating curve with black paper or cloth lining the bottom of the box. It is also instructive to do the experiment with the back and sides of the box wrapped in insulation, such as a large towel or a blanket.

You have now made a solar collector. Real collectors are a little more sophisticated than this, but not much. The main principle is the same: Let sun fall on a good absorber of radiation and trap the resulting heat in a box.

Figure 9.4 shows some data taken with the box. One curve is for when the bottom is lined with white paper, and the other curve is for when the bottom is lined with black paper. Another set of data were taken using a piece of black corduroy cloth lining the bottom of the box. On that day a cold front was

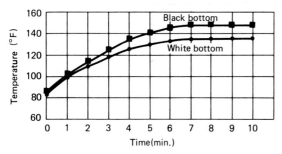

Figure 9.4 Heating curves for a covered box.

approaching, and at about the time the box was put in the sunlight for this trial, clouds started to appear. The result is shown in Figure 9.5. Exactly the same thing happens with real solar collectors—the temperature goes up and down as clouds pass in front of the sun. However, the black cloth proved to be a somewhat better absorber than the black paper; when the sun came out to stay for a while, the temperature of the box reached a maximum of about 165°F. (This was done in April. When repeated the following February, with an ambient temperature of 20°F, the same box soon rose to a temperature of 145°F.)

Flat-Plate, Air-Type Collectors Collectors of solar energy are usually grouped into two types, *flat-plate collectors* and *concentrating collectors*. Flat-plate collectors are much like the simple collector you made from a box; thus they are relatively cheap and durable. The

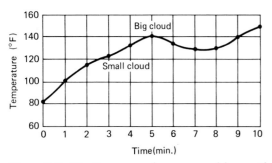

Figure 9.5 Heating curve for a covered box with a cloth-lined bottom.

upper two collectors shown in Figure 9.6 are typical flat-plate collectors. A flat-plate collector is simply an enclosed box, surrounded by insulation on the bottom and sides and covered with a transparent cover sheet, as shown in cross section in Figure 9.7. The cover sheet may be made of glass or clear plastic. Some cover sheets are coated with a thin antireflective film, which minimizes the reflection from the cover and maximizes the transmission of radiation to the absorber. Often a collector has two covers, with an air space between the two sheets, as shown in Figure 9.8, and sometimes there will be three sheets. (These are called *double-glazed* or *triple-glazed* collectors.) The dead air space insulates the collector, just as the air space between a window and a storm window helps to insulate the house.

The absorber may be a layer of black cloth, or a metal sheet covered with a very flat black paint. (*Flat*, in this case, means the opposite of glossy; the paint reflects very little visible light and, it is hoped, very little infrared radiation, as well.) More recently, metal absorber plates have been covered with a *selective surface coating*. A selective coating is designed to make the metal an excellent absorber of solar radiation and a poor emitter of heat. That is, the surface absorbs the visible and short-wavelength infrared radiation well but does not emit the longer wavelength infrared very well. Thus, the incoming radiation warms the absorber, but not much energy is radiated back into the sky. Absorbers that have a selective coating may appear bluish gray or violet, rather than black.

Until recently, it was thought that the "greenhouse effect" is important in solar collectors (and in greenhouses). The greenhouse effect is very important in the energy balance of the Earth, and it works in the following manner. Much of the energy from the sun is in the visible and near infrared regions, and quite a bit of it passes through the atmosphere. It then strikes the surface of the Earth where most of it is absorbed, thus warming

Figure 9.6 Four flat-plate collectors mounted on the roof of Northlake College in Dallas, Texas. The upper two collectors are air types, the lower two are water types.

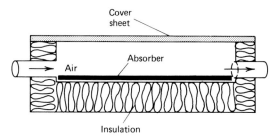

Figure 9.7 Cross section of a flat-plate, air-type collector.

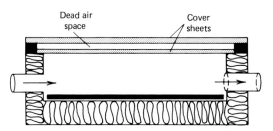

Figure 9.8 Cross section of a double-glazed flat-plate collector.

the Earth. The warm Earth also radiates energy but, since it is much, much cooler than the sun, the energy is in the far infrared, with much longer wavelengths than the incoming energy. This long-wavelength radiation cannot pass through the atmosphere, and is mostly absorbed by it. The atmosphere is then warmed up, and it radiates energy, partly into space, but much of it back to the Earth. In a sense, the energy is "trapped" by the atmosphere, and it cannot readily escape into space.

For many years it was believed that the plain glass on a greenhouse—or a solar collector—served a function similar to that of the atmosphere in trapping the energy radiated from the inside. This is no longer believed to be the case—the effect is there, but it is small. In the atmosphere, a thickness of several miles of air serves to absorb the infrared radiation from the Earth and reradiate some of it back. Plain glass on a collector is too thin for this effect to be very noticeable. However, it is possible to put "selective" coatings on glass or plastic, so that the cover is very transparent to incoming solar radiation but reflects the heat (infrared) radiation from inside the collector, and thus traps it there. Such selective covers are not always used because they are expensive. But, whether plain glass or selectively reflecting glass is used, attention must be paid to the construction of collectors. So that they might function as effectively as possible, they must be well sealed to prevent infiltration of cold air, and they must be well insulated, to minimize the loss of heat by conduction.

EXAMPLE

A flat-plate collector, 8 ft wide and 16 ft long, operating at 110°F above the ambient temperature is calculated to have heat losses through the bottoms, sides, and top totalling 9500 Btu/hr. If solar energy falls on the collector at the rate of 200 Btu/(hr ft^2) and the collector absorbs 70 percent of the incident energy how much usable heat is produced? What is the efficiency of the collector?

Solution

First, we need to find the total insolation on the collector. This is just the rate per square foot times the area:

$$\text{Insolation} = \text{rate} \times \text{area}$$
$$= 200 \, \frac{\text{Btu}}{\text{hr ft}^2} \times 8 \text{ ft} \times 16 \text{ ft}$$
$$= 25,600 \text{ Btu/hr}$$

The rate at which heat is collected is 70 percent of this:

$$P_{in} = .7 \times 25,600 \text{ Btu/hr} = 17,920 \text{ Btu/hr}$$

The net usable power is this amount minus the losses:

$$P_{usable} = 17,920 \text{ Btu/hr} - 9500 \text{ Btu/hr}$$
$$= 8420 \text{ Btu/hr}$$

The efficiency of a collector may be defined as the energy available to heat the house divided by the total energy striking the collector. Thus, in terms of a percentage, the efficiency is

$$Eff = \frac{P_{usable}}{\text{insolation}} \times 100\%$$
$$= \frac{8420 \text{ Btu/hr}}{25,600 \text{ Btu/hr}} \times 100\%$$
$$= 33\%$$

Questions

A flat-plate collector is 4 ft by 8 ft by 6 in. deep. On the bottom and sides, it is insulated to an R-value of 20, and the top is single glazed, with an R-value of 0.9. It is so well sealed that there is no appreciable infiltration. On a day when the ambient temperature outdoors is 30°F, the temperature in the collector is 130°F.

What is the heat loss to the outdoors by conduction through the bottom? Though the sides? Through the top? If the top is changed to a double-glazed cover, with a total *R*-value of 1.7, what is the total heat loss?

At a particular time, solar energy, both direct and diffuse, reaches the double-glazed collector at the rate of 250 Btu/hr per ft². The collector captures 80 percent of the energy, with the other 20 percent being partly reflected from the cover sheets and partly ra-

diated directly into the sky. How much energy is available per hour to heat the house? What is the efficiency of this collector?

Flat-Plate, Liquid-Type Collectors   In many solar systems, a liquid—most often water or water plus antifreeze—is used as the medium to move heat from the collector to the house. This kind of collector includes plumbing of one sort or another, typically copper tubing attached to a flat absorber plate, as shown in Figure 9.9. Thus heat can pass from the ab-

Figure 9.9   A typical flat-plate, liquid-type collector. The absorber has tubes running between two sheets of copper. It is painted black and mounted in the collector, with intake and outlet pipes (manifolds) running across the top and bottom.

sorber through the tubing and into the water. Some absorber plates are made with two layers of sheet metal with the water-carrying tubes built right into them. At the top and bottom of the absorber, there is a *manifold,* a pipe to carry the liquid to and away from all of the pipes in the collector. Figures 9.10 and 9.11 are photographs of two experimental collector installations.

When one hears about problems in maintaining a solar system, it probably involves a system that uses a liquid. But air-type systems also have problems. For example, an improp-erly installed collector may cause roof leaks, or it may be damaged by strong winds. Further, having to move large amounts of air requires a fairly large fan which can be costly to run. However, when plumbing and liquids are involved the chances for something to go wrong are increased. A leak in the system could be disastrous. Sometimes just a bubble of air trapped in one of the pipes can make the whole system work poorly. Perhaps worst of all is the problem of collectors freezing in cold climates. When metal pipes get cold, they contract but, when the water in them freezes,

Figure 9.10 Scientists at the U.S. Department of Agriculture Animal Genetics and Management Laboratory in Beltsville, Maryland, are experimentally heating the wash water and milking parlor with solar energy. Four different types of collectors are being compared. (USDA photo.)

Figure 9.11 The Campbell Soup plant in Sacramento, California, is testing a solar industrial process hot water system. The water is used to wash empty and full soup cans on a production line. Another, nonsolar, production line is being run in parallel to compare efficiency, reliability, and operating costs. (Photo courtesy Acurex Corporation.)

it expands; the resulting pressures crack the pipes and make them useless. There are two ways to get around this problem. One is to install an automatic drain-down system, which senses when the temperature outdoors approaches freezing and drains all of the water out of the collectors. The other is to use a mixture of water and antifreeze.

There are disadvantages to both of these methods. The drain-down system is expensive and, if it fails, damage will be done by a freeze. Using antifreeze avoids the need to drain down, but antifreeze has a lower specific heat than water so, with the same tem-

perature difference between collector and house, more antifreeze will have to be pumped than water, thus requiring a bigger and more expensive pump. Also, when antifreeze is used in a system that produces hot water as well as space heating, some means is needed to keep the antifreeze out of the household water supply. This can be accomplished by a *heat exchanger*, such as the one shown in Figure 9.12, but this adds expense and cuts down on the overall efficiency of the solar system. There is also some chance of a leak in the heat exchanger, which would contaminate the drinking water, and some kinds of antifreeze are

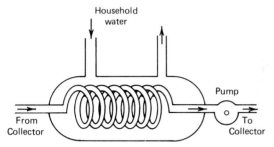

9.12 A heat exchanger keeps two liquids separated while passing heat from one to the other.

toxic. (It is highly recommended that only a nontoxic antifreeze, such as propylene glycol, be used in solar systems.)

Collector Efficiency Discussions of the efficiency of solar collectors are liable to be misleading unless some ground rules are understood from the beginning. For one thing, the efficiency of any given collector will vary greatly, depending upon the conditions. Then, too, if the definition of efficiency involves usable heat, the efficiency of the collector can be zero, or even negative! For example, the little collector you made for Activity 9.1 had an efficiency of zero. It got warm, all right, but none of the heat produced was usable. As the collector warmed up, the heat losses by conduction and radiation became greater and greater. This went on until the losses were equal to the gains, at which point the temperature could rise no more. Because there was no way to extract the heat from the collector, its efficiency was zero.

Zero-efficiency may occur in a real solar installation, but even worse is the "negative-efficiency" situation. This can happen when a system is so poorly designed that it continues to operate when the collector temperature is lower than the indoor temperature. In that case, the unfortunate home owner pumps heat from his home to the outdoors. Such an effort probably contributes little to improving the outdoor climate, but it has a serious effect on the indoor comfort.

Let us define the efficiency of a solar collector as the usable heat divided by the total insolation. Put in terms of a percentage:

$$Eff = \frac{\text{usable heat}}{\text{insolation}} \times 100\%$$

Remember the insolation is measured for a surface that is *perpendicular* to the incoming radiation. It can be quite different in amount from the direct radiation falling upon the surface of the collector. To see this look at Figure 9.13. When the collector is tilted so that the sun's rays come in perpendicular to its surface, the maximum amount of energy is collected. However, when the collector is flat on the ground, much of the energy that previously hit it misses it entirely, and is thus wasted. (When the collector is flat on the ground, it can "see" more of the sky than when it is tilted, and thus more diffuse ra-

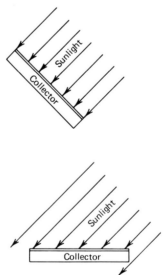

Figure 9.13 The angle at which the sunlight strikes the collector surface affects the amount of energy striking the surface. In the upper picture, all of the energy coming in strikes the surface. In the lower picture, some of it misses.

diation falls upon it. However, the direct radiation lost is greater than the diffuse radiation gained.) Therefore, under otherwise similar conditions, a collector that is tilted toward the sun is more efficient than the same collector lying flat on the ground or mounted on a vertical wall.

Making the angle at which the sunlight enters even more important is the fact that the least reflection from the cover sheet occurs when the sun comes in perpendicular to the surface. When that angle is less than 90°, greater reflection occurs, so there is less energy available to heat the collector. Figure 9.14 is a graph showing how the efficiency of a typical collector changes as the angle at which the sunlight strikes it changes.

As a result, the way to get the maximum energy during the day and year is to continually turn the collector so that it faces the sun at all times during daylight hours. It is possible to install an automatic "tracking" system which causes the collector to follow the sun. However, such a system cannot be justified for flat-plate collectors on the basis of costs; the extra energy gained would not pay for the cost of the tracking system in its lifetime. Looking again at Figure 9.14, you can see that until the angle between the sunlight and the

collector surface gets smaller than 60°, or so, the efficiency does not go down too much.

As a compromise for space-heating solar systems, a flat-plate collector is usually pointed to the south and tilted at the angle that most directly faces the winter sun—in December or January—and left in that position. However, for hot-water heating, the system is used year round, and there is a large change in the position of the sun from December to June. In such systems, the collector is sometimes hinged so that its tilt can be changed about once a month to take the best advantage of the sun during that month. Even with an adjustable mount, the sun will hit the collector at different angles during the course of a day, with the most heat being available at solar noon.

In addition to the type of cover sheet, the coating on the absorber, the insulation of the collector, and the angle at which the sun strikes it, the temperature of the collector affects its efficiency. You saw this with your little collector: When the temperature got high enough, all of the energy gained was simultaneously lost, and the temperature could go no higher. The efficiency of the collector was thus zero. As a general rule, the greater the temperature difference between the collector and the outside air, the greater will be the heat losses and the lower will be the efficiency. Figure 9.15 shows the effect of temperature on efficiency for a particular collector (two cover sheets, nonselective absorber). When the temperature of the collector is just a little above the outdoor temperature, the efficiency is quite high; much of the incoming radiation is turned into useful heat. However, at a temperature of about 160°F above the ambient temperature, the collector loses most of the heat collected and only about 15 percent is useful heat.

One final word about collector efficiency: it is not all-important! A moment's reflection upon Figure 9.15 will convince you that the condition with the highest efficiency may be useless in practice. If the ambient temperature

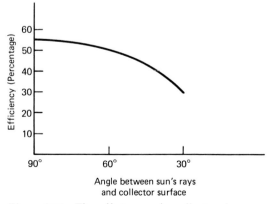

Figure 9.14 The efficiency of a collector decreases as the angle at which the sun's rays hit it goes to less than 90°.

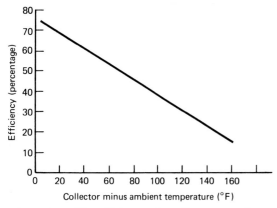

Figure 9.15 The efficiency of a collector goes down as its temperature goes up.

is 30°F, then a collector at 40°F may be theoretically efficient, but it will not put any heat into a house that is at 70°F. In fact, if it is put into operation the collector will remove heat from the house. The trick, then, is to get a collector that has a reasonably high efficiency at a high enough temperature to provide the needed heat.

This discussion of efficiency points out one way in which unethical or unknowledgeable promoters of solar heating equipment may knowingly or unknowingly misrepresent their products. A salesperson may say that the collector is "up to 75 percent efficient" and show data to prove it. The truth may be, however, that at useful temperatures (say 100°F above ambient), the collector is closer to 40 percent efficient, and the collector may provide only half the heat promised.

Concentrating Collectors

Activity 9.2

Find a lens that is thicker in the center than at the edges. The "fatter" it is the better. If you cannot find one anywhere else, you might borrow a pair of strong bifocals from someone over 40 years old (a far-sighted person, that is).

On a bright, sunny day—summer or winter—hold the lens a few inches from your hand and focus an image of the sun on your palm. (It will look like a bright disk.) If, after trying a range of distances from your palm, you cannot form an image of the sun, you probably have the wrong kind of lens—one that spreads out the light rather than focusing it (thinner in the center than at the edges).

Once you have an image move the lens closer to and farther from your palm until the image of the sun becomes as small as you can get it. Ouch! That burns, doesn't it? Don't wait to read the rest of this before you move your palm. Try focusing the sunlight on the bulb of a thermometer and seeing how hot it gets. If you have a really powerful lens, you may be able to make a piece of paper smoke by focusing the sun's image upon it. (Incidentally, even though it may be the size of a pinhead, the image you have on the paper is still that of the sun.) With the right kind of tinder, it is possible to start a fire using just a powerful lens and the sun.

Sunshine falling on your skin may cause it to "burn" after a period of time, but that is different from what you just experienced. A sunburn is caused by ultraviolet radiation destroying skin cells over a period of minutes or hours. When you focus the sun's image on your palm, the burning is immediate and is caused by intense heat, produced by visible and infrared radiation. (In fact, a glass lens will not allow ultraviolet radiation to pass through it.) Can you explain why the radiation is so much more intense under the lens?

The lens takes the energy falling on a relatively large area, and concentrates it into a small area. For example, if the situation is as in Figure 9.16, the diameter of the lens is 16 times the diameter of the spot it produces. Since the area of a circle is proportional to the diameter squared, the area of the lens is 16 × 16 = 256 times the area of the spot. But, except for the part that cannot pass through

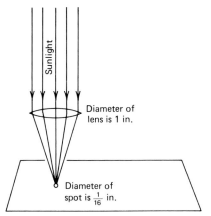

Figure 9.16 Focusing an image of the sun on a piece of paper.

the glass, all of the energy that falls on the lens is concentrated in the little spot. Thus, if 90 percent of the radiation passes through the glass, the intensity on the paper will be 230 times the intensity with no lens. (*Intensity* is a new word: It means power per unit area, such as Btu/(hr ft²) or W/m², which we have already used.) It is that factor of 230 that causes

your palm to feel the intense heat so suddenly. In heating the bulb of a thermometer I discovered that a relatively weak lens, not capable of burning my hand, produced the highest temperature reading. Why? Well, it happened to be a much larger lens than the strongest ones I had available; as long as the image of the sun is not larger than the bulb, more power is concentrated on the bulb by the larger lens.

A *concentrating collector* takes advantage of the ability of a lens or a mirror to concentrate the sun's rays. Figure 9.17 shows one kind of concentrating collector, and Figure 9.18 is a drawing of its cross section. The lens used is long and narrow instead of circular. It concentrates the sunlight on a pipe running through the collector. The pipe is painted black, and water or another liquid is pumped through it to collect the heat and carry it to where it is needed.

Another type of concentrating collector is shown in cross section in Figure 9.19. Instead of using a lens to concentrate the sunlight, this collector does it with a properly

Figure 9.17 A concentrating solar collector array. Each collector in the three banks is aimed to the west in this photo. (Photo courtesy Solar Tech Project.)

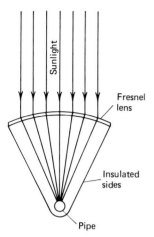

Figure 9.18 Cross section of a concentrating collector.

shaped mirror. Once again, most of the sunlight strikes a black pipe that absorbs the energy, and the energy is distributed by means of a liquid flowing in the pipe.

The main advantage of concentrating collectors is that they are capable of producing high temperatures with reasonable efficiencies. The total amount of collecting area may not be any greater than for a flat-plate collector, so the total energy collected when the sun's rays are perpendicular to the collector will not be greater. However, the higher temperature reached by a concentrating collector may be desirable, for example, in producing very hot water. Further, since concentrating

collectors are usually made to track the sun, the energy collected for a full day will be greater than for a flat-plate collector of the same size.

There are also several disadvantages of concentrating collectors. For one thing, they are more expensive to build and maintain than are flat-plate collectors. Also, lenses can be broken and are costly to replace, and mirrors gradually lose their reflectivity and must be polished or replaced.

Finally, most concentrating collectors must track the sun as it moves across the sky. The reason for this is shown in Figure 9.20. If the sun's rays are perpendicular to the paper, the sun's image hits on a spot we can call the "intended spot." This is like the sun's rays in the collector hitting the pipe carrying the liquid. However, when the sun moves, unless the lens and paper also move, the image moves to a different spot. You could see this by clamping a lens over a piece of paper in the sunlight and watching it for a few minutes. The same thing occurs in a nontracking concentrating collector, as shown in Figure 9.21. The tracking system usually includes an electronic circuit to sense the position of the sun and follow it, a motor controlled by that circuit, and the appropriate gears. There also

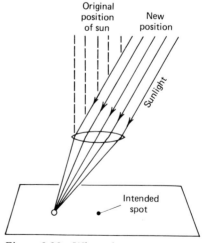

Figure 9.20 When the sun moves, unless the lens and the paper also move, the sun's image drifts away from the spot it is intended to hit.

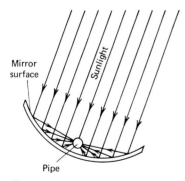

Figure 9.19 Cross section of a reflecting, concentrating collector.

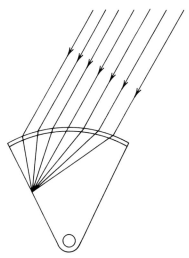

Figure 9.21 If the sun's rays come to a concentrating collector at an angle, the focused light fails to hit the liquid-carrying pipe.

must be an arrangement, usually automatic, for turning the collectors back to the east each evening, so they will be ready for the morning sun.

Heat Storage

The nice thing about the energy in oil or gas is that it is *stored* energy. You can keep it in a tank, ready and waiting, and use it when it is convenient for you. The sun, on the other hand, does not shine at your convenience. You must either use its energy when it is available, or find some way to store it for later use.

Solar hot-water systems usually have a built-in heat storage device: the tank that holds the hot water. The water is heated when the sun shines, and some of it may be used at that time, but some of the heat is stored for later use. In a hot-water system, the hot water itself is used; in a space-heating system, the heat must be extracted from the water or other material in which it is stored.

In solar systems, heat is stored as *sensible heat* or *latent heat*, with the former being the most common so far. The term *sensible heat*

has nothing to do with common sense. It just describes the kind of heat storage we have already discussed, in which heat added to a material causes its temperature to rise. The two most common materials used to store sensible heat are water and rock. As you know, the specific heat of water is 1.000 Btu/(lb °F), meaning that if you raise the temperature of 1 lb of water by 1°F, you have "stored" 1 Btu in it. Under the proper conditions, this Btu may be recovered from the water and used for space heating or other purposes.

Most rocks and stones, on the other hand, have a specific heat of around 0.2 Btu/(lb °F). In order to store 1 Btu of heat in rock, it is necessary to raise the temperature of 1 lb of rock 5°F or raise the temperature of 5 lb of rock 1°F. This is a disadvantage, compared to the heat-storing abilities of water, but there are times when rock storage is preferable.

The simplest storage system consists of an insulated tank, connected to a water-type collector, storing the hot water, as indicated in Figure 9.22. The warm water comes in at the top of the tank, and the cool water leaves at the bottom.

One way to understand such a system is to think of the circulating water as cooling the collector. When the sun first strikes the collector, the collector is cool. If the water is not circulated, the collector will become very hot,

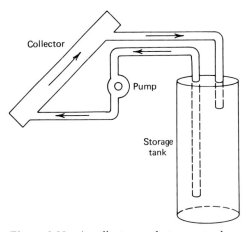

Figure 9.22 A collector and storage tank.

often reaching 200 or even 300°F. Thus, the water must be pumped to cool the collector and warm the water in the tank. One can start the pump just as soon as the collector gets a little warm, or one can wait until the water in the collector is very hot, then start the pump to move the hot water to the tank. Which do you think would be more effective?

If your answer was to start the pump right away, you were correct. Remember that a collector is generally more efficient at transferring heat when it is cooler—closer to the ambient temperature. As it gets very hot, it loses more energy to its surroundings and becomes less efficient at transferring heat to the tank. Thus, it is best to start the pump as soon as the collector temperature is a few degrees higher than the tank temperature and to keep it running as long as the collector is warmer.

Questions

A solar system has a 1000-gal water tank for heat storage. The large bank of collectors that it is connected to is capable of collecting 350,000 Btu on an average January day. On a particular morning, the water in the tank is at 80°F just as the sun starts hitting the collectors. This is a well-insulated house, so it takes 250,000 Btu to heat it for 24 hours at this time of year. What is the temperature of the water in the tank the next morning? At this same rate of energy collection and use, for how many days would the sun have to shine in order to store enough heat to keep the house warm for one day of cloudy weather? With that much stored energy, what would be the temperature of the water in the tank? (Use 75°F as the lowest temperature at which useful heat can be extracted.)

Rock storage of heat is used in systems that run on air rather than on water. Although it takes five times as much rock as water to store the same amount of heat, air-based systems tend to be cheaper and easier to maintain. Also, the rock storage, although bulky, is not expensive or difficult to make. It can be contained in a simple bin in the basement, as illustrated in Figure 9.23. The warm air comes in through a duct at the top of the bin. It then passes through holes in the duct and then through and around the rocks, giving up heat in the process. The cooled air comes out a duct at the bottom of the bin.

The rocks used might be as small as 1 in. or as large as 5 or 6 in. in diameter. The smaller rocks pack more closely together, thus having more weight for a bin of given volume, but it requires a larger fan to blow the air through the smaller spaces between rocks. The rocks also need to be clean; any mud or dirt decreases their effectiveness in storing heat. (I know a family who once washed 18,000 lbs of very muddy rocks by hand.)

Questions

It is desired to size a rock bin so that the rocks are capable of storing at least 300,000 Btu of usable heat. The maximum temperature of the

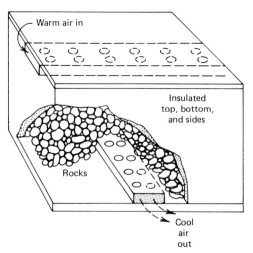

Figure 9.23   A rock storage bin.

rock bin should be 110°F and the distribution system is capable of extracting usable heat until the rock temperature goes below 75°F. The specific heat of the rock is 0.21 Btu/(lb °F), and it weighs 160 lb/ft³. What is the weight of rock required? What volume will that much rock occupy? Assuming that rock occupies 60 percent of the space in the rock bin and air the other 40 percent, what will the volume of the bin have to be? What would be reasonable dimensions for a rock bin of this volume?

Activity 9.3

In this activity you will find approximately how much heat is needed to melt a gram of ice. To begin, fill a styrofoam cup about two-thirds full of cold water. Add some crushed ice and stir with a thermometer until the temperature stops dropping. What is the lowest temperature reached? Now remove all of the remaining ice and make a mark on the inside of the cup at the level of the cold water. Quickly, while the water is still cold, dry some crushed ice, perhaps 10 or 20 g, on a paper towel and add it to the water. Start a stopwatch and, stirring continuously, take a temperature reading every minute. Note carefully the time when the last of the ice melts and continue to take temperature readings every minute for ten minutes beyond that time. Now make a mark on the inside of the cup at the new water level. By pouring the water into a measuring cup (or cylinder) marked in milliliters, find the amount of water in the cup when it is filled to the top mark, then the amount when it is filled to the bottom mark. The difference in the two measurements is the amount of ice you started with. Make a graph of temperature versus time. The fact that the temperature rises slowly after the ice is completely melted indicates that heat is flowing into the water from the surrounding room, which is at a higher temperature. In the 10-min period for which you took read-

ings after the ice melted, how much did the temperature rise? Thus, what was the average rise in temperature per minute? How many grams (milliliters) of water were warming up? Thus, how much heat, in calories, went into the water per minute?

It is reasonable to assume that heat went into melting the ice at roughly the same rate as it went into warming the water after the ice was all melted. How long did it take to melt the ice? Thus, how many calories went into melting the ice? How much ice did you start with? Finally, how many calories per gram does it take to melt ice?

Your graph should look something like Figure 9.24. When you first start timing, and until all of the ice is gone, the temperature of the ice–water mixture stays at 0°C. Even though heat is being added, the temperature does not change, for the energy goes into changing the ice to water. Remember that earlier you learned that solids have molecules that are tightly bound to each other, whereas the binding between molecules in a liquid is not as strong. To separate the molecules enough to change the substance from solid to liquid takes energy. The energy supplied is called the *heat of fusion* (or the *latent heat of fusion*). In the case of water, 80 calories must be added to each gram of ice to cause it to melt. (How close to that did you come?) The temperature of the ice is 0°C just before melting occurs and the temperature of the melt water is also 0°C. To freeze water at 0°C to ice at 0°C requires that 80 cal be *removed* from each gram. For comparison, the melting point of copper is 1083°C and its heat of fusion is 42 cal/g; the melting point of mercury is −39°C and its heat of fusion is 2.8 cal/g.

Activity 9.4

This is similar to Activity 5.3; you will heat water in a styrofoam cup using an immersion

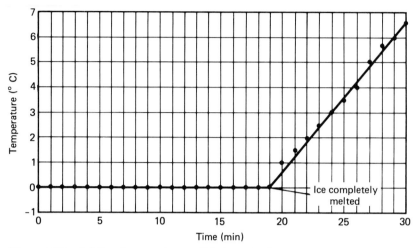

Figure 9.24 Adding heat to ice until it is melted and the water warms. (Start with 24 g of ice in 134 g of water at 0° C.)

heater. The difference is that this time you will continue heating until some of the water boils away.

With the (cool) immersion heater and the thermometer in the cup, fill the cup nearly full of water. Mark the water level and measure the volume of water by pouring it into a measuring cup marked in milliliters. Now measure out an amount of water that is 25 ml less and pour this into the styrofoam cup. Mark the level, also with the heater and the thermometer in the cup, and you will have a cup with two marked levels—one for a nearly full cup and one at the level that is 25 ml less.

With the immersion heater and the thermometer in the cup, fill it with cold water to the top mark. Now—being sure the heater does not touch the sides of the cup—turn on the immersion heater and start timing. Stir the water continuously and read the thermometer about every 30 s. It is very important for the stirring to be continuous and vigorous—so stir!

Watch for boiling and mark the time at which it starts. Keep stirring! After the water starts boiling, continue heating and taking the temperature until the water level reaches the lower mark. (That is, until 25 g of water have boiled away.) Note the time and stop the experiment.

Graph your results—temperature versus time.

Your graph should look something like that of Figure 9.25. As heat is added to the water its temperature rises, of course. When boiling starts, the temperature levels off and stays there. As with the melting of ice, once again the energy added is going into breaking the bonds between molecules and moving them even farther apart. The steam—or vapor—produced is at a temperature of 100°C, the boiling point. The heat added per gram to change water to steam is called the *heat of vaporization* (or *latent heat of vaporization*). The following set of questions will help you to find the heat of vaporization of water from your graph.

Questions

Fix your attention on the part of your graph where the temperature is increasing and pick

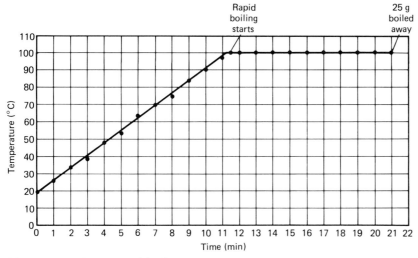

Figure 9.25 Heating and boiling water. (This experiment started with 200 g of water at 19° C. The immersion heater was rated at 100 W.)

out a convenient time span, say 100 seconds. During that time, what was the increase in temperature? What was the total amount of heat added? How much heat was added per second? One way to check this answer is to multiply it by 4.2 to change it to J/S (equals W). Does the answer you get that way match the power rating of the immersion heater? If it is not fairly close, you have made an error somewhere.

How long, in seconds, did it take for 25 g of water to boil away? It is reasonable to assume that heat was added by the immersion heater at the same rate during the boiling period as it was during the heating period. If that is the case, how much total energy went into the water while the 25 g were boiling away? How much energy is this per gram?

So what?

Well, the possibilities for energy storage are tremendous. It requires 80 times more heat to change ice to water to water than to raise the temperature of the same amount of water by 1°C. Since 1°C = 9/5°F, the heat needed is 144 (that is, 9/5 × 80) times that needed to raise the temperature of the same amount of water by 1°F. That is, changing the solid to a liquid (called a *phase change*) stores 144 Btu for each pound of water involved, if only we could find some way to get the stored energy back. This fact is used to advantage in the orange groves. When a freeze is expected, the orange trees are sprayed with water. Before the water can freeze, the atmosphere must remove enough heat to get its temperature down to the freezing point, then an additional 144 Btu/lb. As long as the water doesn't freeze, the blossoms and fruit it covers don't freeze either.

The freezing point (or melting point) of water is too low for it to be used to store latent heat in a solar system, but there are other substances that will work. What is needed is a material that is a solid at room temperature, but melts at a somewhat higher temperature. Some salts do have melting points in the range of 74 to 120°F. One is called *Glauber's salt*. It melts at 90°F, and its heat of fusion is 104 Btu/lb. Further, as a liquid above 90°F, its specific heat is about 0.7 Btu/(lb °F). Therefore, in being heated from a solid at 90°F to liquid at 120°F, 1 lb will store 104 + 21 = 125 Btu.

If the temperature of 1 lb of water is raised through the same limits, it will store only 30 Btu. Thus, 1 lb of Glauber's salt will store as much as 4 lb of water or about 20 lb of rock.

Salts such as this have been used in solar installations to store "latent heat." The problem is that when properly packaged in noncorrosible containers, they tend to be expensive. Also, after one of the salts has gone through a large number of cycles of melting and freezing (solidifying), its character gradually changes in a way that makes it useless for further heat storage. These problems are being attacked by researchers, and some of these salts hold great promise for future use.

SOLAR HOT WATER

The other day I left a garden hose filled with water out in the sun for a couple of hours. When I checked it later, the temperature of the water in the hose was more than 30°F higher than the outside air temperature. This is fairly impressive, considering that the hose is made of a material that is a good insulator. It suggests that solar energy is a good way to provide hot water at little cost.

In fact, this is far from a new idea. Solar water heaters were used in California before 1900. In the 1930's there were hundreds of solar hot-water systems in Florida. But then other forms of energy—electricity and gas—got so cheap that solar systems could not compete. No further systems were built, and the existing ones were gradually replaced with electric or gas water heaters.

Now that electricity and gas are becoming more expensive, new consideration is being given to solar water heating, and new systems are being put on the market almost daily. Here, we shall consider only "domestic" water heating, as opposed to industrial or commercial water heating. Domestic hot water, whether it is used in homes or businesses, is hot water used for baths, cooking, washing,

dishwashers, and other uses found in homes. In many homes, up to 20 percent of all the energy used goes to heat water. It could be even more than that if the water heater is old, inefficient, and poorly insulated and if the water temperature is kept higher than it needs to be. (About 140°F is high enough for residential needs. If there is no dishwasher, or the dishwasher is an "energy saver," which heats the incoming water, 120°F is high enough.) A well-designed and properly installed solar hot water system could save a sizable fraction of the costs of hot water heating and it could pay for itself over a period of time.

One type of solar hot water system is almost as simple as the garden hose left in

Figure 9.26 A simple solar hot-water heater. (Reprinted with permission of *Rodale's New Shelter* magazine. Copyright © 1981, Rodale Press Inc. All rights reserved.)

the sunlight, for it is just a tank of water left in the sunlight. It may be in its own insulated box outdoors on the ground, or it may be in a closet with a south-facing window. A simple tank system is shown in Figure 9.26. This system incorporates a dual reflector, which helps to direct the sunlight to the tank. A system with a reflector of this type is not quite as efficient as a system that tracks the sun, but it is very easy to make and there are no maintenance costs. The tank and reflectors should be pointed to the south and tilted at an angle about equal to the local latitude.

The most straightforward way to install the solar hot water system is shown in Figure 9.27. There are no moving parts, except the water itself. The heated water in the tank rises to the top by convection, and the cooler water sinks to the bottom. When a tap is turned on, the cold water coming in at the bottom forces hot water out the top. The electric heating element takes over during long periods of cloudiness or whenever the solar heat cannot keep up with the demand. (Actually, there is often a second hot water tank indoors, with electric or gas heat, for use when needed.)

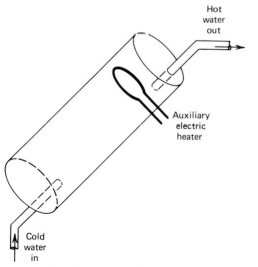

Figure 9.27 The required plumbing for a simple tank heater.

Tests indicate that a very simple passive tank system of this sort can provide about one-third of the hot water needed by a conserving family of four.

EXAMPLE

A tank solar hot water heater in Ohio adds an average of about 11,000 Btu of heat to the water per day, excluding the winter months. The cold water in that area is about 60°F, and it is desired to have the hot water at 120°F. It is used about nine months a year, and it cost $400 to build and install. (It was homemade.) How many gallons of water a day can be heated to the desired temperature? Assuming that the electric power rates remain at $0.06 per kilowatt-hour, how long is the payback period?

Solution

To find the amount of water that can be heated per day, use

$$H = Cm\Delta T$$

or

$$m = \frac{H}{C\Delta T} = \frac{11,000 \text{ Btu}}{1 \dfrac{\text{Btu}}{\text{lb °F}} \times 60°\text{F}}$$

$$= 183 \text{ lb} \approx 22 \text{ gal}$$

To see what the payback time would be, first calculate the total heat provided to the water in a year (nine months):

$$H_{\text{total}} = 11,000 \text{ Btu/day} \times 9 \text{ months} \times 30 \text{ days/mo.}$$

$$= 2,970,000 \text{ Btu} \times \frac{1 \text{ kWh}}{3413 \text{ Btu}}$$

$$\approx 870 \text{ kWh}$$

At $0.06/kWh, the electricity needed to produce this much heat would cost about $52. Thus the payback period would be almost eight years.

There is an important lesson, or two, to be learned from the results of this example. First, even a simple and inexpensive system will require years of trouble-free service to pay for itself in electric power saved. Claims that a solar system will pay for itself in a year or two are generally highly inflated. Also, the amount of hot water produced by the system is not very large. The answer turns out to be about 22 gal/day of hot water. A person who takes very hot, long showers could easily use over half of that in a single shower.

This brings us to a general and very useful rule of thumb which can be applied to all solar systems, whether they are for heating water or space. It is that the installation of a solar system is often a waste of money and effort—unless some conservation measures have been taken first. Generally speaking, conservation measures can save three or four times more energy than a solar system can provide, and the cost is usually far less. *After* conservation has been taken care of, then solar energy can make a lot of sense.

In the case of domestic hot water, the steps taken can include both conservation of water and conservation of energy. If less hot water is used, then less energy is needed to heat it. A nonconserving family can easily use 60 or more gallons per person per day. Of the 60 gallons, 20 or more may be hot water. Some measures to conserve water (which is a worthwhile aim in itself) include installing a pressure-reducing valve where the water enters the house, flow restrictors on the shower head and the faucets, and smaller tanks on the toilets. Stopping all drips would help a great deal, as would changing some habits. For example, filling the washing machine with clothes and the dishwasher with dishes to capacity before each use could save considerable water. To heat the hot water that *is* nec-

essary more efficiently, one can insulate the hot-water tank and put a timer on it so that it does not heat water when it is not needed, such as at night. (After all, heat escapes from the tank at night, but it is rare that one needs much hot water at 3:00 A.M.)

The tank system described is a simple type of *passive* solar hot-water system. Another type of passive solar hot-water heater is the *thermosiphon*. A thermosiphon uses a separate collector and storage tank, arranged in such a way that natural convection carries the hot water to the tank. This is shown in Figure 9.28. When the water is not running out of the tap, natural convection causes the water in the collector to rise to the top of the tank. It is then replaced by cold water from the bottom of the tank, which gets warmed in turn. When a tap is turned on, hot water from the top of the tank is supplied. A major disadvantage of a thermosiphon system is that a properly sized water tank weighs 400 or 500 lb, or more, and supporting that much weight at the top of the attic could be more than the structure of the house could stand.

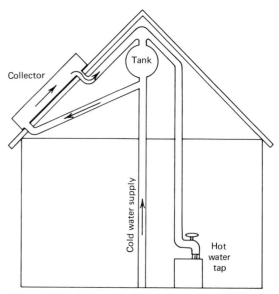

Figure 9.28   A thermosiphon hot-water system.

A disadvantage of both passive systems described is that they will not work in long periods of freezing weather. (The tank heater could be used if it were in an indoor location, such as a closet with a south-facing window.) The water freezes on cold nights and ruins the plumbing. To get around this problem, the collector must either be filled with an antifreeze-water solution, or some way must be arranged to drain it on cold nights. In either of those cases an active system, with pumps to move the water around, is required. An active, drain-back hot-water system, with heat exchanger, is shown in Figure 9.29.

The recent history of solar water heating systems is filled with sad stories of leaks, storm-damaged collectors, tiny amounts of luke-warm water being produced, leaking roofs, pipes bursting because of freezing, drinking water contaminated with antifreeze, and on and on. If you talk to someone who has had a solar system installed, chances are you will hear a sad tale of woe. This need not be the case and, as the fly-by-night installers are gradually weeded out, such stories are becoming less common. A well-designed system, installed by a reputable contractor—or by a competent do-it-yourselfer—should give many years of trouble-free service. As successful experiences with solar systems increase, two cardinal rules have emerged. First, and foremost, before installing any kind of a solar system, be sure that all reasonable conservation measures have been taken. Otherwise, the solar installation will probably not be satisfactory. Second, keep the solar system as simple as possible. For many applications, this will mean installing a passive rather than an active system. However, there are situations for which an active system is preferred, such as when solar hot water is desired year round. Even then, one should resist the American love of gadgets and keep the system simple; the more gadgets there are, the more there is to go wrong.

SOLAR SPACE HEATING

The heating of living and working spaces by means of solar energy is still a relatively young art. However, much has been accomplished in the past few years, and there is much promise for the future. Both active and passive systems have been used for space heating, but the simplicity, lower cost, and lower maintenance of passive systems makes them very attractive.

Active Systems

There are, of course, numerous different kinds of active solar systems. They range from very expensive custom-made systems installed by professionals to home-built systems installed by do-it-yourself homeowners. One type of active solar system is shown in Figure 9.30. This is a liquid-based system that uses a heat exchanger and two pumps. In this type, the fluid circulating in the collector can be water

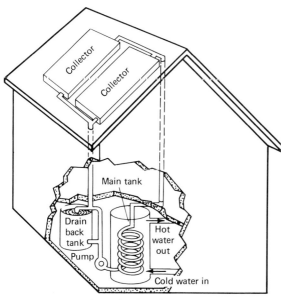

Figure 9.29 A drain-back solar system with heat exchanger. (The electrically operated valves are not shown.)

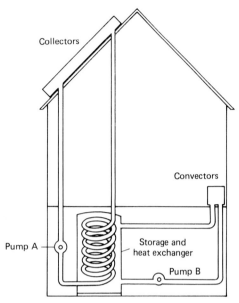

Figure 9.30 An active solar heating system.

and antifreeze, and the system need not be drained down on freezing nights. At night, pump A can be shut down, while pump B continues to circulate water between the warm storage tank and the house.

The many variations on the active-solar-system theme include air-based systems, drain-down systems, and systems that work with heat pumps. In the latter, heat is put into storage by a solar system, and a heat pump uses the stored heat instead of having to extract the heat from the much colder outdoors. The advantage of this is that the storage can be at a relatively low temperature—45 or 50°F—and the heat pump will still be able to use the stored heat with good efficiency. With the storage at these temperatures the solar collectors will also work more efficiently.

Active solar systems have one advantage over passive systems: As you will see in the next section, a passive system usually requires that the house be designed to use it or, at the very least, that there be substantial building or remodeling. An active system can be "retrofit" to existing houses. (That word doesn't appear in my dictionary, but it is widely

used these days. I think it was adopted from the space program.) In principle, retrofitting—putting an active solar system into an existing house— should be straightforward.

But beware! Numerous such systems have been installed and found to have payback times of 100 years, or more. That is, even after extensive insulating and sealing of the house, the fuel saved by the solar system is so little that it would take 100 years for the savings to pay for the cost of the system. Even if the house is still there after 100 years, you won't be! The reason is that good active systems tend to be expensive and, depending on a variety of local conditions, may provide only a fraction of the heating requirements of the house.

Certainly, if a solar installation is being contemplated for a new home, the house should be designed by someone with solar experience. Further, the design of the entire house should be based on the fact that it will have solar heat. In the event that this is done, it would be wise to give very serious consideration to passive, rather than active, solar heat.

Passive Systems

A *passive* solar system is one in which there are no mechanical parts: The collection and distribution of the heat are done by natural convection, conduction, and radiation alone. A passive space-heating solar system is usually an integral part of the house. Figure 9.31 shows a house with several solar features built in. I don't know of any house that has all of these features, but all of them have been used in various houses. (*Caution*: Don't get the idea from this drawing that a solar house can be just thrown together by adding multiple solar features. A solar house must be carefully designed so that all of the features work together. Failure to follow this rule may result in a house that is terribly overheated, terribly cold, or both, in different sections of the house or in different seasons.)

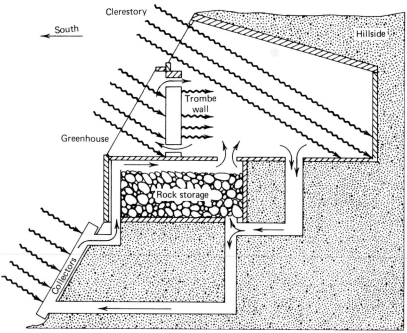

Figure 9.31 A multisolar house. This shows many possible passive solar features, but they would probably never all be built into a single house.

First of all, the house is highly insulated, and the north side is built into a hillside. Both the east and west ends of the building are partly underground. Dirt also covers the roof, and a lawn is growing above. Building into the ground this way can be a good conservation practice. In addition to providing protection from the wind, the ground a few feet down from the surface is usually considerably warmer than the winter air and cooler than the summer air. Thus less heating and cooling are required at the appropriate times.

The second feature of this imaginary house is a set of collectors on the hillside below the house. These work on the thermosiphon principle, which you have seen before in a hot-water system. The air heated in the collectors rises by natural convection into the space above the rock storage bin. If the house is warm enough, the ducts into the house can be closed, and the cool air from the bottom of the storage bin flows down the duct and into the bottom

of the collectors, where it is heated. Then it rises to heat the upper part of the storage bin. If the house needs heating, the ducts into the house are opened, and the warm air flows into the house, with cool air flowing back from the house to the bottom of the collector. At night, the collectors can be shut off, and heat from storage will keep the house warm.

Above the collectors is a solar greenhouse. It can be used for growing plants, and it also contributes heat to the house. The main mechanism by which it heats the house is the "Trombe" wall, named after its inventor. The wall is painted a dark color, and it may consist of any material that has a large *thermal mass*— that is, an ability to store heat. Trombe walls are sometimes made of concrete, or cement blocks, or 5-gal tins of water. While the sun is shining, some of the heat from the wall is transferred to the air just in front of it. That air rises by natural convection and goes through holes near the top of the wall into the main part of the house. Cooler air from the house enters through the holes near the bottom of the wall.

During the day, while the sun shines, the Trombe wall itself gets warmer and warmer. Since it has a large thermal mass, it can store a lot of heat. At night, the holes at the top and bottom of the wall are closed, and the warm wall radiates heat into the house (and into the greenhouse, keeping the plants from freezing.) In climates that are not too severe, this Trombe wall arrangement can provide all of the heat needed by the house. Often, instead of a greenhouse, there is just a vertical glass wall a foot or so away from the Trombe wall, and this works the same way. With this arrangement, overheating can sometimes occur, and some way of venting the space between the wall and the glass is needed.

Getting back to Figure 9.31, above the greenhouse is a *clerestory*, which gives heat to the inside of the house by *direct gain*. If this system is to be used, the back walls and the floor, where the sunshine strikes, must have a large thermal mass and be a dark color.

This is not a complete rundown on passive solar heating, but it will give you some idea of the possibilities. In general, if one is building a new house, passive solar heating can be built in for little extra cost, and with a pleasing and effective design. The experience has been that a well-designed passive house can get most or all of its winter heat from the sun almost anywhere in the United States. Solar homes and other buildings built in Canada have demonstrated that passive solar heating can be effective even in the far north.

SHADING AND SITING

One big problem is associated with a house like the one we've been discussing. Imagine living in that thing in the summertime! All of that sunshine that was so terrific in the winter could raise the summertime temperatures to unbearable highs. Even with ventilation all

around, this could be a problem, and since this house is covered with dirt on three sides, there are few windows to open. The answer is to use a combination of ventilation and proper shading. This is where your knowledge of the sun's path at the different times of the year comes in.

Activity 9.5

The purpose of this activity is to help you to gain a good mental image of how one can shade solar walls—or windows, for that matter—with a properly designed overhang at the roof or window level. To do so, you will use a shading model, like the one shown in Figure 9.32. The vertical part of the model represents the solar wall, and the horizontal section above is the roof overhang. You can make the model by using the template of Figure 9.33. It is a lot like the cutouts children make with the backs of cereal boxes. Reproduce the pattern on very stiff paper or cardboard, cut it out, fold along the dotted lines, and tape the tabs to the proper surfaces.

Now tape the shading model to a globe map of the world so that its front surface is

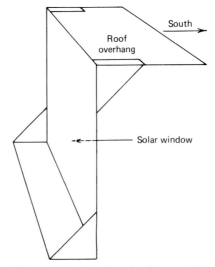

Figure 9.32   A solar shading model.

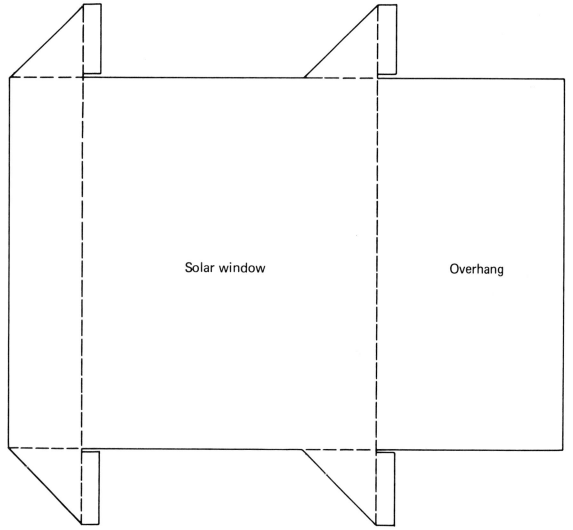

Figure 9.33 A template for making a solar shading model.

at your location and facing directly south on the globe. In a darkened room use a bare light bulb some distance from the globe to represent the sun, and the globe to represent the Earth. Align the axis of the "Earth" so that it is in its June 21 (beginning of summer) position. That is the position with the Earth's axis tilted directly toward the sun (North Pole closer to the sun). It is also the position for

which the sun is the highest in the sky in the Northern Hemisphere. Turn the Earth until it is solar noon at the location of the shading model, and see where the shadow of the overhang falls. The shading model is shown in use in Figure 9.34. By cutting the end back, adjust the length of the overhang so that the lower edge of its shadow hits just at the base of the solar wall. That is, so that the entire

Figure 9.34 The solar shading model in use. The shadow of the overhang is nearly halfway down the "window."

TABLE 9.1 *Shading Chart*

North lattitude (degrees)	Wall height overhang (ratio)	Percentage Shading	
		winter	spring/fall
25	38.2	2.3	5.6
30	8.8	8.4	19.7
35	4.9	12.5	29.1
40	3.4	14.8	35.3
45	2.5	15.5	39.4
50	2.0	14.8	41.8

side in the summer, when the afternoon sun really heats up the house.)

Compare your measurements to Table 9.1, the calculated values for latitudes in the United States. You can find the latitude at your location by using the globe; latitudes and longitudes are marked on it. (The latitude circles are the circles parallel to the equator. The longitude circles cross the equator.) Each figure in Column 2 of the table is the ratio of wall height to the overhang needed if the solar wall is to be completely shaded at noon on June 21.

wall is in shade. Measure the length of the overhang and divide it into the height of the wall. What is that ratio?

Now align the Earth's axis to its December 21 (beginning of winter) position. (That is, the Earth's axis is now tilted away from the sun.) How much of the solar wall is in shade at solar noon? Now find the percentage of the wall that is shaded for spring (March 21) and fall (September 21), when the Earth's axis is tilted sideways to the sun.

If you have the time and the inclination, you could turn the solar wall to the west or the east on the globe and see how the sun strikes it during the day at different times of the year. This will show you how the sun affects windows on the east or west side of the house. (Of greatest interest is the west

EXAMPLE

A solar wall 10 ft high is built into a house that is located at a latitude of 35° north. How much must a roof overhang jut out over the wall to shade it completely June 21? With that much overhang, how much of the wall will be shaded on December 21? On March or September 21?

Solution

A look at Table 9.1 shows that, at the latitude we are interested in, complete shading is provided if the wall height divided by the overhang is 4.9. Thus, the overhang must be about 2.0 ft out from the wall when the sun is high-

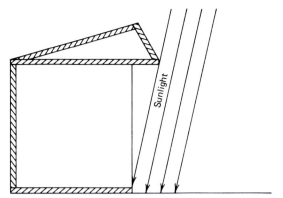

Figure 9.35 Shading at noon at a latitude of 35°
N on June 21.

est in the sky, as shown in Figure 9.35. The
angle between the sun's rays and a vertical
line is 11.5°. That is, it is the latitude minus
the tilt angle of the earth (23.5°). Can you
figure out why? (A drawing will help.)

At noon on December 21, according to
the table, about 12.5 percent of the solar wall
will be shaded. This is a little more than the
top foot of the wall, as shown in Figure 9.36.
The angle the sun's rays now make with a
vertical line is the latitude *plus* the Earth's tilt.

As you see, with an overhang of this type,
the greatest area of solar window is exposed
to the sunlight on December 21. The problem
with this is that the climate lags the sun by a

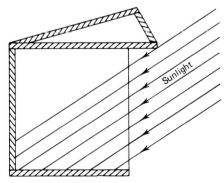

Figure 9.36 Shading of the same house at noon
on December 21.

month or two. That is, although the sun is at
its lowest on December 21, the coldest months
are generally January and February. By February 21, that same solar wall will be 20 percent shaded, meaning that the top 2 ft will be
in shade. Likewise, although complete shading occurs on June 21, by August 21 the wall
will be only 49 percent shaded. Also, March
21 and September 21 get the same amount of
shading, about 29 percent, but the needs are
quite different since March is a cold month
and September is a warm one. The overhang
can be extended further, but then the amount
of shading in the wintertime increases. Some
sort of compromise must be settled upon.

Obviously, also, some kind of protection
in addition to a roof overhang is needed. This
is accomplished in a variety of ways. Some
homes have an overhanging trellis on which
vines grow. In the summer, when shade is
needed, the vines grow in front of the solar
wall and provide shading; in the winter the
leaves fall off, and sunshine can pass through
the vines and the trellis. A slightly different
idea is shown in Figure 9.37. Also, movable
awnings of various types have been used, some
of which roll down directly in front of the
solar wall. Indoor measures, such as venetian
blinds or drapes are not as effective as outdoor sun-blocking devices, but they can help
a lot. One of the most ingenious devices is a
"beadwall," shown in Figure 9.38. The panels
consist of two panes of glass or plastic with
about a 6-in. space between them. At times
when the wall must be well insulated, small
styrofoam beads are blown into the space,
filling it with a very good insulation. This keeps
out the summer heat but is even more useful
for winter nights, when the beads keep the
solar panels from losing the heat gained during the day. At other times the styrofoam beads
are pumped out of the wall and into a holding
container. Other measures include appropriate landscaping, such as well-placed decidu-
ous trees, which provide shade in the summertime but allow the sunshine through in
the winter (at least partially). This is especially

Figure 9.37 An overhanging framework. In the summertime, slats are placed in the framework to shade the solar panels. In the wintertime, they are removed. Incidentally, this home not only has solar heat; it is also partially buried in the ground. The side pictured is the only side above ground. (Reprinted with permission of *Rodale's New Shelter* magazine. Copyright © 1982, Rodale Press Inc. All rights reserved.)

important on the west side, to keep the afternoon summertime sun from overheating the house. Placing a house properly on a lot is also important. The most desirable location is on a south slope, where the house gets the full winter sun but is protected from north winds. If the house is not earth sheltered—dug into the hillside—then the proper location of trees can also help shield the house from unwanted winds. In many locations, the winter wind is often from the northwest, whereas the summer winds are from the west or southwest. Evergreen trees planted to the northwest of the house will help to divert the winter winds away from the house, but will not interfere with the cooling summer winds. In planning such plantings, one should consider not only the prevailing winds in the general area, but the winds at the building site itself. Often the local geography will cause the winds to be different at one building site than they are at another site nearby.

It should be clear by now that casually throwing a collector onto the roof is not going to provide a satisfactory solar dwelling. To "go solar" one needs to take into consideration a large number of factors, including the local climate and winds, the latitude (for solar altitude calculations), the average number of degree-days experienced, the site, the condition and design of an existing house, the habits of the occupants, and so on.

PASSIVE COOLING

When discussing home energy use, heating is only half of the story. In our modern society, we have become accustomed to artificially cooled buildings, and the energy used for air conditioning far exceeds that used for heating in many parts of the country. As the costs of energy get higher, we may either have to do without air conditioning or find some

Figure 9.38 A beadwall prior to installation in a home. When there is no sunlight, styrofoam beads are pumped into the space between the two panes of glass to insulate the wall. When the sun shines, the beads are automatically pumped out. (Beadwall ®, courtesy Zomeworks Corporation, Albuquerque, N.M.)

other means of cooling ourselves. In this section we shall explore *passive cooling*, by which I mean cooling using mostly natural air movements.

As is the case with so many of the "new" ideas we have explored, the art of natural cooling was much better known to our ancestors than it is to us. For example, the adobe dwellings in the Southwest had a large thermal mass. In addition to storing solar energy during winter days and releasing it into the house during the night, the walls and ceiling

had the property of cooling down on summer nights and helping to keep the interior cool during the hot days. In the Southeast, houses were built with "dog runs" to help funnel cooling breezes to the interior. Everywhere, houses were situated to take advantage of the local climate and winds, and trees were left standing for shade. In recent years it has become our practice to bulldoze an area bare of even a blade of grass, plop down a house as near as possible to its neighbors, and turn on the juice. Now, with the need upon us, we are starting to relearn the fine arts of building homes that are naturally comfortable.

As with heating, the first rule of cooling is to stop unnecessary and unwanted heat transfer. This means, especially, that the home should be well insulated. Also, since the summer sun is high and beats upon the roof, a light colored roofing material is preferred. A good selection of site, with natural shading and summer breezes directed toward the house by landscaping and shrubbery can help a good deal. Once these details have been attended to, it can be fruitful to look for other means of keeping cool.

For human comfort, there are actually two aspects of "keeping cool." One is to keep the temperature at a level low enough for comfort. However, you know that at times you can be very comfortable in a temperature of 90°F, and at other times 85°F seems uncomfortably warm. The reason is that the main cooling mechanism of the body is the evaporation of perspiration. Recall that when you boiled water, it required the input of heat in the amount of nearly 1000 Btu to vaporize 1 lb (1 pint) of water. This energy went into the breaking of the bonds between the water molecules in the liquid. When water evaporates, the same process of breaking the bonds occurs. The temperature of the water is determined by the average vibrational energy of the molecules in the water. However, some water molecules have more energy than others, and the most energetic ones have enough energy to break the bonds and escape from the surface of the water. This is indicated in Figure 9.39. Since the most energetic mole-

Figure 9.40   If there are water molecules in the air above the surface, some of them enter the water.

cules leave, the ones left behind have a somewhat lower average energy and therefore they have a lower temperature. Because the process, like boiling, is one of breaking the bonds holding water molecules together, the energy carried away is the same, almost 1000 Btu per pound of water evaporated. When perspiration on the body evaporates, it produces a cooling effect, removing almost 1000 Btu per pound of perspiration. That is why, when you engage in strenuous exercise, you sweat, thus allowing the extra heat generated to be carried away from your body.

However, if there are water molecules already present in the air above the surface of the water, moving about in random directions, some of them will enter the water, as indicated in Figure 9.40. This has the effect of slowing down the evaporation for, although as many molecules leave as before, some of them are immediately replaced in the water. The more water vapor there is in the air, the slower is the rate of evaporation. Finally, sometimes the air is holding as much water as it possibly can. This is called *saturation*. The amount of water that the air can hold depends upon the air temperature, with warmer air being able to hold more water. When the air is saturated, we say that the *relative humidity* is 100 percent. Relative humidity is defined by the following equation:

relative humidity
$$= \frac{\text{amount of water in the air}}{\text{amount of water the air can hold}} \times 100\%$$

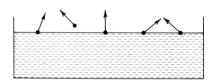

Figure 9.39   In evaporating, water molecules escape from the water surface.

Since the air can hold more water when it is warmer, the actual amount of water in the air for a given relative humidity will depend on the temperature.

When the relative humidity is low, say 50 percent, evaporation of perspiration occurs readily, and considerable cooling results. However, when the relative humidity is 100 percent evaporation cannot occur, for then as many molecules enter the perspiration from the air as leave it. Thus, with 100 percent relative humidity, there is no cooling effect. As the fellow says, "It ain't the heat, it's the humidity!"

Even when the humidity is not 100 percent, if the air is still, air directly above and around a pool of water can become saturated, and no evaporative cooling occurs. If a breeze comes along, it blows away the water vapor, and evaporation can then take place. Thus, although the breeze is just as warm as the still air, it can make you feel cooler by allowing your perspiration to evaporate. This is related to the "wind-chill factor" so often talked about in the wintertime.

Because of the cooling effect of moving air you can be a lot more comfortable in hot weather (unless the relative humidity is 100 percent) if you can find a way to move air past your body, either with a fan or by other means. There is nothing wrong with using fans and, compared to conventional air conditioners, they use little energy. However, if one is going to build a passive solar house anyway, it may cost little more to include some features that can provide good ventilation. Some of the methods we can use were invented in Persia many centuries ago. For example, a *thermal chimney* (also called a *solar stack*) can be very effective in moving air. One version of it is shown in Figure 9.41. Sometimes the chimney will have a wind turbine on its top, as does the one shown here, and any wind that exists then helps the natural convection to move the air.

The action of a thermal chimney is shown

Figure 9.41   A thermal chimney.

in Figure 9.42. The reason it is called a thermal chimney is that its action is similar to that of a fireplace chimney. The air in the chimney is warmed by the sun. The warm air rises, and pulls air from the house with it. This circulation of air will then make the house feel more comfortable, even if the temperature is not changed. If possible, the air can be pulled from a basement, where it is cooled by the ground around it, and cooling will take place in addition to the circulation of air. A variation of this is where the stack is just a solar wall vented at the top and arranged to pull cool air in from a crawl space under the house.

Going one step farther is the use of what have been called *cool tubes* or *earth tubes* to bring cooled air into the house. The means of moving the air may be a thermal chimney,

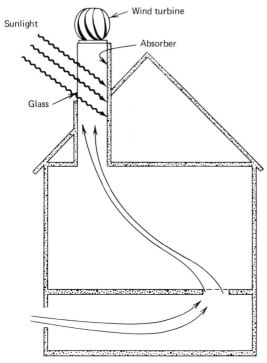

Figure 9.42   The action of a thermal chimney.

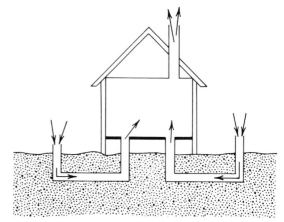

Figure 9.43   The use of "cool tubes" buried in the ground.

wind turbines, or fans, but, in each case, the fresh air comes in through one or more long tubes buried deep in the ground. This is shown in Figure 9.43. At a depth of 8 or 10 ft, the ground temperature remains constant all year long. Depending on the location, it may be at 55 or 60°F. As the incoming air passes through the tubes, it is both cooled and *dehumidified* (moisture is removed).

If you have ever held a cool plate in the steam from a boiling teapot, you know how dehumidification occurs. The cooler the air is, the less water vapor it can hold. As the air passes through the cool tube, the part of it that is close to the walls of the tube is cooled considerably. When that happens, the water vapor *condenses* (forms little droplets) on the tube and, if the tube is properly slanted, it runs out a hole in the far end. This is very much the same as the formation of dew on the ground: When the ground cools (by ra-

diation) at night, the moist air comes in contact with it, is cooled, and gives up some of its water by condensation. Actually, the reason that mechanical air conditioners are so effective is that they do the same thing: As air passes over the cool coils, it is both cooled and dehumidified. Air conditioners have an arrangement for the condensed water to drip away.

In the Southwest, where the summer days are very hot and dry and the nights are quite cool, the *roof pond* is proving to be quite effective. In principle, this is just a pool of water on the roof over part or all of the house, and it works well for both winter heating and summer cooling. (Actually, the water is usually contained in plastic bags, something like a water bed, to prevent evaporation.) Figure 9.44 shows how it works. In the wintertime, the insulation is left off the pool during the day, and the water stores heat from the sun. At night, the insulation is put over the pool, and heat stored in the water heats the house by radiation through the ceiling. In the summertime, the cycle is reversed: During the day, the insulation is kept over the pool, and the water absorbs heat from the house. At night, the insulation is removed and the water radiates its heat to the cool night sky. (Using a

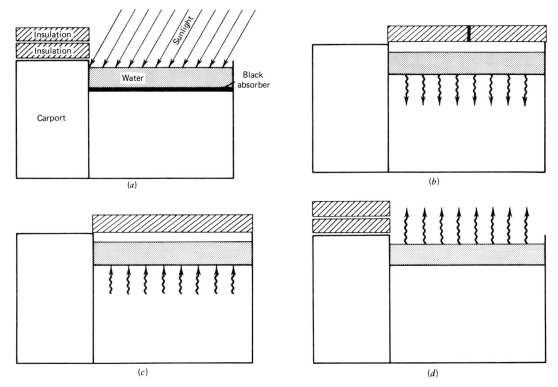

Figure 9.44 A roof pond on a winter day. (*a*) Solar energy is stored in the water. (*b*) On a winter night, heat radiates from the pond down into the house. (*c*) On a summer day, heat from the house is absorbed by the pool. (*d*) On a summer night, heat from the pond radiates to the sky.

system somewhat like this, the ancient Persians were able to make ice on nights when the temperatures were close to, but above, freezing.) Because the climate is so dry in the Southwest, no dehumidification is necessary. A roof pond plus thick walls of adobe or masonry can provide enough thermal mass to heat or cool a house comfortably, without need of a furnace or air conditioner.

SUMMARY

In this chapter we have taken just a surface look at some of the ways in which solar en-

ergy can be used on a small scale in homes and businesses. Solar systems can be either *active*, requiring pumps and/or fans to move heat-carrying fluids about, or *passive*, operating wholly or mostly on natural convection, conduction, and radiation. Active systems generally require separate collectors, pumps or fans, and an electrical control system. Collectors are often of the flat-plate type but, if particularly high water temperatures are required, they may be of the concentrating type. Passive systems may require separate collectors but, more often, they are built right into the structure of the building. Both types of solar systems usually require some way to store extra energy during periods of sunshine for use in nonsunshine periods. The solar system may store latent heat, by melting a solid salt, but more frequently it will store sensible heat. Poorly designed solar systems may have extremely low efficiencies or even negative efficiencies, in which the system causes a net loss of heat from the building. Even well-designed solar systems may have long payback times, and careful design and good sense should go into any system before the investment is made. This is especially true of retrofit systems, built into existing buildings. Passive systems built into a new home, if well designed, may add little to the cost of construction, yet be very effective in reducing or eliminating heating bills. In either case, it is important to be sure that overheating will not occur, both winter and summer. For this and other reasons, shading and siting of the solar installation are of crucial importance. Passive cooling of homes and other buildings in the summertime is a relatively undeveloped art, but several approaches to passive cooling hold promise as being ways to reduce the need for mechanical air conditioning. As with solar heating, the methods used depend strongly on geographic location.

New, more cost-effective materials for solar applications will undoubtedly be developed in future years. However, with the materials and techniques we now have, a good start can be made on using the sun's energy almost anywhere in the country. Most required are a national consciousness of the possibilities and a willingness to change our life-styles slightly. There are also avenues in addition to solar heating and cooling that we can pursue to find solutions to our energy problems. Some of those are explored in later chapters.

ADDITIONAL QUESTIONS AND PROBLEMS

1. When placed in the sunlight, why does a dark-colored object generally heat up faster and to a higher temperature than a light-colored object? If two objects have about the same color to the eye but one heats up substantially faster and to a higher temperature than the other, what can you conclude?

2. Explain briefly, in your own words, the greenhouse effect, and describe its importance in the energy balance of the Earth.

3. What are some of the different types of solar collectors, and how do they work?

4. How is the efficiency of a solar collector defined? What is meant by "negative efficiency"? Why is a solar collector most efficient when its temperature is only a few degrees higher than ambient? Why may it be more effective to run a collector at a lower efficiency than its maximum?

5. Why does a solar collector that is tilted to face the sun generally gain more energy than one that is laid flat on the ground (horizontal) or one that is upright (vertical)? If the sun is coming to the ground at an angle of 45° above the horizon, about how much less direct radiation strikes the collector if it is either horizontal or vertical? (You can estimate this with a careful drawing, something like Figure 9.13.)

6. Suppose, in your present location, you would like to install a solar hot-water system, using flat-plate, liquid-type collectors. You decide to use the system year round, so you hinge the collectors so that their tilt can be changed easily. To get maximum benefit from the sun, at what angle to the horizontal must the collectors be set on March 21 (spring), June 21 (summer), September 21 (fall), and December 21 (winter). (You need to know your latitude and the fact that the tilt of the Earth is 23.5°. A drawing will make it much easier for you to visualize this.)

7. Suppose that a 4 ft by 8 ft flat-plate, liquid-type collector is working at 50 percent efficiency and the insolation, both direct and diffuse, is 300 Btu/hr per ft². What is the total insolation on the collector? At what rate is energy going into the water? If the water is entering the collector at a temperature of 60°F, and leaving at 140°F, how many pounds of water can be heated per hour? What is the maximum flow rate, in gallons per hour? (Remember that 1 gal weighs about 8.3 lb.)

8. What is a heat exchanger? How does it work? Why is a heat exchanger needed in some solar applications?

9. A concentrating collector 10 ft long is made like the one shown in cross section in Figure 9.18. The width of the lens is 16 in. and the diameter of the water-carrying pipe is $\frac{3}{4}$ inch. The direct insolation on a sunny day is 300 Btu/hr per ft². What is the total direct insolation on the collector? Assuming that 95 percent of the incident energy passes through the lens and strikes the pipe, approximately what is the intensity, in Btu/hr per ft², at the pipe? (You can approximate the pipe as being flat and 1 in. wide.)

10. A collector battery is made up of eight concentrating collectors, each 12 ft long and made like the one shown in cross section in Figure 9.19. Each mirror spans a width of 1.5 ft. With direct insolation of 250 Btu/hr per ft², the collectors are operating at 45 percent efficiency. How much heat goes into the water per hour? If the water enters the collectors at 55°F and leaves at 120°F, how much water can be heated in an hour? Thus, what is the maximum flow rate? (1 gal of water weighs about 8.3 lbs.)

11. A house with an active, air-type solar heating system uses a large rock bin for heat storage. There are 19 tons of rock with an average specific heat of 0.23 Btu/(lb °F) and they can be heated to an average temperature of 110°F. If the house loses an average of 25,000 Btu/hr to the outside, will the heat stored in the rock keep the house at 65°F through a winter night? Will it last through the following cloudy day?

12. It has been suggested that, in colder climates, a big block of ice may be frozen in the wintertime to store some "cool." Then, in the summertime, antifreeze may be pumped through pipes running through the ice and used to cool the house. One article I saw said the block of ice would have to have dimensions something like 20 by 35 by 45 ft, a pretty large ice cube! Is that a reasonable estimate? The following will help you to answer.

A cubic foot of ice has a weight of about 57 lbs. If the huge ice cube is melted completely, how many Btus will it absorb? Assuming 50 percent efficiency in using the ice to cool the house (just a guess), how much heat can be extracted from the house in the course of the cooling season? Consider a location with hot, humid summers and about 2000 cooling degree-days annually. I estimate that a large, not-too-well-insulated house with an indoor temperature of 75°F might gain 50,000 Btu/hr from the outdoors for 100 24-hr cooling days. What is the total heat gain? How does it compare to the ability of the ice to extract heat from the house?

Does this entire scheme seem reasonable to you? Why?

13. Draw a schematic diagram of an active, liquid-type space heating solar system, including a heat exchanger.

14. What is the difference between active and passive solar systems? What are some advantages and disadvantages of each?

15. Design and sketch a "multi solar" house, including at least three passive and/or active solar devices, and explain the operation of each.

16. By checking with the local weather bureau, gather information about the average number of sunny and cloudy days in the wintertime in your locality. How will this information affect a decision about solar heat?

17. The cost of heating a home in the colder regions of the United States might be in the range of $600 to $1000 a year, or more. Suppose you could install for $8000 an active solar system that will provide all of the heat you need. Assuming absolutely trouble-free operation of the system for its useful lifetime, what would be the payback time? (Use either $800 per year or, if available, the actual figures for heating a house in your locality.) Suppose, however, that like most of us you cannot afford the $8000 all in one lump, and you must take out a 20-yr mortgage to pay it. Ask your local bank or credit union what such a mortgage would cost. That is, what would be the total cost of monthly payments for a 20-yr period? Now, what is the payback time? (Of course, to be realistic about this, you probably should include an inflation factor for the rising cost of fuel.)

18. Describe the wintertime and summertime operation of a roof pond on a house in the Southwest.

10 Electricity and Magnetism

In our lives electricity is all about us. It powers machinery, lights, radios, telephones, typewriters, televisions, computers, curling irons, automobile accessories, aircraft radar, movie projectors, elevators, calculators, saws, insect killers, razors, heart pacemakers, can openers, air conditioners, lasers, sewing machines, and on, and on, and on. In an industrialized society, when the electric power goes off, many, many routine activities cease completely. We all use electricity in a most familiar way.

Yet many people live their entire lives without having the least understanding of this wonderful servant. Indeed, many people fear electricity and feel completely helpless when something electrical does not work. One should respect electricity, for it can be dangerous, but fear comes from ignorance. Knowledge of the basics of electricity can help one to use it confidently and wisely, and it can lead to an understanding of the problems associated with the production and distribution of electric power, the topic of Chapter 11. In this chapter, you will learn about the behavior of electricity and magnetism and how they can be useful to us.

ELECTRIC CHARGE

Activity 10.1

This activity works best in cold weather. If it is hot and humid, you may have difficulty in getting all or parts of it to work properly.

Using light sewing thread, lightly moisten your fingers and roll one end of the thread into a tight ball, perhaps $\frac{1}{4}$ in. in diameter. Then, using about 12 more inches of the thread, hang the thread ball in midair, attaching it to any convenient place. The activities to follow work best if the thread ball is slightly moist (not soaked). If, after some experimentation, the effects are not as noticeable as they were at first, rub the thread ball between moist fingers to wet it again.

Bring the end of a plastic ruler (or a comb, or something else made of plastic) close to the hanging thread ball, as in Figure 10.1. Does anything happen? Now rub the end of the plastic ruler vigorously with a piece of wool cloth. (If you cannot find wool, cloth that feels like wool should do.) Slowly bring the end of the ruler you have just rubbed closer and closer to the hanging ball. Describe what happens.

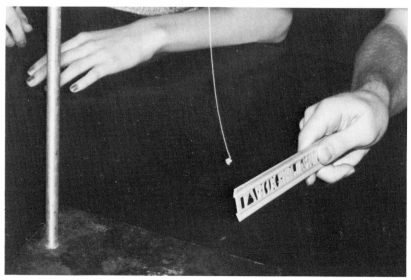

Figure 10.1   Bring a ruler close to the hanging thread ball.

You may have to experiment to find a combination of cloth and plastic that works well, giving very noticeable effects.

Rub the ruler with the cloth again, and this time let the hanging ball touch the ruler. In fact, let it touch repeatedly along the ruler to transfer whatever it is on the ruler to the ball. Now let the ball stop swinging without touching it, rub the ruler with the cloth again, and slowly bring it closer and closer to the ball. What happens this time?

Now find a glass rod or tube, $\frac{1}{4}$ to $\frac{1}{2}$ in. in diameter. In this plastic age that will not be easy, but clear plastic will not do; it must be real glass. You will also need a piece of real silk—not nylon. A silk scarf would be good. When you have found these hard-to-get items, first squeeze the thread ball with your fingers, then repeat the experiments you did with the plastic ruler and the wool cloth. Describe what happens.

Now rub the plastic ruler with the wool cloth and transfer whatever is on the ruler to the hanging ball, as you did before. Then, without wasting any time, rub the glass rod with the silk cloth and slowly bring the glass rod closer and closer to the hanging ball. What happens? To be sure the hanging ball has not changed, bring the plastic ruler near it again. What happens?

Finally, hang two thread balls side by side, so that they are touching. Rub the plastic ruler with the wool cloth, and touch the ruler to the two balls. Then let the thread balls hang side by side again. What happens?

You might want to try variations of these activities.

As you probably know, you have been dealing with *static electricity*. It is closely related to lightning and to the mild shock you get if you touch something metal after shuffling across a rug in the wintertime. Before reading on, try to write down in an orderly way exactly what you have observed concerning static electricity.

As you are aware, when Ben Franklin did this same set of experiments, and others, he concluded that there were two different kinds of *electric charge*. Each kind attracts the other but repels itself. He named the kind of charge found on the plastic ruler *negative* charge (−). The other kind, which you found on a glass rod, he called *positive* charge (+). Figure 10.2 summarizes the results of experiments like the one you just did, using plus signs for positive charge and minus signs for negative charge. In words, we can summarize as follows:

1. Two objects with opposite charges attract one another.

2. Two objects with charges of the same sign repel one another.

In our modern age, we know that matter is made of *atoms* and that atoms consist of *electrons, protons,* and other particles. There is a more accurate description of atoms in chapter 12. For now, Figure 10.3 will do. A *neutral* (normal) atom has a number of protons in a

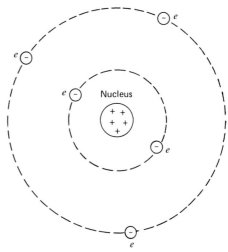

Figure 10.3 A simplified model of an atom having a nucleus with a number of protons and an equal number of electrons in orbits.

tiny core, called tbe *nucleus,* and an equal number of electrons running around the outside. The path of an electron is called an *orbit*. The electrons have the same charge that Franklin called negative, and the protons have positive charge. The size of the positive charge of one proton is the same as the size of the negative charge of one electron.

Atoms are so tiny that a small sample of any material will have an enormous number ot atoms in it. For example, a single gram of iron contains over 10^{22} iron atoms, and each iron atom has 26 protons and 26 electrons. The protons, being in the *nuclei* (plural of *nucleus*), are fixed and cannot move around in the material. Some of the electrons, on the other hand, can move easily from atom to atom, and can even leave the sample of iron completely.

When you rubbed the plastic ruler with a wool cloth, the plastic was no longer *neutral*—it no longer had the same number of electrons and protons. As said before, the plastic had a *negative charge*. That is, it had extra electrons. It got that charge by tearing electrons out of the cloth with frictional forces

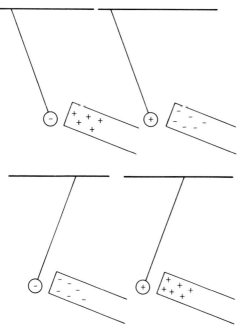

Figure 10.2 Like charges repel; unlike charges attract.

as the cloth rubbed the plastic. The electric charge of a single electron is very tiny. In fact it is so tiny that even the very small charge you were able to produce on the plastic ruler involved the transfer of millions of electrons.

Before you rubbed it with the cloth the plastic ruler was neutral—it had very nearly the same number of protons and electrons in it. That is to say, there is no *net* charge in a piece of neutral material even though there are millions upon millions of electrons in the material; each electron charge is "canceled" by a proton charge. By rubbing the plastic with the cloth, a "few" (a few million, that is) *excess* electrons are transferred to the plastic, and it thus gets a net negative charge. However, those excess electrons were not created; they had to come from the cloth. Thus the cloth must have a net *positive* charge, as indicated in Figure 10.4. Because the charge tends to get spread about on the entire cloth, it is more difficult to detect the net charge on the cloth than that on the plastic. However, on a cold, dry day, it is often possible to detect the charge on the cloth by bringing it close to the hanging thread ball.

If you're so inclined, you might try putting an electric charge on a metal rod. You will find that, try as you might, with every kind of cloth available, you will not be able to detect the slightest charge on the rod. The reason is that some of the electrons in the metal move very easily away from the atoms they are attached to. Since they can move about anywhere on the metal rod very easily, those electrons are called *free electrons*. The metal is a good *conductor* of electricity, whereas plastic and glass are good *insulators* because their electrons are bonded more strongly to the atoms. The concept of conductors and insulators is very similar to the notion of thermal conductors and insulators. (In fact, a good conductor of electricity is usually a good conductor of heat, as well.) It is hard to get an excess charge, either positive or negative, on the metal rod because it can easily accept elec-

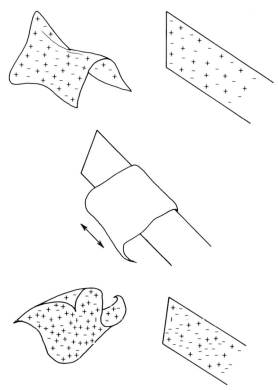

Figure 10.4 Transferring electrons by friction. At the top, the plastic ruler and the cloth are both neutral. After the ruler is rubbed with the cloth (center), the plastic has an excess of negative charge and the cloth an excess of positive charge (bottom).

trons from your body, (or give them to your body). Your body is also a fairly good conductor of electricity, and it is usually electrically connected to the Earth, which is almost an infinite source of electrons. (However, your entire body can get an excess charge, as when shuffling across a carpet on a dry day, which it then gets rid of as a "spark.")

ELECTRIC FORCE

You may not have noticed, but I have not yet explained the very first phenomenon you observed in Activity 10.1. When you brought a charged ruler close to a *neutral* thread ball, the

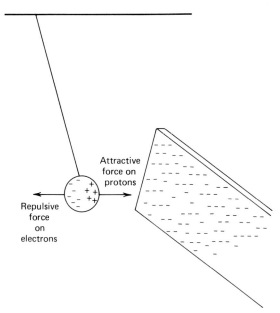

Figure 10.5 When a charged ruler is brought close to a neutral thread ball, there is a net attractive force.

ball was attracted to the ruler. But the neutral ball has an equal number of protons and electrons, and protons are attracted to the ruler whereas electrons are repelled. Thus, one would think that the two opposing forces would be equal and cancel out, so that there would be no net force.

The answer to the puzzle is shown in an exaggerated fashion in Figure 10.5. Although thread is a pretty good insulator, the moisture you put into the thread ball makes it into a fair conductor. (This is why you should not handle plugged-in electrical appliances while you are wet; the moisture makes your skin into a rather good conductor, too!) Thus, some electrons are able to move about in the ball. When the negatively charged ruler comes near, the free electrons "flee" to the far side of the ball. This is a subtle proof that the electric force depends on the distance between the two interacting charges: if it did not, the repelling force on the negative electrons would still be the same as the attracting force on an

equal number of positive protons, and there would be no net force. However, since some of the electrons are slightly farther away from the ruler than are the corresponding protons, the force on the electrons is slightly less. Thus, there is a net attractive force, which you observed. (Incidentally, a charged object will also attract a neutral insulator, in which the electrons cannot leave the atoms they are attached to. Put crudely, the negatively charged object pushes electrons in the insulator toward the far side of the atoms, so the electrons are slightly farther away, on the average, than their corresponding protons. Thus, the attractive force on the protons is slightly greater than the repulsive force on the electrons. You can demonstrate this with little pieces of paper, if you wish, attracting them to the charged ruler.)

If you were to do some further experimenting with a charged ruler and a charged thread ball, you would undoubtedly observe the following:

1. The closer to each other the two charged objects are, the greater is the force between them.
2. The greater the amount of charge is, the greater is the force involved.

The French physicist Coulomb made careful measurements and formulated a law having to do with the forces exerted by two charged bodies on each other. The amount of charge may be given the symbol Q; thus we shall say the first body has a charge of Q_1 and the second Q_2. Then, if the distance between the two is d, the force each exerts on the other is given by

$$F = K\frac{Q_1 Q_2}{d^2} \qquad (10.1)$$

This equation is called *Coulomb's law*. In the SI system of units, the unit of charge is called the *coulomb* (C). The distance between charges is given in meters and the constant K is equal to 9×10^9 (N·m²)/C².

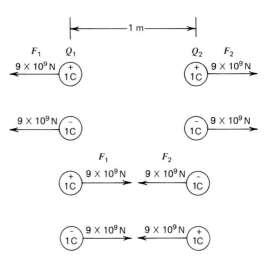

Figure 10.6   The forces acting on two 1 C charges 1 m apart.

EXAMPLE

What are the forces acting on two objects, each having a charge of 1 C, that are 1 m apart? (Note that Newton's third law is valid for charged bodies, too: If Q_1, exerts a force on Q_2, then Q_2 exerts an equal and opposite force on Q_1.)

Solution

Of course, we must use Coulomb's law to find the answer. For example, if each of the objects has a positive charge, then the force acting on either one is given by

$$F = K \frac{Q_1 Q_2}{d^2} = 9 \times 10^9 \frac{\text{Nm}^2}{\text{C}^2} \times \frac{1\,\text{C} \times 1\,\text{C}}{1\,\text{m}^2}$$

or

$$F = 9 \times 10^9 \text{ N}$$

If the two charges are both negative, the result is the same, for $-1 \times -1 = +1$. If one charge is negative and the other is positive, the result is $F = -9 \times 10^9\,\text{N}$. The negative sign implies an *attractive* force, whereas the positive sign means that the force is *repulsive*. Figure 10.6 shows these results. The tremendous force between the two charges tells us that a coulomb is a very large unit of charge. It requires an excess of about 6×10^{18} electrons (or protons) to produce a net charge of 1 C.

ELECTRIC FIELDS

Suppose you have an object with a net positive charge on it. You know that if another charged object is brought nearby, it will feel a force. One way of thinking about this is to imagine that every charged object has surrounding it a *field* or, as they call it in science fiction, a "force field." This imaginary *electric field* is not something you can see, touch, or smell, but you could detect it by bringing a "test charge" into it and measuring the forces on the test charge. The electric field is then defined as the force on a small, positive test charge, divided by the test charge. That is, the electric field E is

$$E = \frac{F}{q} \tag{10.2}$$

where q is the size of the small test charge in coulombs. The units for an electric field are force divided by charge, or N/C. The electric field not only has a size, given by the Equation 10.2, but also has a direction, defined as the direction in which the force on the test charge acts. Figure 10.7 indicates how to find the electric field at any given location. Note that the only charged object in the picture is the small test charge. There is no large charged object shown. Yet, an electric field must come from objects with excesses of either positive

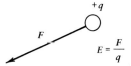

$$E = \frac{F}{q}$$

Figure 10.7 The electric field at any location is equal to the force on a small positive test charge divided by the charge. The field is in the same direction as the force.

or negative charges. However, we do not need to be able to see those objects to specify the electric field they produce. We can just merrily use our test charge out in space, and find the field at every location, regardless of the field's sources.

Suppose we were to allow the test charge in Figure 10.7 to move very slightly in the direction in which the electric field pushes it. Then, at the new location, we would be able to measure the new electric field and draw a line connecting the two points. From the new point we could again let the test charge move slightly in the direction it is being pushed—that is in the direction of the electric field—and do the same thing. By repeating this process a number of times, we would be able to produce an *electric field line* (also called a *line of force*), as shown in Figure 10.8. That line would tell us the direction of the force acting on a test charge located anywhere on the line, as in Figure 10.9. By moving a test charge around in space, we could draw a kind of map of the electric field lines surrounding any charged object.

In general, the electric field lines can be very complicated. However, certain simple cases are instructive to examine. For example, a positively charged sphere produces the electric field lines shown in Figure 10.10. What this map tells us is that, everywhere in the

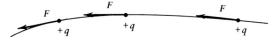

Figure 10.9 At any point, the direction of the force on a positive test charge is the same as that of the line of force.

space around the sphere, a positive test charge will be pushed directly away from the sphere. Further, the electric force, and thus the electric field, will be greater near the sphere than farther away. The field-line map tells us this, as well: Where the lines are close together the field is strong; where they are farther apart the field is weaker.

Figure 10.11 shows the same kind of map for a negatively charged sphere. It is identical to Figure 10.10, except that the arrows point inward. That is, the force on a positive test charge will be in toward the charged sphere everywhere in the space around it. Figure 10.12 is a cross-sectional view of part of an infinitely large charged flat plate. (A good approximation is a large flat plate, just as long as we stay away from the edges.) Here the field lines

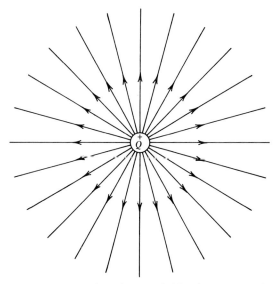

Figure 10.10 The electric field of a positively charged sphere.

Figure 10.8 An electric field line (line of force).

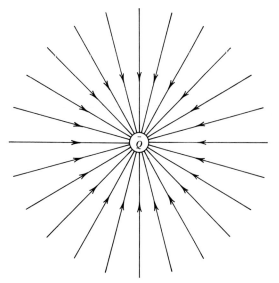

Figure 10.11 The electric field of a negatively charged sphere.

are all parallel and equally spaced. That means that the electric field is the same size everywhere in the space surrounding the plate. What would the field map look like for a negatively charged flat plate? (Note that, if we consider all of space, every field line that leaves a positive charge must end up on a negative charge

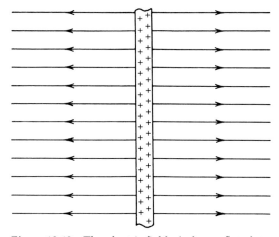

Figure 10.12 The electric field of a large, flat plate. (Cross-sectional view. The plate is perpendicular to the paper.)

somewhere. However, we won't need to use that fact in what follows.)

MAGNETS AND MAGNETIC FIELDS

Activity 10.2

For this activity you will need a couple of small magnets. Small, bar-shaped magnets are easier to analyze than are the disk-shaped ceramic magnets. However, if disk-shaped magnets are all you have, go ahead and use them; the principles involved are the same.

Hang one magnet horizontally from a piece of thread, as in Figure 10.13(*a*), and wait until it stops swinging and turning. In which direction does it point? (That is, is it along an east–west line, a north–south line?) Mark one end for identification, and let the magnet hang again. Does the same end point in the same direction? Now repeat the whole thing with the second magnet, and any others you might have. If you are using disk magnets, you will have to hang them so the disk is vertical, as in Figure 10.13(*b*). Mark a face of the disk.

Bring one end of a magnet close to a hanging magnet. Whoops—not too close! Describe carefully what happens. Try this a few times to see if the results are consistent. Now try the other end of the magnet in your hand. How does the result differ from the first try?

Figure 10.13 Hanging a small magnet.

If you have additional magnets, try it with each of them. Try every variation you can think of and record your observations.

Questions

In Figure 10.14, two magnets are shown in various positions. On each, the end marked with a dot pointed north when the magnet was hung from a thread and left free to turn. Using the knowledge of magnetic forces you have gained from the previous activity, show the direction of the force(s) on each of the magnets.

No doubt, you noticed that the magnets tend to point north and south. The end that points north is called the *north pole* or the *north-seeking pole* of the magnet. (That figures, doesn't it?) The other end is, naturally, the *south pole*. You also must have discovered that north poles repel north poles, south poles repel south poles, and north and south poles attract each other. This is summarized in Figure 10.15.

This is similar to the situation with electric charge, in which like charges repel and unlike charges attract: like poles repel and un-

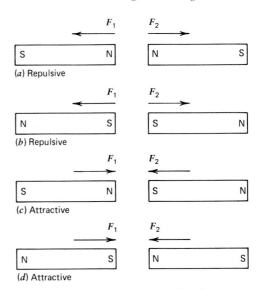

Figure 10.15 A summary of the forces between two magnets.

like poles attract. There is one very big difference, however. An electric charge, either positive or negative, can exist all by itself on a given object. A magnetic pole cannot! Whenever you find a north pole, there must be a south pole on the *same* magnet.

We can also invent magnetic fields, as we did electric fields. It is just a little harder, because we cannot produce a single magnetic pole to use as a test pole. Since the field lines exist only in our imaginations, anyway, just imagine a very, very lo-o-ong but thin magnet, such that its north pole can be in the range of influence of the magnet we are testing, but its south pole is far away. Then, the magnetic field is defined by the force on this test north pole. If we imagine probing the space around a magnet with a test north pole this way, we can map out magnetic lines of force. For example, for a bar magnet, the test north pole will be pushed away from the north pole of the bigger magnet and pulled toward the south pole. Mapping out the entire area gives a set of magnetic field lines like those shown in Figure 10.16. Every line of force that

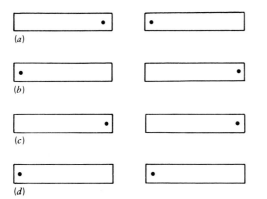

Figure 10.14 Show the directions in which the forces act in each of the four cases.

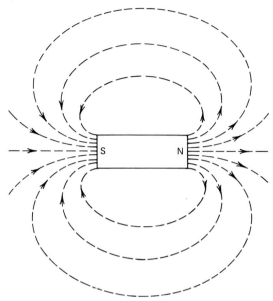

Figure 10.16   The magnetic field lines of a bar magnet. You can imagine that even the ends of the straight line in the center might curve around in the universe and eventually meet.

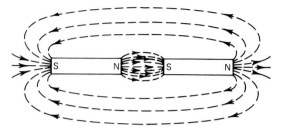

Figure 10.18   A more complete view of the magnetic field produced by two bar magnets.

ure 10.18 should help. There, a greater field of view is shown, and you can see that a line of force that starts at the north pole of the right-hand magnet goes to the south pole of the left-hand magnet, passes *through* the magnet, comes out at the north pole of the left-hand magnet, and goes to the south pole of the magnet it started from. (From there, it passes through the right-hand magnet, back to the point from which we started tracing it. Thus, this is just like a longer single magnet and, if you allow the two poles in the middle to come together, it *will be* a single magnet. Similarly, if you break a single magnet in half, you get two complete smaller magnets.)

starts out from the north pole of a magnet must end up on the south pole of the same magnet.

The magnetic field between two bar magnets, with the north pole of one near the south pole of the other, is shown in Figure 10.17. A test north pole placed midway between the two magnets would feel a force in the same direction as it would feel with either magnet alone, but the force would be double.

But wait a minute! I said earlier that a line of force starting on the north pole of the magnet had to end up on the south pole of the *same* magnet, and that does not seem to be the case in Figure 10.17. Relax. A look at Fig-

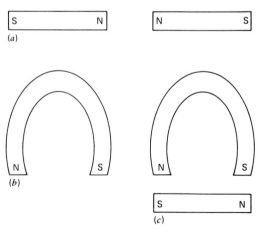

Figure 10.19   By imagining the force on a small test north pole at each location, draw the magnetic lines of force for these three cases.

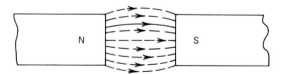

Figure 10.17   The field between two bar magnets.

Questions

By imagining the force exerted on a test north pole at any location, draw the magnetic lines of force for each of the three examples shown in Figure 10.19.

ELECTRIC CURRENT

Activity 10.3

For this activity you will need a length (say 10 feet) of insulated wire, a magnetic compass, and two 1.5-volt dry cells (usually called batteries) with contact posts and knurled knobs at the top. Size "D" flashlight batteries will do, but it is a little hard to use them because you will have to touch the wires to their ends by hand. (But it can be done.) Holders for flashlight batteries would be helpful. You can find the dry cells at a hardware store and the battery holders at an electronics supply store.

Note: Some people are afraid of electricity and, when dealing with the 120-volt supply found at your home outlets, being afraid to mess with it is a very good idea. Electricity from those outlets can hurt you badly, or even kill you, if you do not know what you're doing. On the other hand, the electricity provided by the dry cells that you will be using in the activities in this chapter cannot hurt you. You can touch any terminals or wires without hurting yourself at all. But be sure not to place a short, small diameter wire directly across the posts of one of the dry cells; the wire might get hot enough to give you a burn. Also, it is not wise to work with anything electrical with soaking wet hands, so keep your hands reasonably dry. Otherwise, feel free to try out anything that seems interesting.

With that preamble, we are ready to start. Place a section of the wire on the compass, running north and south. Be sure the compass is level and the pointer is free to move.

The wire should now be in the same direction as the pointer. Now touch the two ends of the wire to the terminals of a dry cell. (If you are using a flashlight battery, the terminals are the top center and the bottom of it.) What happens to the compass? This is shown in Figure 10.20.

Try the same thing, with some variations. For example, switch the ends of the wire when touching them to the dry cell.

Now connect the two dry cells together by means of a single wire going to one terminal on each dry cell. Then touch the ends of the wire to the two free terminals and see what happens to the compass. I am deliberately not telling you which battery terminals to connect together. You have three choices: you can connect + and +, − and −, or + and −. (The + terminal is the center one on the dry cell with terminals and the top one on a flashlight battery.) Try it all three ways and see what happens.

Now wrap several turns of the wire around the compass, as indicated in Figure 10.21, and try the same things. Do you detect any difference in what happens to the compass pointer?

With the wire and the compass, you have made a simple *galvanometer*, an instrument for measuring electric current. The fact that the compass pointer deflects when contact is made with the dry cell is an indication that there is a current flowing in the wire.

An *electric current* consists of a flow of electrons: There are electrons moving through the wire. There is no way you could have deduced this from the experiments you have done; it originally took years of experimentation and thought by a number of people to come to that conclusion. However, the deflection of the compass pointer *does* indicate that there is an electric current in the wire, and the current *does* result from electrons flowing through the wire.

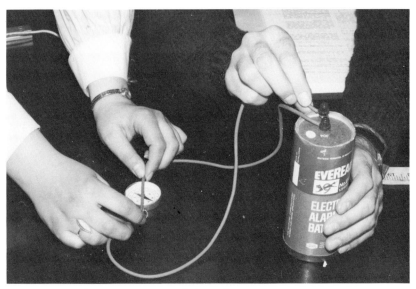

Figure 10.20 Compass, wire, and dry cell. A current passing through the wire deflects the compass pointer.

What causes the electrons to flow? Well, in a sense, the dry cell pushes them. To see how this can come about, first look at Figure 10.22. It shows the space between two flat plates, one with a positive charge and one with an equal negative charge. There is therefore an electric field E in the space between the two, and we can imagine putting a small charge q in that region. Since the charge is in an electric field, there is a force acting on it.

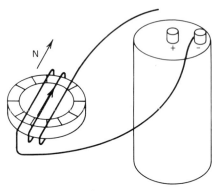

Figure 10.21 The compass, with a few turns of wire coiled around it.

From Equation 10.2, that force is

$$F = qE$$

Then, if the force is qE and the distance between the plates is d, the work done on the electric charge by the field, as the charge moves from the left plate to the right one, is

$$W = \text{force} \times \text{distance}$$
$$= qEd$$

Of course, the charge bangs into the right-hand plate because, as the force is acting on it, it gains kinetic energy. Saying this slightly differently, the work done on the charged object is converted into kinetic energy of the object.

EXAMPLE

Suppose the two charged plates are 0.1 m apart and the electric field between them is 10 N/C. If the charge put in the space near

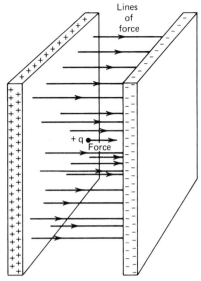

Figure 10.22 A charged object between two charged plates experiences a force.

the left-hand plate is 10^{-6} C (a *microcoulomb*), how much work is done on the charge as it moves through the space from left to right? When you use these quantities, does the unit for work turn out right? What is that unit?

Solution

The work is given by the equation

$$W = qED = 10^{-6}\,C \times 10\,N/C \times 0.1\,m$$
$$= 10^{-6}\,N{\cdot}m = 10^{-6}\,J$$

The unit turns out to be joules, as it must, since we are calculating work in SI units.

Another way of looking at this situation is to note that, when the charge is near the left-hand plate, it has a certain potential energy. As it moves to the right, the potential energy decreases and the kinetic energy increases, with the sum of the two staying constant. That is, potential plus kinetic energy is conserved. This is much like a ball falling in a gravitational field, only now a charged object "falls" in an electric field. If the potential

energy at the right-hand plate is taken to be zero, what is the potential energy of the charge when it is at the left-hand plate? It is, of course, equal to the work that must be done *on* the charge to move it from the right-hand plate to the left-hand one. That is: $PE = qEd$.

Actually, as with other kinds of potential energy, we are really more interested in the *difference* in potential energy (ΔPE) between two different positions. In this case, the difference in potential energy between the right-hand and left-hand plates is simply

$$\Delta PE = PE_{left} - PE_{right} = qEd - 0$$
$$= qEd$$

EXAMPLE

Often it is advantageous to know the difference in potential energy *per unit charge* between two positions. That is, given ΔPE for a charge q, we want $\Delta PE/q$. This is called the *potential difference* between the two points, and it is usually given the symbol V. In the case we are talking about, how large is V? What are the units of V?

Solution

$$V = \frac{\Delta PE}{q}$$
$$= \frac{qEd}{q}$$
$$= Ed = 10\,\frac{N}{C} \times 0.1\,m = 1\,\frac{N}{C}\cdot m$$

Thus, the potential difference does not depend upon the size of the test charge. It depends only upon the size of the electric field and the distance between the two points being considered. The units for potential difference are joules per coulomb:

$$\text{units} = \frac{N}{C} \times m = \frac{N{\cdot}m}{C} = \frac{J}{C}$$

This idea of *potential difference* is very important. The unit joule per coulomb, J/C, is called a *volt* (V). This is the familiar "volt" used by electricians. Also, the potential difference is often called the *voltage*, a more familiar term. The potential difference between two points—or the voltage, if you wish—tells how much work the electric field can do on a charge. For example, suppose that a total charge Q is pushed through the potential difference V; the work done on it is

$$W = QV \qquad (10.3)$$

Now, we are ready to return to the dry cell which is causing current to flow in the wire. First of all, remember that the wire is made of metal (probably copper), a good conductor of electricity. That means that in the wire there are millions upon millions of free electrons, ready to move about at the slightest nudge. The dry cell—through a type of chemistry magic that we need not go into here—is able to maintain a potential difference between its two terminals. The potential difference is small, only 1.5 V, but it is enough to make electrons move. When the wire is connected to the two terminals of the dry cell, an electric field is set up inside the wire, and that is what pushes the electrons through the wire. The amount of work done by the dry cell is equal to the voltage times the total charge pushed through the wire.

Before we proceed to new topics, there are a couple of things that remain to be explained about electric current. First, when all of the free electrons originally in the wire have been pushed out of one end, does the current stop? Well no; you can leave the dry cell connected for a long time, and the current will continue to flow. (Until the dry cell "runs down.") The electrons must leave the wire and go into the dry cell, and the dry cell must be providing new electrons. In fact, the same electrons run around through the dry cell and back into the wire. In order for this to happen—in order for a current to flow at all—there must be a *complete circuit*, as indicated in Figure 10.23. The electrons are moving around the circuit in a clockwise manner. However, because of a historical quirk going back to Ben Franklin, it is customary to speak of the *current* as being in the other direction. That is, as shown in Figure 10.24, the current flows in the direction positive charges would move in the wire. Of course, the positive charges that are in the wire are protons, and they cannot move. This should not be a problem for you, but you do need to remember that it is really the electrons that move.

The other point of interest is the actual motion of the electrons in the wire. If a charged object is placed between two charged plates, as in Figure 10.22, it is accelerated, moving faster and faster until it bangs into one of the plates. In the wire, the electrons must undergo an acceleration to get them moving from rest when the battery is connected, but they cannot continue to accelerate for long. If the current flows for a long time and the electrons were to continue to accelerate, they would be going impossibly fast after a while and the current would become greater and greater. This does not happen. A pinball machine gives a fairly good example of what is happening in the wire. In a pinball machine, a ball rolls down an incline. If the incline were clear, the ball would accelerate all the way down the

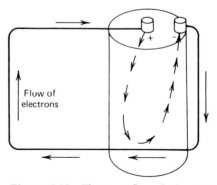

Figure 10.23 Electrons flow clockwise in this circuit.

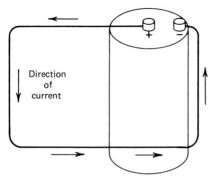

Direction of current

Figure 10.24 The *current* flows in the wire from the positive to the negative terminal of the battery.

slope. However, there are bumpers and flippers along the way. Every time the ball hits one of them it bounces off, giving up some of its energy, and then it starts down the incline again. Depending on the skill of the operator, and the energy put back into the ball by the flippers, the ball may bounce around for some time, but eventually it gets to the bottom of the incline.

Electrons in a wire, with a potential difference pushing them, also continually bump into things. The "things" they bump into are the atoms of the wire. When an electron hits an atom, it bounces off, giving up some of its energy to the atom in the process, and then it is accelerated again. The bumps are so frequent that the electrons move along quite slowly. In fact, the motion is often referred to as *electron drift* because it is so slow. To give you some notion of how slow it is, I just did a calculation of the speed of the electrons in the house wiring when I plug my toaster in. It turns out that the speed is such that it would take an electron over 40 min to travel 1 foot in the wire.

If the electrons move so terribly slowly, how is it possible for them to accomplish anything? As with Egyptian slaves building the pyramids by hand, the key is that there are so very many of them. In 1 ft of copper wire there are about 1.6×10^{23} free electrons (two

for each copper atom). This means that, even though they are traveling so very slowly, about 6×10^{19} electrons pass a given point per second. Since each electron carries a charge of about 1.6×10^{-19} C, the current is about 10 C/s.

Well, the secret is out: the unit of current is the coulomb per second (C/s). This often-used unit is named the *ampere* (A), commonly called "amps." My toaster happens to require approximately 10 A (10 amps) to operate.

OHM'S LAW

Activity 10.4

For this activity you will need the compass and some lengths of wire, two 1.5-V dry cells, and two 3-V flashlight bulbs. This is the type of bulb used in flashlights that require two "D" batteries. It would also help to have two sockets to screw the bulbs into and to which the wires can be attached. (If you have enough hands to hold wires to batteries and bulbs, you can do without the sockets, but the sockets make it easier. They are sold in electronics supply stores.)

Using one battery and some wire, light one of the bulbs. Remember that you must have a complete circuit for a current to flow.

In case you have not done this before, here are a few tips on how to get the bulb lit. As shown in Figure 10.25, the two contacts of the bulb are on its base. One is around the side of the base, and the other is at the bottom. If you don't have a socket for the bulb, you can wrap one wire around the base, as shown, and touch the other to the bottom of the bulb. Be sure to strip the insulation off the ends of all the wires you use, so that bare metal is touching bare metal for all contacts. Remember that there has to be a complete circuit. That is, you should be able to trace the path of the current from the positive side of the battery, through a wire to the bulb,

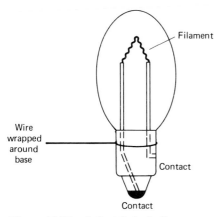

Figure 10.25 A flashlight bulb.

through the bulb and out of its second contact, through the second wire, and back to the negative side of the battery. If you have serious problems lighting the bulb, look ahead to Figure 10.28, which shows how it is done.

Once you have the bulb lit and have noted how brightly it burns, put the compass galvanometer in the circuit. That is, wrap two or three turns of one of the circuit wires around the compass, as you did before, and be sure that the current has to flow through this piece of wire to get to the light bulb. (Try to do it on your own but, if you have trouble, Figure 10.29 will show you how.) Adjust the number of turns around the compass so that when the current is flowing, the pointer is deflected about half-way through the full 90° it can go. (That is, so it is pointing northeast or northwest.) This deflection will be a measure of the current flowing in the wire; if more current flows, the deflection will be greater; if less current flows, the deflection will be less. There is one oddity about this that you should be aware of: If twice as much current flows, the deflection of the compass pointer will not be twice as great. That is, the deflection is not proportional to the current. However, the galvanometer, carefully read, will tell you whether more or less current is flowing. (In a good, commercially made galvanometer, the deflec-

tion of the pointer will be proportional to the current that flows in the circuit.)

Light the bulb using both batteries. This can be done in two ways: by putting the batteries in "parallel" and by putting them in "series". Both parallel and series connections are shown in Figure 10.26. What do you note concerning the brightness of the bulb and the reading of the compass in the two kinds of connections?

Now light both bulbs using one battery. Try the bulbs in series and in parallel, as shown in Figure 10.27. Note the brightness and galvanometer readings in each connection. You might also try the batteries both in series and in parallel, giving a total of four cases:

1. Bulbs in series, batteries in series.
2. Bulbs in series, batteries in parallel.

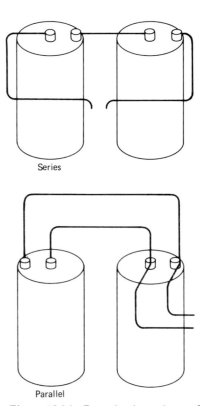

Figure 10.26 Batteries in series and in parallel.

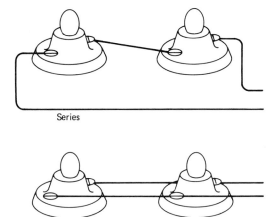

Series

Parallel

Figure 10.27 Bulbs in series and in parallel.

3. Bulbs in parallel, batteries in parallel.
4. Bulbs in parallel, batteries in series.

___

Now let us summarize your findings in this activity. First, it requires a complete circuit to light a single bulb (or to do anything useful with electricity). Figure 10.28 shows a circuit for lighting the bulb, drawn in two different ways: in a "picture," with drawings of the parts; and in a *circuit diagram*, or *schematic diagram*, with electrical symbols representing the parts. When the galvanometer is put into the circuit, as in Figure 10.29, it does not change anything; it simply gives us a way of measuring current. (Incidentally, using modifications of the simple galvanometer, *voltmeters* for measuring voltages and *ammeters* for measuring currents can be made. If you have access to voltmeters and ammeters, you can do the experiments in Activity 10.4 with more numerical precision.)

In addition to learning that a complete circuit is necessary for electric current to flow usefully, you should have observed the following:

1. When a single bulb is lit with a single battery, it glows dimly. When the single

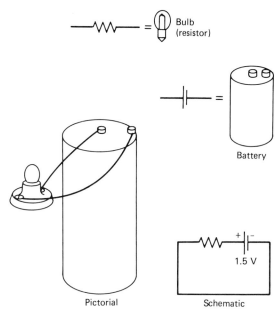

Pictorial   Schematic

Figure 10.28 Pictorial and schematic circuit diagrams for lighting a flashlight bulb.

bulb is lit with two batteries *in series*, it glows brightly and more current flows than with one battery. When it is lit with two batteries *in parallel*, the glow and the current are the same as for one battery.

2. When two bulbs are in parallel, the observations made are exactly the same as

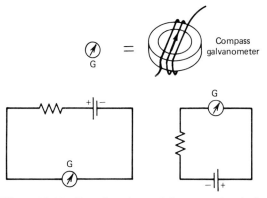

Figure 10.29 Two drawings of the same circuit. It doesn't matter where the galvanometer is placed in this circuit.

those made for a single bulb, except that the total current is greater in each case. (With two bulbs in parallel, the current supplied by the batteries is just double what it was for one bulb. Did you measure the total current with your compass galvanometer, or just the current through one bulb? Figure 10.30 shows the various possibilities.)

3. When two bulbs are in series, they glow more dimly for either way of wiring the batteries, and less total current flows than with a single bulb.

4. If the two batteries are wired in series or in parallel in the "wrong" way, strange things happen.

Let us start with the last observation first. Figure 10.31 shows the results of wiring together two batteries in various ways. In each case, parallel or series, the two batteries must be wired so that they help each other. If they are opposed, no useful voltage will be produced. Wired properly, two batteries in parallel still give a voltage of only 1.5 V, but they have twice the current-producing capacity of one battery alone. Two batteries in series, on the other hand, will give 3 V, but the maximum current from them will be the same as for a single battery. (It figures—in the series wiring, all of the current in the circuit has to pass through each of the batteries; in the parallel wiring, each of the batteries can contribute half of the current.)

With this information, you now know whether the voltage was 1.5 or 3 V for each method of wiring, and the rest of your observations can be summarized in two statements:

1. Increasing the voltage across a bulb causes an increase in the current through the bulb.

2. Compared to the current through a single bulb, for a given voltage, putting two bulbs

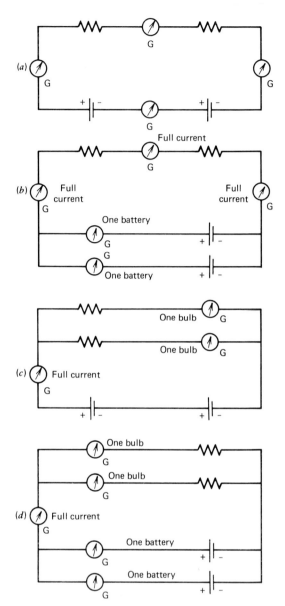

Figure 10.30 Measuring current in the various ways possible. (*a*) Bulbs in series and batteries in series. Same current everywhere in circuit. (*b*) Bulbs in series and batteries in parallel. Galvanometers measure full or half. (*c*) Bulbs in parallel and batteries in series. Galvanometer measures either full current or current through a single bulb. (*d*) Bulbs in parallel and batteries in parallel.

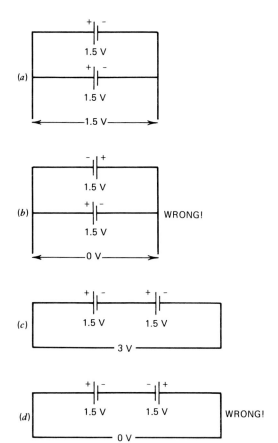

Figure 10.31 Two batteries wired in parallel and in series. (*a*) Parallel: The two batteries help each other. (*b*) Parallel: The batteries oppose; although a large current runs through the batteries, no voltage is available to light a bulb. (*c*) Series: Batteries help each other. (*d*) Series: Batteries are opposed; no current flows and no voltage is available to light a bulb

in series decreases the current, and putting two bulbs in parallel increases the current.

Believe it or not, we can summarize this in one simple equation, if we first come to grips with the notion of *resistance*. Remember that when electrons move through a wire, they continually bump into atoms and get slowed

down. If this did not happen, the current would soon become infinite, which is an impossibility. We say that the wire *has resistance* or that it is a *resistor*. Light bulbs are also resistors, having much more resistance than a wire. (In fact, for the circuits under investigation here, you can treat the wire as if it had no resistance at all, with all of the resistance being in the light bulbs.) Careful measurements, using voltmeters and ammeters, give the following relationship:

$$V = IR \qquad (10.4)$$

This equation is called *Ohm's law,* and it is another example of a "law" that is the result of many experimental observations. The V stands for the voltage (potential difference) across the two ends of the resistor, and the I is current through the resistor.

The quantity R in Ohm's law is the resistance. By writing the equation as $R = V/I$ and putting in the proper units, you can see that the units of resistance are volts per ampere. This unit is called an *ohm* (symbolized by Ω, the Greek letter *omega*). That is:

$$1\ \Omega = 1\ \frac{V}{A}$$

It is worth noting that light bulbs do not have a constant resistance. As the voltage across the filament increases, and thus as the filament gets hotter, its resistance also increases. However, many resistors have very nearly a constant resistance as the voltage is changed and, in what follows, we shall treat them all as if they behaved that way.

EXAMPLE

Suppose you have two resistors, such as those used in electronic circuits, rated at 3 Ω and

6 Ω, respectively, and two batteries, one a 1.5-V dry cell and the other a 6-V lantern battery. Using only one battery and one resistor at a time, what currents can be produced?

Solution

Ohm's law can be used for each connection of battery and resistor. The form in which it is useful here is

$$I = \frac{V}{R}$$

When the 1.5-V dry cell is connected to the 3-Ω resistor, the current produced is

$$I = \frac{1.5 \text{ V}}{3 \text{ }\Omega} = 0.5 \text{ A}$$

Similarly, for the other three combinations, the currents produced are

$$I = \frac{1.5 \text{ V}}{6 \text{ }\Omega} = 0.25 \text{ A}$$
$$I = \frac{6 \text{ V}}{3 \text{ }\Omega} = 2 \text{ A}$$
$$I = \frac{6 \text{ V}}{6 \text{ }\Omega} = 1 \text{ A}$$

Questions

An automobile headlight has a resistance of 1 Ω, and it is connected to a battery that provides 12 V. What is the current through the headlight?

Suppose two 1-Ω headlights are connected in series, and the battery is used to light them (rather dimly). What current will flow through the headlights? If a voltage of 12 V were to cause that same current to go through a single resistor, what would its resistance be? (Use Ohm's law to find out.) Then, what can you conclude about the total resistance if two resistors are connected in series?

Usually, two headlights are connected in parallel in an automobile. (Otherwise, if one burns out, the other would go out as well, like some strings of Christmas tree lights.) If the headlights are connected in parallel, and the battery is used to light them, what is the current through each? (Each one gets the full 12 V.) What is the total current? Then, what can you conclude about the total resistance if two resistors are connected in parallel?

No doubt, you concluded that the two bulbs (resistors) in series provide more resistance (twice as much) than does one bulb. Thus, less current flows. However, the two bulbs in parallel offer two routes for current to flow, thus the total current is twice as great using the same battery. That means the resistance is less (half as much) than for one bulb. What you have concluded about two resistors in series and in parallel can be extended to any number of resistors. For example, if a number of resistors are connected in series, as in Figure 10.32, their total resistance is the sum of the individual resistances. If you were to electrically test a box with two wires sticking out of it, as in Figure 10.33, you would not be able to tell whether there is one resistor with a resistance of R_{total} or a number of resistors in series whose sum is R_{total}. Writing this as an equation, for a *series* connection:

$$R_{total} = R_1 + R_2 + R_3 + R_4 \ldots$$

When two or more resistors are in parallel (Figure 10.34), the situation is a little more complex. The two identical resistors in parallel have a resistance equal to *half* the resistance of either one alone. In general, if there are a number of resistors in *parallel*, the total resistance is given by

$$\frac{1}{R_{total}} = \frac{1}{R_1} + \frac{1}{R_2} + \frac{1}{R_3} + \frac{1}{R_4} + \ldots$$

Figure 10.32 Several resistors in series.

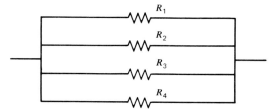

Figure 10.34 Resistors in parallel.

Again, if the resistors are all in a box with just two wires sticking out, as in Figure 10.33, there is no way to tell how the resistors are arranged inside without opening the box. You can only measure the total resistance.

In your home, all of the appliances and lamps are wired in parallel, so that each one gets the full voltage available.

ELECTRIC POWER

What is it that lights the flashlight bulb, anyway? Obviously, energy in the form of light and heat comes *from* the bulb. Then, the battery must be providing energy *to* the bulb, for "you can't get something for nothing."

The explanation is really quite simple. Recall the discussion of the motion of electrons through a wire. The electric field in the wire—produced by the voltage—did work on the electrons. If the electrons were in empty space, that work would be converted to kinetic energy, with the electrons going faster and faster as they moved in the electric field. In the wire, however, the electrons continually bump into atoms, giving up some of their energy. The atoms, as a result of gaining the extra energy, vibrate more rapidly than

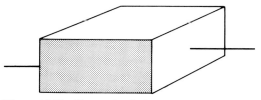

Figure 10.33 If you check the wires emerging from a box, you can measure only the total resistance inside the box.

they did before. That is to say, their *temperature* increases.

In the wire, the resistance is very small, and large currents flow easily. The tiny coil of wire in a flashlight bulb (the *filament*) is made to have a larger resistance, but it is still small enough to permit a large current to flow. When current flows, enough energy is transferred to the filament so that it becomes white hot and gives off light and heat.

It is easy to compute the amount of energy given to the light bulb. Let us ignore the resistance of the connecting wire, because it is so small, and assume that all of the energy lost by the electrons in the circuit goes into the filament. This is a good approximation to the actual situation.

Suppose that a bulb is wired to a battery and remains connected until a total charge Q flows through it. Then, that charge has done the work of lighting the bulb. The potential difference across the filament in the bulb is V, the voltage provided by the battery. The amount of work the charge Q does in passing through a potential difference V was given earlier in Equation 10.3:

$$W = QV$$

That is, the work is the total charge times the voltage.

However, we are often more interested in *power*, the rate of doing work, than we are in the work done. Suppose the bulb must burn for a time t for the total charge Q to flow through the bulb. Then, the power is the work divided by the time:

$$P = \frac{W}{t} = \frac{QV}{t} = \frac{Q}{t}V$$

But the charge divided by the time—the coulombs per second—is just the current. Thus, where I is electric current, we can write

$$P = IV \tag{10.5}$$

This is a very important result, and we shall use it later.

Questions

The wires in a toaster have a total resistance of 10 Ω, and the voltage available at the outlet is 115 V. When the toaster is plugged in, how much current flows through it? What is the power used by the toaster?

Remember that current has the units C/s and voltage has the units J/C. What are the units for electric power?

_____

Were you surprised that the units for *electric* power turned out to be watts? Of course not!

In the previous questions, you used Ohm's law to find current, then you used IV to find power. This process can often be shortened by writing a couple of variations of Equation 10.5. Start with

$$P = IV$$

Then, substitute for V the product IR (because $V = IR$). This gives

$$P = IV = I(IR)$$

or

$$P = I^2R \tag{10.5a}$$

Likewise, you can substitute V/R for I and get

$$P = \frac{V^2}{R} \tag{10.5b}$$

Both of these expressions are useful.

ELECTRIC MOTORS

Activity 10.5

For this activity you will need only the magnetic compass, a long piece of wire, and a battery.

Once again, make a few turns of wire around the compass and place the compass where it will be level. Touch the two ends of the wire to the terminals of the battery, and observe the motion of the compass pointer. Now reverse the two ends of the wire, touch them to the terminals, and observe the motion. How does the motion in the two cases compare?

This part will take some coordination, which you can develop with a little practice. With the same setup of compass and wire, attach one end of the wire to one battery terminal, and just tap the other terminal with the other end of the wire. The compass pointer will probably make several rapid revolutions. Now try tapping repeatedly, at different rates, and see if you can keep the pointer turning in the same direction. Faster! Faster!

With practice, you should be able to make the pointer turn very rapidly in either direction.

_____

What you have done is to make a simple electric motor. Although a real motor is somewhat more sophisticated, it works on exactly the same principle.

What is the principle, anyway? Remember that the compass pointer is a small magnet. The thing that exerts force on a magnet is a *magnetic field*. You saw that earlier, with the magnetic field coming from a second magnet. Now the magnetic field must be coming from the current in the wire.

Activity 10.5 has shown that an electric current (or *any* moving electric charges) produces a magnetic field, an important result. For example, in making a coil with loops of wire and passing current through it, you made a magnet that has lines of force very similar to those of a bar magnet, as is shown in Figure 10.35. The magnetic field from each loop of wire reinforces the field from each other loop, so the coil produces a magnetic field much stronger than that produced by a single straight wire or a single loop. Wrapping the coil around some iron makes the magnet even stronger: When the current is turned on, the tiny "magnets" in the iron all line up, and the iron temporarily becomes a magnet. When the electric current is turned off, the little magnets all relax into random directions, and the iron is not a magnet any longer. This kind of magnet is called an *electromagnet.* Using the right kind of alloy of iron and other metals, one can also line up the little magnets permanently and make a permanent magnet, like those you have been using.

So now you know that you can make an electromagnet, which is a magnet only when the current is on. Also, a motor can be made by mounting a magnet so that it can rotate and by applying a changing (on and off) magnetic field to it with an electromagnet. A smoother motor can be made with several coils that are switched just at the right times for each to give the rotating magnet a push as it goes by. You can see the sets of coils in the photo of the inside of an electric motor, Figure 10.36. In addition to the outer coils, very often the rotating magnet itself is an electromagnet. The switching of the current usually occurs in the turning coil, and is done by a contact ring—called a *commutator*— mounted on the shaft on which the coil turns. The commutator is split into a number of segments, depending on the number of coils in the motor.

Electric motors are used in a wide variety of applications, and they range from the tiny ones that power electric wristwatches to the huge monsters that power industrial equipment. Figure 10.37 shows a massive dragline, used to work surface coal mines, that is powered by very large electric motors.

ELECTRIC GENERATORS

Activity 10.6

For this activity you will need 10 or 15 ft of wire, the compass, and a strong magnet. The type of magnet I have in mind is usually bent into a horseshoe shape. The stronger it is, the better the activity will work. The experiment will work with almost any wire, but does better if the wire is of a fairly heavy gauge, say number 16 or number 14. (Number 14 is bigger than number 16.) The heavy-gauge wire works better because it has less resistance than a smaller-diameter wire; thus current flows more readily in it.

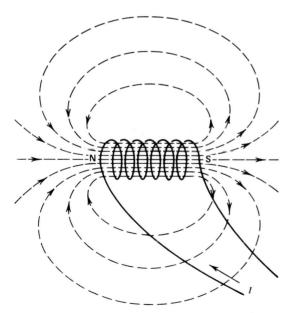

Figure 10.35 The magnetic field of a coil of wire resembles that of a bar magnet.

Figure 10.36 An electric motor has numerous coils wrapped around iron to produce magnetic fields.

Wrap a number of turns of the heavy wire around the compass, depending on how much wire you have. About 3 ft from the compass, wrap another coil of as many turns as possible using the same piece of wire. Then, keeping the second coil about 3 ft from the compass,

join the two free ends of the wire to make a complete circuit. The second coil should be of a size such that one side of it can fit between the poles of the horseshoe magnet, as shown in Figure 10.38. Note that this is a complete circuit, but one with no battery in it. (If the equipment is made available, this activity can also be done easily by connecting a ready-made coil of wire to a commercial galvanometer and using a bar magnet. Some students are shown doing this in Figure 10.39.)

Now, hold the second coil between the poles of the horseshoe magnet. Is the compass deflected away from north? Then, suddenly jerk the magnet away from the coil. The more sudden the movement, the better. Do you note a deflection of the compass pointer now? Does it stay deflected, or come back to north? Now—suddenly again—bring the magnet back to the coil. Which way is the deflection this time? Does the compass stay deflected? Now try moving the coil instead of the magnet.

Figure 10.37 This huge dragline, working a surface coal mine, is powered by electric motors.

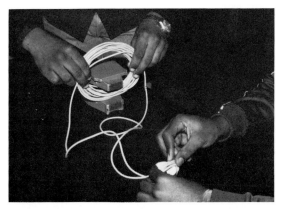

Figure 10.38 One leg of the horseshoe magnet is in the second coil.

Earlier you discovered that a current in a wire produces a magnetic field. What you have seen here is that a *changing* magnetic field produces a current. When the magnet is suddenly removed from the coil, the magnetic field in the region goes suddenly from a strong field to none. This causes electrons to flow in the wire, and the compass pointer is de-flected. When the magnet is suddenly brought back, current flows again, but in the opposite direction. When the magnet is stationary, no matter where it is, there is no current. In order for current to flow, the magnetic field must be *changing.* The strongest magnetic field, if it is not changing in the region of the wire, will produce no current. The change of magnetic field near the wire may be accomplished by actually changing the field strength or, as here, by moving the wire in and out of the magnetic field region.

The second coil and the moving magnet comprise a primitive *electric generator.* In fact, the "generator" and the "motor" (the compass with the coil around it) are much like the generator–motor combination that we use in our homes and in industry. In that case, a large generator at the power station produces electricity, and the motor at the other end of the transmission lines changes the electric energy into work.

A real generator is built very much like a motor. (In fact, some motors can be run as

Figure 10.39 A ready-made coil and a commercial galvanometer can be used in Activity 10.6.

generators and *vice-versa.*) There are rotating coils surrounded by another set of coils. When operating, either the rotating coils (the *rotor)* or the stationary coils (the *stator*) have a current applied, making them into electromagnets. When one set of coils feels a changing magnetic field as the other set goes whizzing by, a current is produced in it.

There is one important aspect of generators and motors that has only been mentioned in passing so far. It is that conversion of energy from one form to another is involved. To run the generator requires energy. Work must be done to turn the shaft of the generator. I am not talking about the work needed to overcome friction; in a well-made generator the friction will be very small. Most of the work goes into producing the electricity. That is to say, mechanical energy (work) is converted into electrical energy. When that electrical energy reaches a motor, it will be converted into mechanical energy again to run a machine of some sort. (Or, the electrical energy can just be converted into heat, as in a toaster.)

Of course, in the real world, a conversion of energy into another form is usually not 100 percent efficient, and the losses that result are something we will have to look at later.

TRANSFORMERS AND ALTERNATING CURRENT

Activity 10.7

For this you will need the compass, the same long wire you used in the last activity, another piece of wire 10 or 15 ft long, and two batteries.

The compass is wired as before, with several turns of wire wound on it. This time, the second coil, made with the same piece of wire, should be wound tightly on a large bolt. Wrap as many turns as you can. We will call this coil the secondary coil, or just the secondary. Now, with the other piece of wire, wrap another coil on the bolt, over the first one and with its ends sticking out. Again, wrap as many times as possible. This coil is called the primary. Be sure that there is no electrical contact between the two coils on the bolt. The

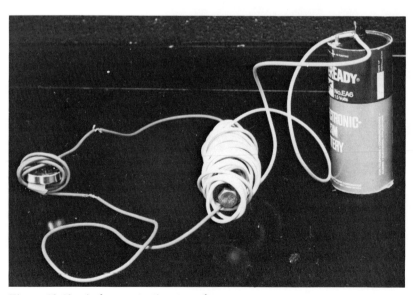

Figure 10.40 A demonstration transformer.

finished product will look something like Figure 10.40. (Figure 10.41 shows how to produce the same effects using two large coils and a galvanometer.)

Now, using the two batteries in series, put a current through the primary. Note the compass deflection as the current starts. Does the deflection continue, or does the pointer swing back to north? Now stop the current and note the deflection. By alternately turning the current on—off, on—off, see how great a swing of the compass pointer you can get.

By now, I am sure you know what is happening. The secondary coil can produce a current if it is subjected to a changing magnetic field. The primary coil provides that changing field. When the batteries are not attached, no current flows in the primary coil, and there is no magnetic field. When the batteries are attached, a current suddenly flows and a

magnetic field suddenly builds up around the primary. This field is also building up in the region of the secondary coil, and a current is produced in the secondary as the field changes. With a steady current in the primary, the field is present but unchanging. Thus no current flows in the secondary. However, when the primary current suddenly stops, the field dies, and again a changing field produces a current in the secondary. The current in the secondary is in opposite directions in the two cases.

This is an example of a *transformer*. Transformers are very useful in power transmission, for they can change (or transform) the voltage. However, in order for a transformer to work, the current in the primary coil must keep changing. A steady current would do no good at all. Figure 10.42 shows real transformers of the type used in electronic circuits.

I suppose every transformer could have a little gnome in it to open and close the pri-

Figure 10.41 A demonstration transformer using two ready-made coils, side by side, and a commercial galvanometer.

Figure 10.42 Real transformers of the type used in electronic circuits.

mary circuit, but people have found an easier way to change the field. So far, we have been talking about *direct current* (dc), a current that always flows in one direction. To make a transformer work, we need *alternating current* (ac). An alternating current flows first in one direction through the circuit, then in the other. This happens very rapidly. In the U.S. power grid, the standard rate of switching alternating current is 120 changes of direction per second. That is 60 complete cycles per second. Figure 10.43 shows graphs of direct current and alternating current. A graph of the voltage applied across the ends of a circuit to produce an alternating current would look exactly the same. So we could, if we wished, speak of "alternating voltage," but it is customary to talk about alternating current. This is abbreviated ac, and the alternating current found in your home is called "60 cycle, ac," or "60 hertz, ac."

As I mentioned, the importance of transformers is that they can change the voltage. If the secondary of a transformer has 10 times as many turns as the primary, then the outgoing voltage will be 10 times as great as the incoming voltage. Similarly, if the primary has

twice the turns, then the secondary voltage will be only half of the primary voltage. You will see in Chapter 11 that this can be very useful in power transmission.

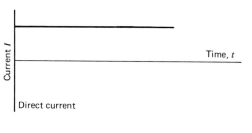

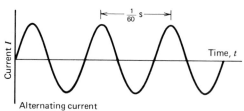

Figure 10.43 Graphs of direct and alternating current. When the alternating current is below the time axis, it is traveling in the direction opposite to the direction of its flow when it is above the axis.

SUMMARY

Electrical effects responsible for all of the benefits of the various uses of electricity arise from the charges resident on electrons and protons. A study of static (nonmoving) charges shows that there are two kinds of charge, usually labeled "positive" and "negative." Two objects with like charges repel each other, whereas unlike charges attract each other. The force between two charged objects is given by Coulomb's law.

When an object with a net charge exists in space, it will exert forces on any other charges in the area. These forces can be mapped by imagining lines of force, which show the direction of the force that would be exerted on a small test charge in a given location. The electric field at that location is defined as the force acting on the test charge divided by the size of the test charge.

Magnets have no net electric charge, but they exert forces on one another and on some metal objects. Each magnet has a north pole and a south pole. Like poles repel and unlike poles attract one another. Each magnet produces a magnetic field, and magnetic lines of force can be drawn by imagining the force on a tiny north pole at every location in the surrounding space.

To get useful work from electricity, electric charges must flow to produce a current. Usually, a current is produced by huge numbers of electrons flowing through a conductor, such as a copper wire. Electric currents can be caused to flow by voltages produced by dry cells ("batteries") and generators. When an electric current is made to flow through a resistor some electric energy is changed to heat and, in the case of a light bulb, light. The work done by an electric current in a circuit element, such as a resistor, is given by $W = QV$, where Q is the total charge that flows through the resistor, and V is the potential difference (voltage) across it. Ohm's law is the equation $V = IR$, where I is electric current and R is resistance. Two or more resistors in series add up to a greater total resistance; two or more resistors in parallel have less total resistance than any one of them.

The power used in a circuit is given by $P = IV$, or its variants, $P = I^2R$ and $P = V^2/R$. In a resistor, the power goes into heat and sometimes light. In an electric motor, the electric power is converted to mechanical power. Electric motors use the principle that currents in wires produce magnetic fields, and those fields can be used to exert forces and thus do work. Electric generators use the principle that a changing magnetic field in the vicinity of a wire will produce an electric current. Using permanent magnets or electromagnets, generators produce electric power by moving coils of wire through magnetic fields, and they change mechanical energy into electric energy.

Transformers use both principles mentioned in the preceding paragraph to increase or decrease voltages. A primary coil has an alternating current (ac) passing through it, thus producing a constantly changing magnetic field. A secondary coil, which is not attached to the primary but is near it, and thus is "coupled" magnetically to it, has an alternating current produced in it by that changing magnetic field. The voltage in the secondary will be greater than or less than the voltage in the primary, depending on whether the secondary coil has more or fewer windings.

ADDITIONAL QUESTIONS AND PROBLEMS

1. Summarize, in your own words, the rules of attraction and repulsion of electrically charged objects.

2. Describe the submicroscopic action that occurs whenever one rubs a piece of plastic with a piece of wool cloth.

3. What evidence can you cite from your own experience that static electricity occurs in nature?

4. Suppose you could remove all of the free electrons (about 1.6×10^{23} of them) from a 1-ft-long length of copper wire, leaving the same number of extra protons in the wire. Suppose, further, that you could move the electrons, all in one bunch, 1000 km away from the wire. What would then be the force of attraction between the free electrons and the extra protons in the wire? (Each electron has a charge of 1.6×10^{-19}C.) Why would actually doing this be quite impossible, even though the electrons move freely in the wire?

5. Imagine a sphere in outer space with a charge of $+1$ C. If a small test charge of $+1$ picocoulomb (10^{-12} C) is brought to a distance of 10 m from the large charge, what will be the force acting on the small charge? In what direction will this force act? What will be the size of the electric field at that point? How would the preceding answers be changed if the large charge were -1 C?

6. A flashlight, with 2 "C" cells (1.5 V each) in series is turned on for 1 min. In that time, 120 C of charge flows through the bulb. Using $W = QV$, find the work done by the batteries in forcing the charge through the bulb. How great is the current? What is the resistance of the bulb? Using $P = I^2R$, what is the power supplied by the batteries? Does power times time equal the work you just calculated?

7. Batteries are often rated in ampere hours (A h). A rating of 1 A h would mean that a battery could provide a current of 1 A for 1 hr, or 2 A for $\frac{1}{2}$ hr and so on. I have a flashlight that uses three "D" cell (1.5 V each) in series and a bulb with a resistance of 0.9Ω. The dry cells have a rating of 3.5 A h If the flashlight is loaded with new dry cells, turned on, and left on, how long should the light burn? (It won't actually happen this way because the

light doesn't burn at full strength, then suddenly go out; it gets dimmer as time goes on.)

8. Sketch a circuit for connecting four 1.5-V dry cells in series to light two bulbs in parallel. What is the voltage across each bulb? If each bulb has a resistance of 1 Ω, what is the current through each bulb? What is the resistance of the two bulbs in parallel? What is the total current? How much power do the batteries provide?

9. Have you ever tried to start an automobile and heard only a sickening "click"? Terrible, isn't it? The problem could be a "dead" battery—one whose charge has been lost. However, this frequently happens with a perfectly healthy battery. The reason is that a starter needs a very large current, sometimes as much as 150 to 200 A. In order for that much current to be provided by a 12-V battery, the resistance in the circuit must be small. However, sometimes corrosion occurs between the battery posts and the connecting cables. This corrosion has a relatively high resistance. Assuming that the cables have no resistance, what is the greatest resistance the corrosion may have if the battery is to produce a current of 150 A? If this were to happen to you, how would you fix the problem?

10. What is the resistance of the filament of a 100 W light bulb designed for 120 V operation? What is the resistance of 10 such bulbs in a home? (Remember, they are all in parallel.)

11. The lights and other appliances in your home are all wired in parallel, so that each one will get the full voltage it needs (120 V, ac). If a 100-W lamp is turned on, what current flows? A single "circuit" in a home consists of all of the outlets that connect together in parallel with the current going through a single fuse or circuit breaker. In a 15-amp circuit, the fuse or circuit breaker is made to "blow" if the total current exceeds 15 A by very much. How many 100-W lamps can be operated safely on a single 15-amp circuit? What is the total power consumed if they are all on at the same time?

12. Look at the power pole supporting the power lines to your house or another building. You should see a transformer on it. It is there because the power coming on the trans-mission lines is at too high a voltage to be used in your home. In such a transformer, a "step-down transformer," the number of turns in the primary may be about 25 times the number in the secondary winding. If the voltage at the house is 220 V, what is the voltage at the transmission line?

11 *Generation and Transmission of Electricity*

In our industrial society, electricity is so plentiful, so cheap and, above all, so dependable that we rarely think about it. That is, we rarely think about electricity until something goes wrong; then we find out how utterly dependent upon it we are.

On November 9 and 10, 1965, much of the heavily populated northeastern part of the United States was plunged into darkness when a minor occurrence cascaded in a sort of "chain reaction" throughout the entire northeast power grid. About 30 million people suddenly and unexpectedly found themselves without electric power. Some of them were stuck in elevators on the thirty-fifth floor. Some were stranded in subway cars under the streets of Boston and New York. Everywhere people were without heat. Dinners could not be cooked in many homes, and restaurants were shut down. Lights went out in operating rooms, and emergency power had to be found and used to complete surgical operations. Electric toothbrushes and electric blankets were useless. In the morning, there was no toast. Computers were silenced. Commerce and industry ground to a standstill. Perhaps the worst cut of all, there was no television viewing!

That experience was painful, and the electric utilities have taken all possible steps to see that it does not happen again. However, it was a good object lesson on how dependent we are on electric power. The electric power industry accounts for the consumption of about 25 percent of all the energy used in the United States, and it is one of the biggest industries, if not the biggest industry, in the country and the world. In general, that industry has done well in meeting the demands of the society. In this chapter we shall explore some of the ways in which it does so and some possibilities for the future.

STEAM-DRIVEN POWER PLANTS

Whenever electricity is produced or used, energy is *converted* from one form to another. In 1983 in this country, close to 90 percent of all the electricity generated came from steam-powered turbines. (Most of the rest came from hydroelectric plants and wind power made a small contribution.) A *turbine* is a machine with a shaft on which there are numerous blades against which high-pressure steam is directed. As the steam hits the blades, they force the shaft to turn, much like the blades of a windmill cause a shaft to turn. In recent years, new steam turbines have been manufactured at the rate of adding over 25,000 megawatts (1 MW = 1 million W) capacity per year. The turbines have also become very large, with the largest single unit rated at about 1300 MW.

Figure 11.1 shows a typical large steam turbine in a nuclear power plant. It operates at 3600 rpm (revolutions per minute), and the shaft is connected directly to the generator which it drives. One important part of all turbines is a "governor" of some sort, to regulate the speed. If the generator should lose its load (that is, if a cable should break, so that the generator is no longer producing power) the sudden decrease of resistance could cause the turbine to "run away" to very great speeds and destroy itself. The governor prevents this from happening.

In fact, a modern power plant is literally filled with devices to control and regulate the machinery so that it will run efficiently and safely. Generally, the entire plant will be run from a central control panel, and that panel will be attached to sensors of all types to detect such diverse data as the rate of fuel flow, the generator voltage, the temperature of important bearings, and the condition of valves, as well as many other functions. The scanning of these devices and the control of the operation of the plant is usually computer assisted. This degree of automation ensures that there will be early detection of problems, that the plant will run at top efficiency, and that it will be highly reliable. Figure 11.2 shows part of the control room of a power plant.

Fossil-Fuel Plants

The only major difference in the various kinds of steam power plants is in the source of the steam to power the turbines. *Fossil fuels* are fuels that have developed from the remains of plants and animals that lived millions of years ago. Or, more explicitly, the major fossil fuels are coal, oil, and natural gas. The future availability of these resources is discussed in Chapter 12. In the recent past, it has been very tempting to use oil and natural gas and, as a country, we have not resisted the temptation to any great degree. Since the Arab oil embargo in 1973, however, it has been brought

Figure 11.1 A large steam turbine and generator in a nuclear power plant. (Baltimore Gas and Electric Company, from U.S. DOE.)

Figure 11.2   Part of the control room of a modern power plant. (Rochester Gas and Electric Company.)

home to us that the supplies of oil and natural gas are not infinite, and that alternatives must be sought.

This sounds good in principle. It seems reasonable that we should just replace oil and natural gas with coal and/or nuclear power (depending on one's convictions). However, that may not be as easy as it sounds. Even a switch to coal presents problems. For one thing, oil and natural gas, being fluids, are much easier to transport and handle. They can be pumped easily through pipes and handled automatically. Many power plants, especially those on the heavily populated East Coast, are far from the coal fields but have good access to foreign and domestic oil. Then, too, oil and gas are "clean" fuels, offering much less of a pollution problem and needing less special equipment to burn than coal. In fact, if the Environmental Protection Agency (EPA) were to enforce the air pollution regulations strictly, more than 70 coal-burning power plants in the United States would be subject to stiff fines for every day of operation, and many of them might be forced to close down.

At any rate, whatever fossil fuel is burned, the result in a power plant is the same: Large quantities of steam (as much as 10 million pounds per hour!) are produced to run the steam turbines. You recall that, when water is boiled, its temperature is 212°F (100°C), and the steam that is produced is also at that temperature. More accurately, the water boils and the steam is produced at that temperature when the pressure is *atmospheric pressure* (about 15 lb/in^2 or 1×10^5 Pa). But if, like me, you have tried to cook spaghetti in the high mountains, you know that the boiling point of water depends upon the pressure. At an altitude of 7000 ft, where the atmospheric pressure is low, the boiling point of water is about 92°C, and the spaghetti takes much longer to cook. Similarly, when the pressure is higher, the boiling point of water is higher.

In the case of steam turbines, huge quantities of steam at very high pressures are needed to provide the required power. In some fossil-fuel plants the steam pressure is about 3500 lb/in^2 (over 200 times atmospheric pressure) at a temperature of about 1050°F (566°C). This is called *superheated steam*, and it is produced by raising the temperature of the steam in a closed container until the desired pressure is reached. A few of the newest plants run at a pressure of 5000 lb/in². Approximately 300 miles of pipe must be used in a power plant along with perhaps 50,000 welded joints, so you can imagine the amount of care that must be taken in building and maintaining the plumbing so that it will not explode or leak at this very high pressure!

Remember that a heat engine of any type must give up waste heat in the process of converting heat to work. The steam turbines in a power plant are no exception. After the steam has done its work in the turbine, it must be changed to liquid water and pumped back to the boiler. For this purpose a *condenser* is used. A condenser is just a heat exchanger that gives up heat from the exhausted steam to water in a stream, lake or cooling pond, or directly to the air. In the process, the steam is condensed back into water. In a typical installation, the temperature at the heat exchanger might be about 100°F (38°C). Figure 11.3 is a schematic diagram of a typical fossil-fuel power plant.

EXAMPLE

In the case of the modern power plant just described, what would be the efficiency of the turbine if it were an ideal heat engine?

Solution

You may recall from Chapter 5 that the efficiency of an ideal heat engine is

$$Eff = \frac{T_H - T_C}{T_H} \times 100\%$$

where T_H is the absolute temperature at the hot reservoir and T_C is the temperature at the cold reservoir. In this case $T_H = 566°C = 839$ K and $T_C = 38°C = 311$ K. Thus

$$Eff = \frac{839 \text{ K} - 311 \text{ K}}{839 \text{ K}} \times 100\% = 63\%$$

Because of practical limitations of steam engines and losses in the turbine, the actual

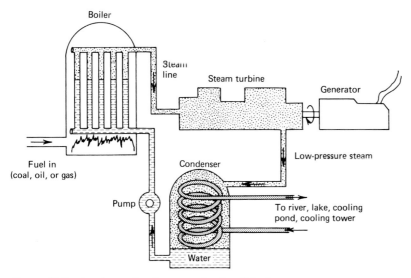

Figure 11.3 A schematic diagram of a fossil-fuel-fired steam turbine power plant.

efficiency turns out to be about 47 percent for the best of the modern steam turbines. In addition, the other components of the power generating system are not 100 percent efficient. For example, the boilers that are now used in power plants can convert about 90 percent of the heating value of the fuel into heat, and the generators are about 99 percent efficient in converting the mechanical energy from the turbine into electrical energy. From this information we can calculate the overall efficiency of a modern power plant in converting the energy stored in a fuel to electric energy. To see how, first consider the following.

Suppose there are 100 bass in a small lake belonging to a friend, and you aim to catch all of them and cook them for a huge neighborhood fish fry. You fish for a long time, but only catch 50 of the bass. Thus, your "efficiency" at catching bass is 50 percent. However, the friend who owns the lake wants to have her own fish fry, so you give her 20 of the fish. Thus your efficiency at getting the caught fish home is 60%. Your overall efficiency, then, at getting the fish from the lake to your frying pan is .50 × .60 = .30, or 30 percent. That is, 30 out of the original 100 fish got there. (This, believe me, is *much* better than my efficiency at catching bass.)

In the case of the power plant, the overall efficiency can also be found by multiplying the efficiencies at each stage of the energy conversion. That is:

overall efficiency = .47 × .99 × .90 = .42

That is, the overall efficiency of the best existing systems is about 42 percent.

Questions

If the power plant being discussed is a 1000 MW plant and it produces at full capacity, how many kilowatt-hours are generated in 1

hr? In 1 day? Given the overall efficiency just computed, how much energy must be supplied by the fuel in 1 day? What is this in Btus? (1 kWh = 3413 Btu.) The heating value of the coal commonly used in power plants varies quite a bit, but it is somewhere in the neighborhood of 13,500 Btu/lb. How many tons of coal are required daily to fuel this power plant?

Although fossil-fuel power plants have many advantages, each type of fuel also has some serious disadvantages. All of the fuels produce air pollution, to varying degrees, and thermal pollution. Oil is a very convenient fuel, but we reached the point long ago where the oil produced in the United States could not meet the demands for electric power generation, transportation, the chemical industry, and other needs. We must, therefore, depend heavily upon foreign oil, in large part from the Mideast. Not only has the cost of Mideastern oil been rising very rapidly in recent years, but the supply depends on the world political situation; it could be cut off next week! Obviously, since we can no longer provide our own oil, it would be silly to build very many more oil-burning plants.

Natural gas is also an excellent fuel; it is easly transported from where it is produced to where it is used; it is easily handled and stored; and it is a relatively clean fuel. However, we have already used more natural gas than remains in *known* reserves in this country. There may be much more natural gas available than we have thought until recently, but this is yet to be proved. In any event, this resource is not limitless, and it should be used thoughtfully.

Coal is the fossil fuel we have in greatest abundance. However, it is much harder to transport, more equipment is required to burn it, and it's use produces more pollution than any other fuel. Much of the coal now available does not meet the standards for clean burning, and large amounts of pollutants are released when it is burned in power plants. We need better means of cleaning up the smoke, and that may very well mean that we will have to make *coal gasification* (changing the coal to a gas similar to natural gas) economical. But even if the efforts to perfect gasification are successful, all of the problems will not be answered. The reserve of coal is very large but it is not infinite; some thought must be given now to the resources left for future generations.

Nuclear Plants

In later chapters we shall examine the theory of nuclear power and some of the effects on society of the use of nuclear power. At this point, we are interested only in nuclear power as a source of fuel for a steam-powered plant.

A nuclear plant is very similar to a coal-fired or oil-fired plant, except that the source of heat to produce the steam is different. Figure 11.4 is a schematic diagram of the essen-tial parts of one type of nuclear plant, and Figure 11.5 is a picture of an active plant. The containment building is designed to keep radioactive materials out of the environment and the rest of the power plant. There are actually four different types of reactors used in various countries. Some of them *(boiling water reactors)* have no heat exchanger, and the water that goes through the core of the reactor also goes through the steam turbine. That water must be made extremely pure, for any impurities become radioactive as they pass through the core. Some reactors *(liquid sodium reactors)* have two heat exchangers between the reactor core and the turbine, and some *(pressurized water reactors* and *gas-cooled reactors)* have just one (Figure 11.4). Using nuclear power affords some real advantages in that there is practically no pollution of the environment (except for thermal pollution), the fuel is easily transported, and there is potentially a lot of fuel available.

However, there is some danger—however slight—that an accident could release

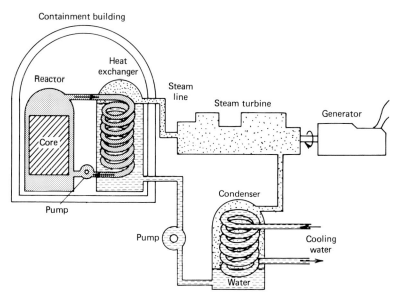

Figure 11.4 A schematic diagram of a pressurized water nuclear power plant.

Figure 11.5   A nuclear power plant. The Millstone Unit Number 2, at Waterford, Connecticut, began full-power operation in November, 1975. It can produce 830 MW. (Combustion Engineering Power Systems photo.)

dangerous amounts of radioactive materials to the environment. A more routine problem is associated with the disposal of all of the waste material from nuclear power plants; it is highly radioactive and will stay that way for many years. Also, the amount of nuclear fuel available is not infinite, and turning to nuclear power will not solve the long-range problems. Finally, there has grown a consid-erable amount of public resistance to nuclear power due to fear of the problems associated with its use. Although some of the fears may be based on a lack of understanding, they are there, nonetheless, and they cannot be dis-counted, as is obvious from the fact that the nuclear power program in the United States, which seemed to be growing rapidly just a few years ago, has come to a virtual standstill.

An understanding of nuclear power is very important, so Chapters 13 and 14 are devoted to it entirely.

Geothermal Power

Geothermal power is derived from thermal energy naturally occurring in the Earth. Although much of that heat is down deep below the surface, in some places we can see evidence of it at the surface. For example, in Yellowstone National Park there is a variety of marvelous bubbling cauldrons, geysers, and steam and smoke holes (called *fumaroles*). Other examples are Mount St. Helens, Mount Etna, and other volcanoes, where the molten rock has come so close to the surface that it occasionally escapes in explosive bursts of energy.

The geothermal energy trapped in hot and molten rocks under the surface of the Earth can sometimes be used to produce useful power. The energy comes from two sources: one is the flow of heat from the deep crust and mantle, by means of which heat is conducted to the rocks above. The other is from the radioactive atoms in the rock which give off energy to the surroundings. In some places where hot rocks are close to the surface, we have learned how to use a little of the energy available. Where the hot rock is porous (somewhat like a sponge), water seeps in and is heated. Often, in fact, the water is superheated, sometimes up to 350°C, producing fairly high-pressure steam (at 350°C—660°F—the steam pressure is about 2400 lb/in^2). Sometimes, by simply drilling a hole down to it, the steam can be used to power a generating plant.

Geothermal energy, like solar and wind energy, is "free." That is, once the power plant has been built, the "fuel" is free. The only large-scale geothermal electric power plant in the United States is at the Geysers in northern California, shown in Figure 11.6. It has several medium-size (about 100 MW) generators,

one of which is shown schematically in Figure 11.7. In 1980 the Geysers had an installed generating capacity of over 900 MW, and an additional capacity of about 350 MW was under construction. A total generating capacity of about 1500 MW is expected before 1990.

Despite vast total resources of geothermal energy, the total worldwide geothermal electric power production is presently only equal to the output of two or three large coal-fired plants. The reason is that, although the energy is free, dry steam sources (with no water mixed in), such as the one at the Geysers, are extremely rare. Modern technology has not yet solved many of the problems associated with extracting heat from some of the other formations, although research is under way. For example, in some locations there are hot, dry rocks near the surface. The difficulty in getting the energy from the rocks results from the lack of suitable heat exchanger. Some experiments have been done to *fracture* (crack) the rocks, pump cold water down through the fractures, and get steam or hot water in return.

However the problems are quite formidable. Even the drilling of exploratory holes is very expensive; because of the high rock temperatures involved it costs as much as four times the cost of drilling for oil. When a field of suitably high temperature is found, it rarely provides dry steam and is more likely to provide "brine," hot water that is loaded with dissolved salts and minerals comprising up to 30 percent of its weight. When the hot water is cooled, as a result of extracting power from it, those materials will no longer stay dissolved, and they foul the machinery. Further, they are very corrosive, and they "eat" away any metals that could economically be used for the pipes. There is also the problem of what to do with the many tons of solids pumped up.

It has been estimated that there is enough heat in the Earth to satisfy the energy needs

Figure 11.6 The Geysers. Over 100 wells have been drilled in this area, with the deepest at 8000 ft. The steam comes from a mass of rocks that are at a temperature of about 255°C and that cover an area of 100 square miles. (Photo courtesy DOE.)

of the world for more than 1 billion years. However, very little of that energy is readily available. A more interesting estimate is that the known available resources in the United States could provide about 50 times the energy of the oil we used in 1981, and the guessed-at reserves could extend that to 125 years' worth. This is a large and valuable resource, but many problems remain to be solved before we can use it effectively.

Questions

One of the problems associated with using geothermal energy is that it adds a great deal of thermal pollution to the environment for the amount of energy gained. This is because of the low efficiency of geothermal plants, resulting from the low-temperature steam used. The efficiency of a geothermal plant may be as low as 10 percent, whereas a modern coal-

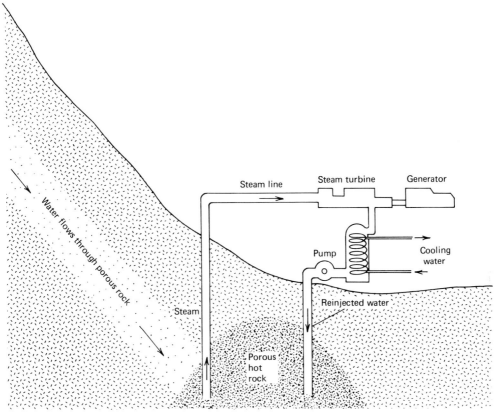

Figure 11.7 A schematic diagram of a geothermal power plant.

fired plant may be 40 percent efficient. For a plant of the same size (say 1000 MW), how much more heat does a geothermal plant throw away compared to a coal-fired plant? (*Hint*: The "obvious" answer of 90%/60% = 1.5 is not correct. The 1000 MW is the electricity produced, taking into account the efficiency. So how much total heat energy does each plant need? How much of this is wasted?) Can you think of other possible uses for this waste heat?

Trash Burners

There are a few cities, such as Saugus, Massachusetts, Harrisburg, Pennsylvania, St. Louis, and Chicago, that have been burning trash and rubbish for several years to produce electric power. In parts of Europe this source of power has been used for about 50 years. The trash-burning power stations generally use steam turbines like those used in fossil-fuel plants.

The trash used is just what your garbage collector picks up, but it requires considerable treatment before it is suitable for burning in a power plant. First, it has to be ground up and dried out, since trash often is 30 percent or more water. Then, the "recoverables" are removed. These include iron and other metals and glass. Recovering them from trash is one way in which these valuable resources can be recycled. Finally, after further grinding, compacting, and drying, a fuel is produced that can be burned in power plants. There are some variations of this approach. For example, one alternative is to subject the trash to *pyrolysis,* a method of heating without complete combustion to drive off burnable gaseous or liquid fuels. These fuels can easily be stored and used later, possibly at a distance from their source. Another alternative combines sludge from sewage disposal plants with the trash and burns the dried-out combination. Still another method uses sewage sludge and trash to make fuel in the form of methane gas, which can be compressed and stored in tanks.

Whatever method is used, properly treated trash will yield an average of about 5000 Btu/lb. This is little more than one-third the heating value of coal, but the trash is free. It is estimated that trash burning can provide up to 10 percent of the electric energy needs of the city that produces the refuse. This estimate may be too high, but burning the trash this way also solves the problem of trash disposal. More and more cities are running out of space for landfills and dumps, and disposing of trash can be a serious and expensive problem. Certainly, this means of partially solving two problems at once deserves further investigation and trial.

Cogeneration of Steam

In the preceding text, mention has been made numerous times of the energy that is wasted because power plants, being heat engines, need a cold reservoir to exhaust heat to. In fact, the average power-plant efficiency of about 30 percent nationwide means that we throw away over twice as much energy in generating electricity as we use. On the other hand, almost 20 percent of the national fuel consumption goes to the production of steam and hot water for industry. Steam and hot water are needed in a variety of industries, such as in chemical plants and in paper manufacturing. At present, there are two almost

completely independent producers in the United States of large quantities of steam: the electric utilities and all of the industries requiring *process steam.* (*Process steam* is the name given to steam used for various industrial processes.)

Cogeneration is a word that was coined when President Carter's energy plan was announced. It means using steam to generate electricity, then using the "waste" heat from that effort to produce industrial process steam and hot water. This is not really a new concept: In 1950, about 15 percent of all the electric power produced in the United States was generated in industrial plants, which then used the electric power to run the plant and the leftover heat to produce industrial steam. In 1970 that figure was down to about 4 percent as a result of low electric rates and policies of the electric utilities that discouraged industry from producing their own electricity. Perhaps now the trend will move back toward cogeneration. The attraction of cogeneration is that the overall efficiency in the use of fuel can be as high as 70 to 80 percent, thus eliminating some of the wasteful duplication of effort.

As with all partial solutions of the energy problem, there are difficulties with this one. In this case, the main problem is to build the electric power plants and the industrial plants close enough together so that they can both benefit from the same boilers. This does not seem to be an insurmountable problem, and perhaps it will be pursued more in the near future. Another helpful thing we can do is to use the "waste" heat from power plants for the heating of buildings in cities. (Of course, the objection to doing this is that the power plant would then be too close to the city—a good point.)

Peaking Units

The need for peaking units in power plants has been mentioned earlier in this book. Now we are in a position to understand that need more fully. Figure 11.8 is a graph of the load curve for a winter day of a power plant that serves a northern city. The base-load units at such a plant will run continuously at full or nearly full capacity. However, remember that the utility companies are required to provide as much electric power as is needed and at very nearly a constant voltage. That means that they have no control over how much power is used; as people turn on more appliances, lamps, and machinery at certain times of the day, the load demand goes up. That is, the voltage stays the same, the total resistance is decreased as more gadgets go on the line in parallel, and thus the current increases. But the power (current times voltage) that the generators can produce has a limit. If the demand is more than the maximum the generators can produce, then the voltage must drop. However, if it drops more than a small percentage, many of the motors, electronic devices, and other appliances on the line can be badly damaged. Thus when the demand is greater than the capacity of the base-load units, the power company has only two choices: cut off some customers, or provide more power.

Naturally, rather than cut off customers, the power companies choose to have *peaking units* on hand to meet peak-load demands. When the base-load units cannot meet the demand, the peak-load units can be quickly brought into operation to provide the needed extra power. Peaking generators may be powered by gas turbines, diesel engines, or stored water. In some cases, they are made from older, small base-load units that have been converted. Peaking units generally have two characteristics in common: They can be brought on-line quickly, and they are inefficient. Some of the units can be brought up to full power in under 10 minutes, but they also have efficiencies under 10 percent. This low efficiency is the reason that there is such a need for a way to store electric power during periods of low demand so that it can be used in periods of peak demand. Even stored-water systems, which lose about 35 percent of the electric energy originally generated, are considerably more efficient than many of the peaking units now used.

Questions

Suppose that a 1000 MW plant is producing power at full capacity and that all of the users are using that power at 120 V. (They don't; even in your home some of the power consumption is at 240 V if you have an electric range or central air conditioning. However, this assumption will do for the moment.) Ignoring any losses that might occur in the transmission lines, what is the total current drawn by all of the users?

Suppose that at the peak-load time the current drawn increases over the figure you just calculated by 15 percent, but that the power plant has no peaking units. That is, the power output cannot go above 1000 MW, but the current drain increases. If all of the customers stay on line, what happens to the voltage? (This is too low for satisfactory operation of

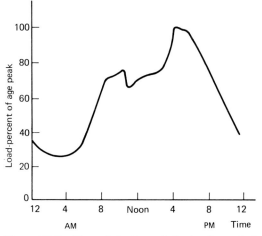

Figure 11.8 The wintertime daily load graph for a northern, urban power plant.

many appliances, and considerable damage would probably result.)

GENERATORS

As you saw in Chapter 10, when a wire moves in a magnetic field, a current is caused to flow. All conventional electric power generators, small and large, work on this principle. The tiny generator on a bicycle works in exactly the same way as the huge generators in a power plant.

The fundamental physical fact is that when a charged particle, such as an electron or a proton, moves through a magnetic field, it experiences a force. This is shown in Figure 11.9 for both a negative and a positive charge. (You can see this happen with a strong magnet and a television set. I would do this with a small, old television set rather than with Mom's $1000 color TV, just on general principle: You wouldn't want to accidentally put the magnet through the picture tube! With the set on, bring the magnet close to the face of the picture tube. As you do so, you will see a distortion of the picture. This distortion occurs because the picture is produced by electrons which hit the inside of the picture tube at high speed. When an extra magnetic field is introduced, it produces a new force

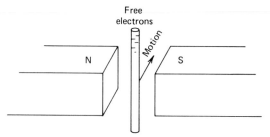

Figure 11.10   When a piece of wire moves through the magnetic field, the force on the free electrons pushes them to the top end of the wire.

on the electrons, causing them to hit the screen at places different than the ones they were originally headed for.)

When a wire is moved through a magnetic field, both the electrons and the protons experience forces. The protons, of course, cannot move within the wire, so they stay where they are. However, in a metal wire, many of the electrons are free to move, and they are pushed to one end of the wire, as indicated in Figure 11.10. If there is a complete circuit, as in Figure 11.11, a current flows. (Remember that the direction of the current is opposite to the direction of the electron flow.)

However, if the entire loop is moving in the magnetic field, as in Figures 11.12 and 11.13, there is no current. In those pictures,

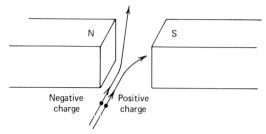

Figure 11.9   Charged bodies passing through a magnetic field. The path of the negative charge is curved upward by the force as it passes through the field. The path of the positive charge is curved in the opposite direction.

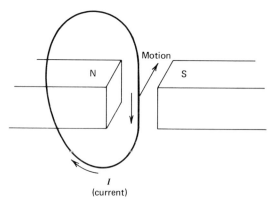

Figure 11.11   When a part of a wire loop (a complete circuit) moves in a magnetic field, a current is produced.

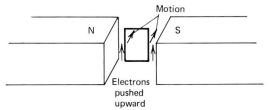

Figure 11.12 When the entire moving loop is in the field, the electrons are pushed upward in both legs. The two oppose each other and no current can flow, so electrons gather at the top.

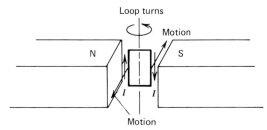

Figure 11.14 When the loop is rotated, electrons rush *up* in the right leg and *down* in the left leg. Thus a current flows in the loop. (Opposite to the electron flow.)

the electrons are pushed upward in both vertical legs of the loop and the two motions oppose each other. Note that when the loop is completely in the field no current will flow no matter what the orientation of the loop. But, if the loop is rotated instead of moved, as in Figure 11.14, then one vertical leg is moving in one direction and the other leg is moving in the opposite direction. This means that electrons are forced upward in one leg and downward in the other, and therefore a current flows in the loop. Note that, when the two legs have interchanged positions (that is, when the loop has turned 180°), the current will flow in the other direction. In other words, if the loop continues to turn, the current will flow first in one direction, then in the other, then in the first, and so forth. Voila: alternating current!

This, then is an easy way to produce continuous alternating current. Figure 11.15 shows how to get the current out of the loop to where

it can be used. There are two rings, called *slip rings*, on the rotating shaft, and each has the wire from one end of the loop attached to it. Touching the slip rings at all times are pieces of metal (or carbon) called *brushes*. The current then flows through the slip rings, into the brushes, and to the outside world. Thus, the current can flow and the loop can turn without getting the wires all twisted up.

A large power generator works on exactly the same principle, except that there are many loops (coils) of wire, and the magnetic field is produced by electromagnets instead of permanent magnets. Those generators also produce alternating current, which (as you will see) is fortunate. Direct current can be produced by a generator by using just one ring, cut in half, with the two ends of the coil attached to the halves of the ring. This is shown in Figure 11.16, and it is called a *commutator*, or *split-ring commutator*. Every time the coil turns 180° and the current flips to the other

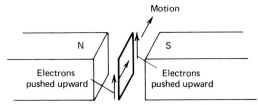

Figure 11.13 Again, with the whole loop in the field, the forces on the electrons in the two vertical legs are opposed and no current flows.

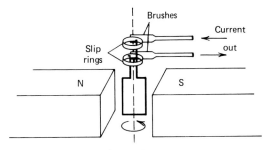

Figure 11.15 An alternating-current generator.

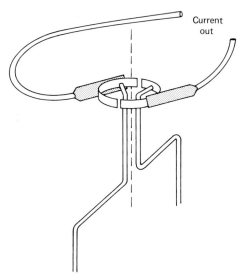

Figure 11.16 The commutator changes alternating current to direct current.

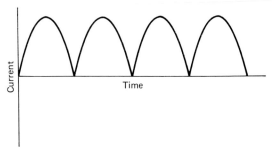

Figure 11.17 Graph of the current from a dc generator.

direction, the commutator also turns 180° and switches the two wires. Thus, there will be alternating current in the coil, but direct current in the outside circuit. The direct current will have the strange appearance of the graph of Figure 11.17, but it can be smoothed out electronically. (The *alternator* in your car is a generator that produces alternating current. The ac is then converted electronically to dc.)

TRANSMISSION OF ELECTRIC POWER

Suppose that a 1000-MW power station sends out power on 10 transmission lines, with each line carrying 100 MW. Typically, each line would have a resistance of 1 Ω, or so. Since the transmission line has a resistance, it is heated up when current passes through it. That is, there is a loss of power in the line equal to $P = I^2 R$. (Naturally, this is called an "I-squared R loss.") Let's take a look at how much power might be lost in the transmission lines.

EXAMPLE

The usual household voltage is about 120 V. Suppose that the power company generated and transmitted the power so that this were the voltage supplied at the ends of the main transmission line, which has a resistance of about 1 Ω. How much power would be lost in the transmission line?

Solution

Assume there are to be 10 main transmission lines from the 1000-MW plant. Each line would have to deliver 100 MW. The current required would then be given by $P = IV$. That is

$$I = \frac{P}{V} = \frac{10^8 \text{ W}}{120 \text{ V}} \approx 8.3 \times 10^5 \text{ A}$$

In other words, to provide 100 MW (10^8 W) of power at 120 V at the consumers' end, the line would have to carry a current of about 830,000 amps! With such a large current, the $I^2 R$ loss in the transmission line would be

$$\text{loss} - I^2 R \sim (8.3 \times 10^5 \text{ A})^2 \times 1 \text{ Ω}$$
$$\approx 6.9 \times 10^{11} \text{ W}$$
$$\approx 7 \times 10^5 \text{ MW}$$

Wow! With a current that high, the losses in a single line would be 700 times the power that the entire plant can supply. Obviously this is not possible. (Furthermore, such huge currents would just melt the wire.)

Since we cannot transmit sufficient power at 120 V, we must find another way to get the job done. The most reasonable thing to do would be to decrease the current. But, if the same amount of power is to be transmitted, the voltage must then go up (because $P = IV$). In fact, power plants generate power at considerably higher voltages than 120 V. One common voltage is 24 kilovolts (1kV = 1000 V). If this voltage were to be delivered at the ends of the transmission line, the current needed to transmit 100 MW would be

$$I = \frac{P}{V} = \frac{10^8 \text{ W}}{24,000 \text{ V}} \approx 4200 \text{ A}$$

This is a little more like it. The I^2R loss is now

$$I^2R = (4200 \text{ A})^2 \times 1 \, \Omega \approx 1.8 \times 10^7 \text{ W}$$

Well, this is much better, but it is still a loss of about 18 percent of the transmitted power, and only 82 percent would be available for consumers. We can take a good thing a step further and go to really high voltages for transmission. One standard voltage used is 345 kV (345,000 V). Why don't you try this one?

Questions

If a delivered voltage of 345 kV is used, what is the current necessary to carry 100 MW in the line? For a transmission line with a resistance of 1 Ω, what is the power loss in the line? What percentage of the power is lost? Some transmission lines are operated at voltages up to 765 kV. What is the advantage of that?

The much greater efficiency in delivering usable power when the voltage is high ac-

counts for the many "high tension" lines strung about the countryside. Generally, the transmission is in the form of *three-phase power*. This term is explained by Figure 11.18. There are three power lines (wires), and the return is through the ground. (Yes, the ground is a pretty good conductor; when a wire is electrically connected to the ground, it is said to be grounded.) The voltage, measured from each of the lines to ground, will be the same, and each graph has the same shape. However, when the three graphs are compared on the same time scale, as in Figure 11.18, they have their peak and zero values at different times. They are said to have different phases, thus the term *three-phase*.

Figure 11.19 shows typical high tension towers and the lines they support. The cur-

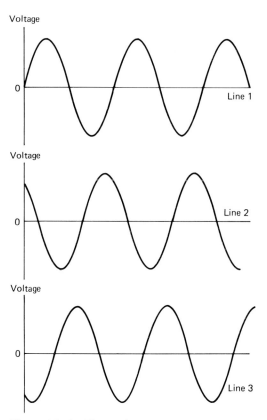

Figure 11.18 Three-phase power.

Ground lines

Figure 11.19 High tension lines. These towers carry three complete circuits of three lines each. The two ground lines are placed on top for protection against lightning.

rent-carrying wires are supported by insulators hanging from the tower. The two "ground" wires above are connected to the ground at each tower. Their function is to intercept bolts of lightning before they can get to the power lines and carry them harmlessly to the ground. The higher the voltage in the high tension lines, the higher above the ground and the farther apart the lines have to be. This is to prevent arcing across from one line to the other, and it means that the larger towers for higher voltage are more expensive. Up to a point, the lower line losses at higher voltage will more than make up for the extra expense of the tower. However, as the voltage gets greater, other losses enter in. You have prob-

ably heard the crackling sound that is given off by very high voltage lines. This is caused by electric charge leaking off the wires and causing many little "sparks" in the air. This is called a corona discharge, and it represents a power loss. (I sometimes do some fishing on a lake that has a high tension line overhead. As you paddle beneath it, not only is the crackling, sizzling, hissing sound eerie, but you can feel your hair trying to stand on end. This is not due to fear—I keep telling myself—but to the electric charge in the air.)

Even with the lines widely separated on the tower, arcing sometimes occurs between two of the wires. The arc, a stream of electrons through the air, may be started by nearby lightning or it may start for other reasons. During arcing, electrons are stripped off the air molecules in the arc, and the air becomes a very good conductor. Thus, once started, the arcing will continue through the air. The only way to stop it is to shut off the power momentarily, then turn it back on when the arc has stopped. The power companies have equipment to do this automatically.

TRANSFORMERS

Power is transmitted on high-tension lines at anywhere from 69 to 765 kV. However, it is produced in power plants at much lower voltages, ranging from 11 to 30 kV. Even those lower voltages would be highly dangerous in your home, so you get power at 120 or 240 V. The device that handily transforms voltage from one value to another is called, naturally, a *transformer*.

Transformers are all much like the simple one you wound, as described in Chapter 10. Each has a *primary coil* and one or more *secondary coils*. One common way of making a transformer is shown in Figure 11.20, with both coils wound on a single iron core. When the current in the primary changes, the magnetic field it produces also changes. That field

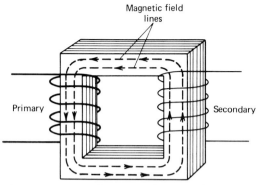

Figure 11.20 A diagram of a transformer. The magnetic field in the core is changing continually.

is carried around through the secondary by the iron core. When the field passing through the secondary coil changes, it produces a voltage in the secondary. This is why it is so fortunate that generators produce alternating current. A transformer needs changing voltages to operate, and an ac voltage is continually changing. When an alternating current passes through the primary, it produces a magnetic field that also is continually changing. That magnetic field, in turn, produces an alternating voltage in the secondary and, if there is a complete circuit in the secondary, an alternating current will flow. If a steady direct current is put through the primary of a transformer, there is no output in the secondary, even though there is a magnetic field passing through the secondary coil.

In addition to the alternating voltage produced in the secondary, the alternating magnetic field in the iron core of a transformer produces currents (called *eddy currents*) in the core. Since the iron has resistance, the eddy currents produce heat and some energy is lost in the transformer as a result. For this reason, the core is usually *laminated*, made of thin sheets of iron separated by insulation. This cuts down on the eddy currents. As a result, a good transformer is almost 100 percent efficient in transferring power, and we shall ignore the small losses involved.

Ignoring the losses means that whatever changing magnetic field passes through the primary coil also passes through the secondary coil. It happens that the voltage in each turn of both coils is the same. The result is that, if there are 10 turns of wire in the primary and 100 in the secondary, the voltage produced across the secondary will be 10 times the voltage across the primary. Within limits, the voltage can be either "stepped up" or "stepped down" to whatever voltage is desired in the secondary by winding the correct number of turns. If N_1 is the number of turns in the primary, N_2 the turns in the secondary, and V_1 and V_2 are the respective voltages, as shown in Figure 11.21, this can be expressed as an equation:

$$\frac{V_1}{N_1} = \frac{V_2}{N_2}$$

or, slightly rearranged:

$$\frac{V_1}{V_2} = \frac{N_1}{N_2} \tag{11.1}$$

However, if there are no losses in the transformer, then the power output from the secondary must be equal to the power input to the primary.

$$P_1 = P_2$$

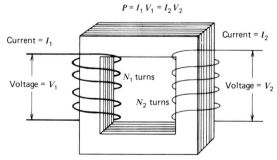

Figure 11.21 The voltage and current in the primary and secondary of a transformer.

or

$$I_1 V_1 = I_2 V_2$$

Combining this and Equation 11.1 gives the result:

$$\frac{I_1}{I_2} = \frac{N_2}{N_1} \qquad (11.2)$$

So, if we know the current or the voltage in the primary and the number of turns in the two coils, we can calculate the current or voltage in the secondary.

Questions

Undoubtedly, you have played with a toy electric train at some time. The little box that you plug in is a transformer and it changes the household voltage of 120 V to something much smaller, depending on the setting of the transformer.

Have you ever noticed that the transformer gets warm when it is plugged in and not attached to the train? (If you can find an electric train set, you might try this.) Since it is not attached to the train, does any current flow in the secondary? In that case, does any current flow in the primary? (Be careful with this one. Remember, it does get warm.) What can you then conclude about the efficiency of this transformer?

EXAMPLE

In the toy train type of transformer, the secondary voltage is varied by varying the effective number of turns in the secondary (that is, by using only part of the secondary coil). Suppose the transformer is set to produce 12 V at the secondary and it is used to power a 12-V automobile lamp with a resistance of 1 Ω. What current flows in the secondary? What is the power output? The house voltage is 120

V. What are the current and the power input to the primary? If the primary coil has 200 turns, how many effective turns does the secondary have when it is producing 12 V?

Solution

First, use Ohm's law to find current in the secondary:

$$I = \frac{V}{R} = \frac{12 \text{ V}}{1 \text{ }\Omega} = 12 \text{ A}$$

The power output is then

$$P = IV = 12 \text{ A} \times 12 \text{ V} = 144 \text{ W}$$

(You could also find the power output by using V^2/R and not finding the current explicitly.)

Since, disregarding losses, the power in is equal to the power out, we can find the current in the primary easily:

$$P_{in} = I_1 V_1 = P_{out} = I_2 V_2$$

Solving for the input current, I_1, this becomes

$$
\begin{aligned}
I_1 &= \frac{I_2 V_2}{V_1} \\
&= \frac{12 \text{ A} \times 12 \text{ V}}{120 \text{ V}} \\
&= 1.2 \text{ A}
\end{aligned}
$$

The input power is 144 W, of course, when calculated either by setting it equal to the output power or from $P = IV$.

Finally, we can find the number of turns in the secondary from the ratio of either voltages or currents Using the former, it is

$$
\begin{aligned}
N_2 &= \frac{V_2}{V_1} \times N_1 \\
&= \frac{12 \text{ V}}{120 \text{ V}} \times 200 \text{ turns} \\
&= 20 \text{ turns}
\end{aligned}
$$

There is one additional lesson hidden in this example. The transformer might well be *rated* at 50 W. This means that 50 W is the maximum power that the transformer can handle safely. As the calculations show, the transformer in question would have to provide 144 W to power the lamp. What happens when an *overload* of this type is imposed? Well, the transformer does its best but, unless some current-limiting device is built into it, it will produce more current than it can handle. It will do so for a short time, that is. Very soon the whole thing will get very hot and "burn up." That is, the insulation on the wires will burn, and the wires will then melt in places, making the whole thing useless. If you have ever smelled a burned transformer, you will probably recognize one the next time. Any electrical device that is subjected to overloads or "short circuits" may have similar problems.

POWER DISTRIBUTION

Figure 11.22 shows schematically the transmission and distribution of electric power from the power plant to the final consumers. The key to being able to distribute the power safely and economically is the transformer. At the power station end, there will be a step-up transformer to boost the voltage, as discussed earlier. However, those high voltages could not be safely used in your home, so there has to be some means of stepping the voltage down. Again, transformers are the answer. The necessary reduction in voltage is done in several steps.

When the power leaves the step-up transformers at the power plant, it may travel some distance at very high voltages, say 100 miles at 345 kV to reach the city where it is to be used. The high-voltage transmission lines

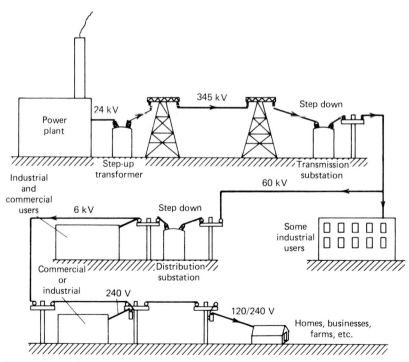

Figure 11.22 The transmission and distribution of electricity.

from numerous power plants in the same geo-graphical region are connected to each other in parallel, and it is now possible to link over 90 percent of all the power generation in the United States and Canada (not including Alaska and Hawaii). Thus, very often during peak-load times one part of the country may buy power from another region that has some to spare at that moment. Later, the buying may go in the other direction.

You have probably seen power *substations,* such as the one pictured in Figure 11.23. The substation is just a collection of trans-formers and some control equipment. At a substation the voltage is stepped down. The first voltage step-down occurs at a *transmission substation,* and it will be stepped down to somewhere between 20 and 140 kV, still a very high voltage. Power at this voltage may be used directly by an industrial plant or a subway system that does its own transform-ing to lower voltages. The *distribution substation,* which you see around town, drops the voltage further to somewhere between 4 and 35 kV. These still rather high voltages are sent out to your neighborhood. (The fact that sev-eral thousand volts are on those lines is a *very* good reason not to climb power poles or oth-erwise mess with the power lines.) Finally, on the pole from which power is distributed to several homes or businesses, there will be the transformer that steps the voltage down for household use. One of these is shown in Figure 11.24.

Figure 11.23 A power distribution substation. Each cylindrical "can" is a transformer.

Figure 11.24 A neighborhood transformer, which drops the voltage to 120/240 V for use in homes.

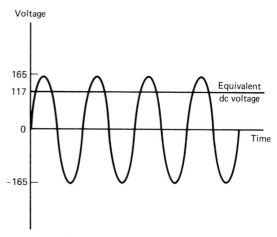

Figure 11.25 Rms voltage. The straight line on the graph is the dc voltage that would produce the same heating of a resistor as the ac voltage shown.

A little arithmetic will show that the peak voltage is then about 165 V for a 117 V, rms, standard.

Well, most people now call it 120 V, and that is close enough. Actually, in our homes there will usually be both 120 V and 240 V available. This is produced by the transformer on the power pole, shown schematically in Figure 11.26. It is just the same as the other transformers we have examined, except that the secondary coil has a wire attached to it at its center. That wire, which is grounded, has the effect of dividing the secondary into two equal coils. Each half produces 120 V, measured between the end of the coil and the

What about that voltage for household use? You will hear it variously referred to as 110 volt, 120 volt, and, sometimes 115 volt. Which is it? The present standard is actually 117 V, rms *(root mean square)*. The term rms is used with alternating current. It means that it is the ac voltage that will produce the same power output in an appliance such as a toaster as a dc source of 117 V will. An ac voltage varies and reaches its peak only twice (once in each direction) in each cycle. Thus, it will not have the same effect as a dc voltage which is as large as the ac peak. This is indicated in Figure 11.25. The factor used turns out to be the square root of 2 (1.414). That is, the rms voltage is the peak voltage divided by 1.414.

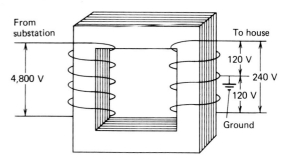

Figure 11.26 A center-tapped transformer, like the one on the power pole.

ground wire, and the voltage across the two ends is 240 V. In your home many circuits, such as those for toasters, lamps, and refrigerators, will use 120 V from one side of the transformer. Each circuit in the house has its own fuse or circuit breaker. The two 120 V halves coming in are divided about equally into the circuits in the house. A few appliances, such as a large air conditioner and an electric clothes dryer, use the full 240 V.

Incidentally, having a ground wire is very important. Newer homes, those with three-prong outlets, have two ground wires in every 120 V circuit. One of them is used as the return wire in the circuit itself; the other is strictly for safety. An appliance that has a three-prong plug has the case connected to the third prong (the round one). We say that the case is "grounded," which means that the voltage between the case and the ground is zero. In an appliance that is not grounded, a malfunction can occur and the case can touch the "hot" wire. This means that the case is at 120 V with respect to the ground. Since you are more-or-less well connected to the ground because you are standing on it or on something which is resting on it, if you touch that appliance, you may get 120 V across your body. That hurts! If you happen to be well grounded, such an event could be fatal. Shocking! (Bad puns aside, it takes only 35 or 40 *thousandths* of an amp through the heart to stop it—permanently. A fairly common occurence is for persons to get injured or killed by touching a faulty, ungrounded refrigerator and a kitchen sink, simultaneously. The refrigerator case provides 120 V and the sink provides a good ground, with the unfortunate person completing the circuit.)

SOME FUTURE POSSIBILITIES

Thus far, this chapter has examined the present production and distribution of electric power. It seems clear that, even if we adopt better energy conservation practices, there will be an increasing demand for power. It is unlikely that we will find *the* solution for this demand. Rather, it is more likely that many smaller efforts, put together, will help us to move in the direction of clean and affordable energy. An increased use of geothermal and wind power are good examples.

Let's take a look at some of the other possibilities in the near future for producing more electric power or using the power we produce more efficiently. I'll leave breeder reactors and nuclear fusion for a more thorough treatment in later chapters.

Direct Current Transmission

When electricity was first used for lighting, most of the distribution systems were direct current systems. Thomas Edison believed that dc transmission was the better way, but lost out to ac when transformers were used to boost the voltage and thus lower the line losses. Nonetheless, some dc distribution was found in this country until 1970, or so.

Now, with the development of "solid-state" converters, dc is starting to look attractive again. (A transistor is a solid-state device. In converting ac to dc, a related device called a thyristor is used.) The reasons for wanting to go to dc are economic in nature. For one thing, a dc transmission line needs only two wires, as opposed to three for three-phase ac. (This is not counting the overhead ground wires that are necessary for both ac and dc.) The wires needed for dc can also be lighter than those needed for an equal amount of ac power. This represents a savings that makes it economical to use converters to produce dc if the distance from the power plant to the users is over 600 miles. In underground or underwater cables, where the wires are much closer together, ac losses are particularly high because the changing magnetic field produced by each wire interacts with the others, and dc transmission is more efficient. This will become an increasingly important factor as the power plants are forced to be farther from the cities and rights of way for overhead lines become more expensive and harder to get. Direct current is also more efficient in superconducting wires.

There is considerable ongoing research into improving dc transmission, and we shall probably see more use of dc in the near future.

Superconductivity

These days, we see the word *super* attached as a prefix to a lot of other words: *supersale*, *supersensitive*, *superneat*, and even *supermouse*. In the case of *superconductivity*, the *super* is very appropriate.

When an electrical conductor (a metal) is heated, its resistance increases. This happens, for example, when the filament of a light bulb gets hot. On the other hand, when the conductor is cooled, its resistance goes down. Near the beginning of the century, an interesting fact was discovered. Solid (frozen) mercury was being investigated at very low temperatures, not much above absolute zero. As the temperature of the mercury was lowered, its resistance became less and less. Suddenly, at a temperature of about 4 K, its resistance dropped to zero. Understand that the resistance did not just become very small—it became zero! Mercury became a *superconductor* at the "critical temperature" of about 4 K. The phenomenon of superconductivity was not understood by physicists until the 1950s.

Now there are some alloys, such as niobium and tin or niobium and titanium, which have critical temperatures as high as 20 K, and they are thus somewhat easier to use as superconductors. Even so, the only material capable of producing such low temperatures is liquid helium, which boils at about 4.2 K. (The next lowest boiling point, that of hydrogen at about 21 K, is too high to make any known material superconducting.) The helium can be liquified through a process of compression and

expansion and, at atmospheric pressure, liquid helium has the very low temperature of 4.2 K until it boils away.

The notion of a material that has no electrical resistance brings to mind some interesting possibilities, and considerable research is being done to develop practical applications of such materials. One such application is the storage of electricity, which, as you have seen, is very important. Once a current is started in a superconducting loop, it will go on forever—as long as the loop is kept cold enough. The only energy cost for storing the electric energy in the loop is thus the cost of refrigeration. You may recall that pumped water storage could be about 60 percent efficient. If the technology is developed, the efficiency of storing electric energy in superconducting coils may turn out to be as high as 90 percent.

Another possibility is to use a superconductor for the coils that produce the large magnetic fields needed in power generators. Much research must to be done yet to make this practical, but it could lead to more efficient generators.

Finally, imagine using transmission lines that have no power losses. This would improve the overall efficiency of power production a bit more. (Actually, it turns out that there would continue to be small power losses for alternating current, but none for direct current.) The only energy cost of transmission then would be the cost of refrigeration. Even with the task of keeping the wires so cold, it is believed that this form of transmission could be less expensive that transmission through conventional lines in some instances. This is particularly true if the transmission has to be of very large amounts of power in underground cables. An experimental cooled conducting cable is shown in Figure 11.27.

Some or all of these new technologies may prove practical soon. The biggest problem is in finding a continuing source of helium for the distant future. Helium is a precious resource that exists in deep underground wells

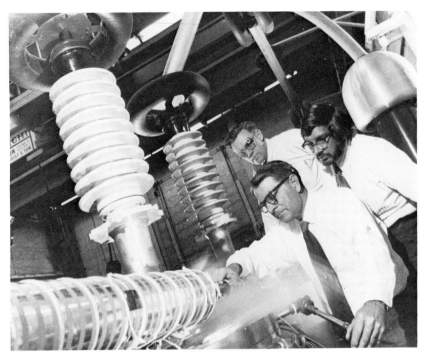

Figure 11.27 An experimental conducting cable being cooled by liquid nitrogen (77 K). (Courtesy General Electric Company and DOE.)

in only a few places in the world. There is no presently known way to replace it in large quantities. Furthermore, because it is such a light atom, when helium is released into the air it gradually rises to the top of the atmosphere and escapes the Earth entirely. Thus, there is no way to reclaim helium from the air. Perhaps some new superconducting materials will be developed that can be used at liquid hydrogen temperatures. Hydrogen is a very plentiful element.

OTEC

OTEC stands for *Ocean Thermal Energy Conversion*. One way of looking at ocean energy is that, like wind energy, it is a form of solar energy. The oceans comprise about 70 percent of the total surface area of the Earth. When the sun shines on the ocean, considerable thermal energy is absorbed and stored by the water. However, warmer water floats on colder water, so the warmed water is primarily near the surface, whereas the water down deep is quite cold, and there is little mixing. But, a warm heat source combined with a cold heat sink is just what is needed to run a heat engine. Because the ocean is so huge, tremendous amounts of heat are stored, and it is available night or day, rain or shine.

Of course, the problem is that the heat available is what we have been calling low-grade heat. In semitropical areas, the temperature of the water near the surface may be 75° to 85°F, and down deep, at 2000 to 4000 ft below the surface, it may be 35° to 40°F. With these relatively small differences of temperature available, the efficiency of an ideal heat engine would be about 6 to 7 percent. (Try it; you know how to do the calculation.)

For a real engine, the overall efficiency would probably be more like 2 percent, as compared to about 40 percent for a modern steam power plant.

Nonetheless, there is so much "free" energy available that it might still be profitable to try to get it. One way that has been proposed and tried on a relatively small scale is to build a generating plant that has a turbine and generator, much like the ones you are already familiar with. However, instead of running on steam, the turbine will run on a "secondary fluid," such as ammonia or propane. A sketch of how such a plant might work is shown in Figure 11.28, and Figure 11.29 is an artist's conception. The ammonia is vaporized in the evaporator, which is just a mammoth heat exchanger, and it powers the turbine, which runs the generator. The condenser, which changes the ammonia back to liquid form, is another huge heat exchanger. The cycle is similar to that for a steam turbine, but it runs at much lower temperatures. Another possible use for ocean thermal energy is illustrated in Figure 11.30.

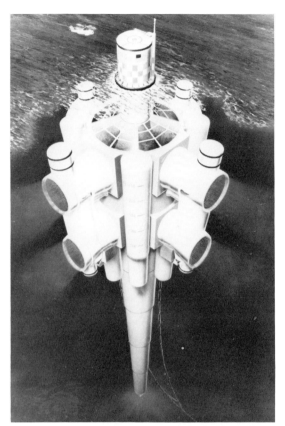

Figure 11.29 An artist's conception of an OTEC power station. (Lockheed Missiles and Space Company, Inc.)

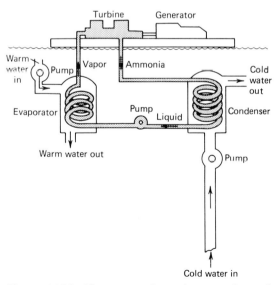

Figure 11.28 The conversion of ocean thermal energy.

Power plants such as the one described here have been tried experimentally, and they may be economically possible in the next few years. For example, in 1982 the Japanese operated a 100-kW demonstration OTEC plant on the island of Nauru. There are a few problems, however. Because the efficiency in extracting heat from sea water is so low, tremendous amounts of it have to be pumped. One estimate is that to operate a 100 MW power plant an amount of water about 2.5 times the average flow of the Potomac River at Washington, D.C., would have to be pumped. That is a relatively small plant but a huge flow of water. Pumping that much water would require a sizable part of the gen-

Figure 11.30   An ocean thermal conversion plant ship. This proposed floating factory could make 586,000 tons of ammonia a year using energy from the warm tropical seas. (DOE.)

erated energy just to run the pumps. (You might think about this for a minute: How would such a plant get started into operation?) Also, the sea water tends to be corrosive, and there are problems with finding the right materials for the huge heat exchangers.

The buildup effects of tiny marine plants and animals on the heat exchangers is also a problem (called biofouling). Finally, although the environmental impact of such a power plant seems as if it would be small, we do not yet know what it might be.

Fuel Cells

When fuels are burned, there is a release of *chemical energy* that was stored in them. For example, one chemical process that releases energy in burning is

$$C + O_2 \rightarrow CO_2$$

That is, carbon (from coal or oil, for example), combines with oxygen to form carbon dioxide. Of course, when fuel is burned, many other chemical reactions take place, and some of them produce unwanted pollutants.

In either a storage battery or a dry-cell battery, energy is *stored* in the form of chemical energy, to be released when needed. In both types, the conversion of energy from chemical to another form comes from a rearrangement of the electrons surrounding atoms and molecules. As an example, in the chemical equation given above, one molecule of carbon dioxide has slightly less total energy than the combined energy of one atom of carbon and one molecule of oxygen. Thus, there is energy lost in the chemical reaction, and this energy becomes heat.

A *fuel cell* is a device for converting the chemical energy stored in a fuel *directly* into electricity. The fuel need not be burned. A very simple fuel cell is shown in Figure 11.31. Two rods made of carbon with a very small amount of platinum are placed in a water and sulfuric acid mixture. (The platinum is a *catalyst* which does not take part in the chemical reaction but helps it to occur.) When hydrogen is bubbled along the surface of one rod and oxygen along the surface of the other, an electric current is produced in a circuit attached to the carbon rods. The potential difference produced by such a cell is about one volt. The chemical reaction at the negative electrode is the following:

$$2H_2 + 4OH^- \rightarrow 4H_2O + 4e^-$$

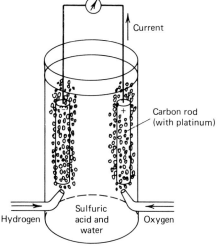

Figure 11.31   A simple fuel cell.

The interpretation of this equation is that two molecules of hydrogen combine with four charged partial molecules (called *ions*). The four OH^- ions each have one *extra* electron, and they exist in the water–sulfuric acid mixture. The result, on the right side of the equation, is that four molecules of water are produced, along with four extra electrons. The four electrons can flow through the wire to the positive electrode—the one with oxygen bubbles around it.

At the positive electrode the electrons that come through the wire combine with oxygen and water, as follows:

$$O_2 + 2H_2O + 4e^- \rightarrow 4OH^-$$

Thus, the reaction at the positive electrode not only uses up the extra electrons; it also produces the OH^- ions needed by the reaction at the negative electrode on a continuing basis.

If we were to look at the reaction from outside the container, without worrying about the details inside, the result of combining the above two reactions is

$$2H_2 + O_2 \rightarrow 2H_2O$$

That is, hydrogen and oxygen are used up and water is produced. In the process, as the electrons travel around the outside circuit, they do work. Electric power is produced.

This is a very sketchy introduction to chemistry, and you don't need to understand it in depth. The important point is that when hydrogen and oxygen are combined in this way, the *only* products are electric power, very pure water, and waste heat. Real fuel cells are somewhat more complex, but they work on the same principle. The major research challenges are to reduce cost and produce fuel cells that will run for a long time before they need to be replaced.

The research in fuel-cell technology is now to the point that it may not be very long at all before useful power plants are working. In fact, fuel cells like the one shown in Figure 11.32 have been used very successfully in the space program. The main barrier to their use to produce electric power commercially is cost, which may soon be competitive with the rising cost of oil, particularly if a good fuel can be made cheaply from coal. Fuel cells could be used in new power plants placed right in the cities, where the power is needed. As transmission line right-of-ways become harder to get and more expensive, transporting the energy in the form of a gas, which is then converted into electric energy in the city, could solve much of the transmission problem. Unlike conventional power plants, a fuel-cell plant would not contribute significant pollution, so it could be located in a population center without harmful effects. Or, a large building might generate its own power using fuel cells and natural gas. The fuel cells may be 40 to 55 percent efficient in converting the fuel to electricity for the building to use. A real advantage is that the "waste" energy can then be used to heat the building, giving the unit very nearly 100 percent efficiency, overall. It is estimated that, using fuel cells in this way, a large building might generate both its heat

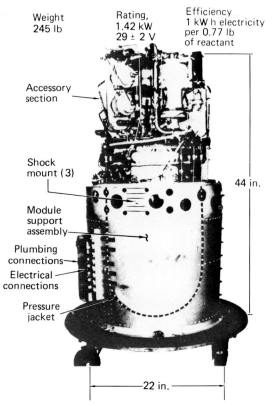

Figure 11.32 A drawing of a fuel cell used in the space program. The pure water produced by the fuel cells is used by the astronauts. (Drawing courtesy NASA.)

and its electric power for not much more fuel than is now used to produce its heat alone.

There are other promising uses for fuel cells. For example, they would be ideal for use as peaking units in a power plant because they have the advantage of being capable of starting and stopping simply by turning the flow of fuel on or off. Unlike big boilers, they run about as efficiently at partial capacity as they do at full capacity. Small units can be made in a factory and installed in a power plant, as needed, to build up the capacity. To add capacity in a conventional plant, a large unit must be constructed in the plant itself, an expensive and time-consuming process.

Another possible use of fuel cells relates to energy storage. When a direct current passes through water that has a little acid in it, the water "decomposes" into hydrogen and oxygen, as indicated in Figure 11.33. This is called *electrolysis*. The hydrogen can then be used as the fuel for fuel cells. Therefore, when the load on a power plant is low, its extra capacity could be used to produce hydrogen, which is easy to store. At peak-load times the hydrogen thus produced could be used to power fuel cells. With more research, this process could be about as efficient as pumped-water storage. Whereas pumped-water storage can work only in certain locations with the proper terrain, this method of energy storage could work anywhere.

There seem to be many advantages associated with fuel cells, and we shall soon see whether they will work well in actual practice.

Magnetohydrodynamic Power Production

Now, that is a mouthful! For short, it is usually called *MHD*. In an MHD generator, the energy of a moving fluid is converted directly

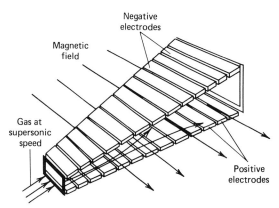

Figure 11.34 MHD power generation. The negatively charged particles are pushed into the upper electrodes and the positively charge particles are pushed into the lower electrodes as they travel through the magnetic field.

into electric energy. In the MHD generators under research investigation, the moving "fluid" is a very hot gas resulting from the burning of coal or other fossil fuels. Figure 11.34 is a diagram of how it works. The gas is *ionized* (that is, many of its particles have an electric charge) by the high temperature and it passes through a magnetic field. A small amount of "seed" material, such as potassium, is added to the gas to increase the number of particles that are ionized. Without the seed material, impossibly high temperatures of 8000 K or more would be necessary to make the hot gas sufficiently conducting to produce the desired effect. With the seed material present, working temperatures of about 2500 K are possible. The magnetic force causes the path of the positively charged particles to curve in one direction and the negatively charged ones to curve in the opposite direction. One set of electrodes then picks up the negative charges and another set picks up the positive charges; the resulting electrons flowing in a circuit provide the electric power. The MHD generator is quite simple in that the only moving part is the hot gas passing through it.

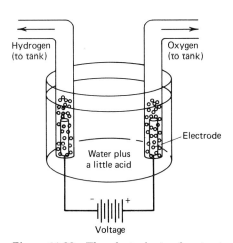

Figure 11.33 The electrolysis of water to produce hydrogen and oxygen. The acid in the water helps the process take place, but it is not used up.

Potential advantages of the MHD generator are efficiency and cleanliness. Also, an MHD generator could be fired up from a cold start to full power in a very short time, probably less than a minute. Thus, such generators could provide a high-efficiency alternative for present peaking units. The efficiency of the first commercial MHD power plants is expected to be about 50 percent and, as experience is gained in operating them, 60 percent is likely. By using the hot exhaust gases to produce steam, the efficiency could be pushed even higher. That is, the steam produced would be used to power conventional steam turbines or for industrial process steam. The high efficiency would result in considerable saving of fuel.

However, practical difficulties remain. One relates to the fact that coal would be the preferred fuel for MHD because of its ready availability. If coal were used in a sizable MHD generator, however, handling the large amounts of coal ash that would be produced would present problems of its own. On the other hand, MHD generators would burn coal more cleanly than conventional boilers because the sulfur in the coal tends to combine chemically with the seed material. The seed material is too valuable to throw away and it must be recycled. When the seed material is collected for recycling, the sulfur can also be discarded. But the seed material and other materials tend to deposit on the electrodes and foul them rather quickly, thus reducing their efficiency, so keeping the electrodes clean is another unsolved problem. One additional complication arises because of the very large magnetic fields needed for a big installation. To produce such fields economically will probably require electromagnets with superconducting wire coils. The economics of making superconducting magnets on a large scale is still uncertain.

The research on MHD continues. The first sizable MHD generator (about 60 kW) has been operating in the Soviet Union for some years, and there is hope that a commercial MHD power plant might be built in the United States in the near future. Figure 11.35 shows the innards of an actual MHD generator.

Figure 11.35 The working inner section of an MHD generator, showing the electrodes on the top surface. (Photo courtesy Westinghouse Electric Corporation.)

Questions

An MHD generator may be thought of as a heat engine, with electromagnetic forces changing thermal energy into electric energy. Typically, one may run with hot gases at 3000 K. It will probably be followed by a steam generator system running on the exhaust gases, and the temperature at which waste heat is rejected might be about 300 K. Treating the entire system as an ideal heat engine, find its efficiency in converting thermal energy into electric energy. This theoretically high efficiency is a major reason that MHD power generation seems so attractive.

Solar Electric Power Generation

Meanwhile, good old Sol is still up there pouring down energy upon us. As you have seen earlier, we are learning to take advantage of that free energy in home heating and other applications. We may also use solar energy indirectly by extracting energy from the wind and waves. But we have not yet managed to extract large-scale electric energy directly from the sun's energy. However, since the solar energy striking the Earth each year is about 20,000 times our present annual use of energy for all purposes, learning how to use more of it directly seems worthwhile. There are a number of ways under investigation by which it might be feasible to produce electric power on a large scale from the sun's energy. A few of them will be touched upon here.

Tower Power One very straightforward way to use the sun's energy to produce electricity is simply to use the sun as the heat source for a conventional steam-powered generating plant. Since the power level of sunlight falling on the Earth is so low, the only way to achieve the temperature and power needed for efficient operation is to concentrate the sun's light. There are several ways in which this could be done, including using concentrating reflec-

tors like those we examined earlier. A promising method involves the use of "tower power." This concept is illustrated in Figure 11.36. The central tower absorbs the sunlight reflected by numerous mirrors and uses the absorbed energy to produce high pressure steam, which then runs a conventional turbine–generator. There may be thousands of mirrors mounted at ground level, each one aimed so that it directs the sun's rays to the tower. (These movable mirrors are called *heliostats*.) As the sun moves across the sky, each mirror has to be moved so that it is always aimed in the right direction. The tracking has to be rather precise, for some of the mirrors may be a kilometer away from the tower. The control of the tracking can be handled easily by a computer.

Several models have been proposed and used for the energy absorbers in the tower itself. For example, Figure 11.37 shows an absorber that approximates a blackbody cavity. The light enters the cavity from below and is trapped among a network of black pipes lining the inside. The pipes may carry steam or, alternatively, a substance such as sodium

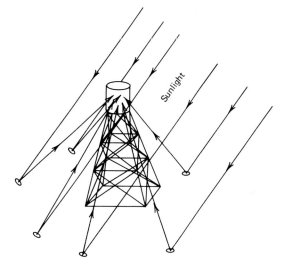

Figure 11.36 Tower power. Thousands of mirrors direct sunlight to a boiler atop a tower.

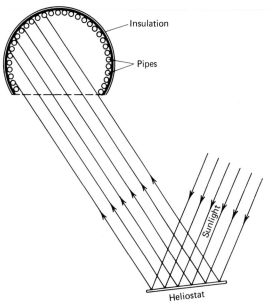

Figure 11.37 One design of a power tower has a cavity approximating a black body. The inside of the cavity is lined with black pipes shaped in such a way as to minimize reflections that could go back out of the opening.

metal, which is liquid at temperatures over 98°C. In the latter event, a heat exchanger would be used to generate steam using heat in the sodium. Figure 11.38 is a photograph of part of an experimental power tower system.

EXAMPLE

If the "concentration factor" is great enough— if, for example, the intensity of the sunlight at the tower is 1000 times the intensity on the ground—a solar tower may be able to provide superheated steam at about 540°C (1000°F). In that case, the net efficiency of the generating process could be about 40 percent. In preliminary experiments, the efficiency of the light collecting, reflecting, and absorbing processes has turned out to be about 25 percent. What would be the overall efficiency of the plant?

If the plant is to produce 1000 MW of electric power, how much sun power must fall upon the heliostats? If the plant is in the desert where the average insolation is 1000 W/m², what must be the total area of the mirrors, in square meters? If each mirror has an area of 20 m², how many mirrors are needed?

Solution

As you have seen earlier, the overall efficiency is just the product of the efficiencies at each stage:

$$eff = 0.4 \times 0.25 = .1$$

or 10 percent. To produce 1000 MW, then, the sun must provide 10 times that, or 10,000 MW. To find the area of the mirrors use

$$total\ power = total\ area \times insolation$$

or

$$
\begin{aligned}
total\ area &= \frac{total\ power}{insolation} \\
&= \frac{10,000\ MW}{1000\ W/m^2} \\
&= \frac{10^{10}\ W}{10^3\ W/m^2} \\
&= 10^7\ m^2
\end{aligned}
$$

If each mirror has an area of 20 m², the number of mirrors needed is

$$N = \frac{10^7\ m^2}{20\ m^2} = 500,000$$

In the answers to these questions, you see some of the problems associated with tower power. First, the only places such towers are likely to be economical is in the desert or high in the mountains, far from where most of the electricity will be needed. Then, a "standard"

Figure 11.38 An experimental tower near Bar-stow, California. There are 1818 sun-tracking mir-rors (heliostats), and the plant will produce about 10 MW of electricity during peak daylight hours. (DOE photo from L.A. DWP, by Cecil Riley.)

size large power plant will require many thousands of large mirrors for its operation. Each of the mirrors will have to be mounted on a very rigid support, for aiming purposes and to be able to withstand high winds, and each will have to have motors to aim it quite exactly.

Also, if a steam-turbine plant is to operate, there must be some way to cool the steam. If the plant is in the desert, there is not likely to be a source of cooling water nearby, and some kind of air-cooling towers will be necessary. This may not cause too great an effect on the environment, but if there are very large numbers of these plants the cumulative environmental effects could be significant. Desert sand, on the average, absorbs about 60 percent of the solar energy falling upon it and reflects 40 percent. A tower power complex may reflect into space only about 10 percent of the energy falling on the mirrors. The extra energy absorbed will show up as heat, partly at the tower site and partly where the electricity is used, thus contributing to thermal pollution.

Finally, as with all solar energy, there has to be some arrangement for providing power when the sun is not shining. This will require energy storage of some kind. Among the substances that may be used for heat storage are some stable salts that are liquid at the desired temperatures, or metals, such as sodium. In fact, if liquid sodium were used as the primary fluid circulating in the pipes in the tower, then heat storage could be provided by a large reservoir of liquid sodium containing a heat exchanger to generate steam. Also, conventional fossil-fuel burners may be used for backup purposes. Then a means of getting fuel to the desert will have to be found.

Obviously, tower power has some problems yet to be solved, including a very high cost for construction, but perhaps it will become usable in the near future.

Solar Cells   Another approach involves the use of *photovoltaic cells* (often called *solar cells*) to convert solar energy directly into electric energy. We have seen these used widely to provide power in spacecraft. Most solar cells are made of silicon (chemical symbol: *Si*), but other materials can be used. One advantage of silicon is that it is the most abundant element on Earth, and thus it is one of the cheapest elements. Sand is largely silicon. However, the processing required to make silicon into solar cells is expensive. In the space program this expense does not matter, but if they are to be usable sources of electric power on a large scale, the cost of solar cells will have to come down a great deal.

To understand how a solar cell works, we first have to look at the structure of the silicon atom. Silicon has 14 protons in the nucleus and 14 electrons in shells surrounding the nucleus. The electrons are distributed as shown in Figure 11.39, with two electrons in the shell closest to the nucleus, eight in the next shell, and four in the outermost shell. The inner two shells are "filled." That is, they will not accept more electrons. The outer shell could hold eight electrons, but there are only four there. These outermost four electrons are

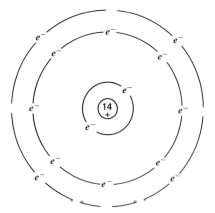

Figure 11.39   A silicon atom. The third electron shell has four electrons, but there is "room" for eight. (This is a schematic representation. The distances are not drawn to scale, and the electron shells are more complex than this.)

the ones that participate in chemical reactions, and they are called the *valence electrons*.

In a perfect crystal of pure silicon each atom is surrounded by four other atoms. This is difficult to show, for the crystal is three-dimensional; Figure 11.40 is a two-dimensional representation of it. Each atom in the crystal shares two electrons with each of its four closest neighbors, and this has the effect of "filling" the outermost electron shell of each atom. That is, because of the four extra electrons it shares with its neighbors, each atom behaves as if its third shell were filled. This is a very stable situation, and in a perfect crystal of silicon there would be no free electrons at low temperatures. However, the addition

of a small amount of energy to the crystal would free some electrons to move about.

Further, every real silicon crystal has some imperfections and impurities, making even more free electrons available to conduct electricity. As a result of the moderate number of free electrons available in the crystal, silicon is a *semiconductor*, being neither a good conductor nor a good insulator. To make semiconductors even better conductors of electricity, so they can be used for transistors, solar cells, and the like, impurities are deliberately and carefully added. This process is called "doping" the crystal. A material often used for doping is phosphorus (P), and it will usually constitute less than 1 percent of the crys-

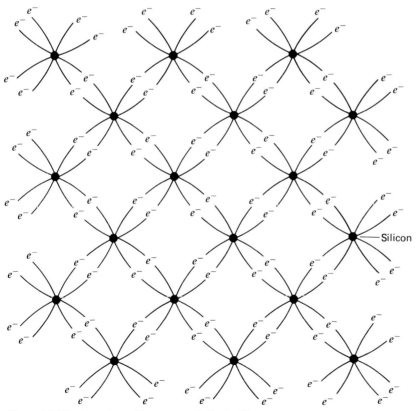

Figure 11.40   A perfect silicon crystal. Each silicon atom shares two electrons with each of four other atoms. Thus, each atom has a "full" third shell.

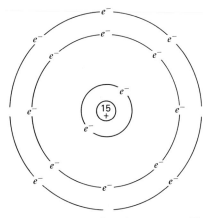

Figure 11.41   A phosphorus atom. Five of the eight places in the third shell are filled.

tal. It has 15 protons and 15 electrons, so its structure is like silicon, except that it has *five* electrons in the third shell instead of four.

(Figure 11.41) The result of doping a silicon crystal with phosphorus is shown in Figure 11.42: Wherever there is a phosphorus atom in the crystal, the outermost electron shells are filled with eight electrons and there is an extra free electron floating around, not attached to anything. The crystal still has no net charge, but there are now more free electrons and the electric conductivity is greater. A crystal doped this way is called an *n-type semiconductor*.

Alternatively, silicon can be doped with boron (B). Boron has only three electrons in its outermost shell, which is not filled until it has eight electrons. Figure 11.43 shows a boron atom, and Figure 11.44 shows what happens when silicon is doped with boron. At each site where there is a boron, there is now a *lack* of one electron. This is called a *hole*. A

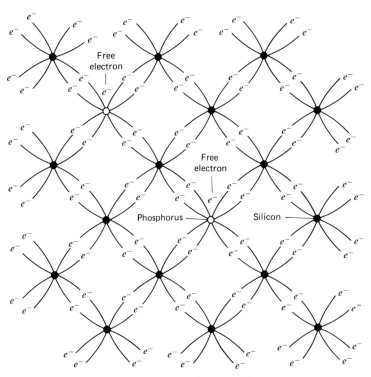

Figure 11.42   An *n*-type semiconductor, silicon doped with phosphorus. Each phosphorus atom contributes a free electron.

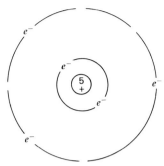

Figure 11.43 A boron atom. Only three electrons are in the second shell, where there is room for eight.

hole at the site of a boron atom can attract an electron from a neighboring atom, getting filled but leaving a hole elsewhere. Thus holes can move in the atom, just as electrons can, and they can be thought of as positive charge carriers. If an electric field were put across the crystal, electrons would move in one direction

and holes in the other. This kind of crystal is called a *p-type semiconductor*.

The payoff comes when an *n*-type and a *p*-type crystal are put together to form a *p-n junction*. Until now, we have been talking about *neutral* crystals, having the same number of electrons and protons. Now, with the two types joined together, the extra electrons in the *n*-type crystal that are near the boundary drift over into the holes in the *p*-type crystal. As indicated in Figure 11.45, this produces an area with excess positive charge in the *n*-type side (which the electrons have left), and an area with excess negative charge in the *p*-type side (where the electrons have gone). As a result, an electric field is permanently set up at the junction. If free electrons wander over to the area near the junction, they will be pushed by the electric field over to the *n*-type side.

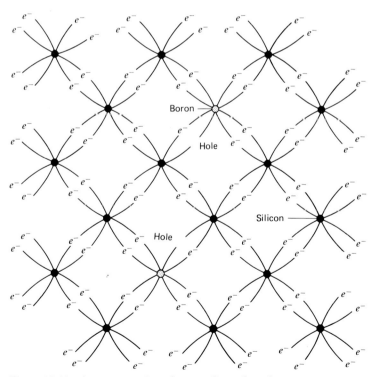

Figure 11.44 A *p*-type semiconductor; silicon doped with boron. Each boron atom contributes one hole.

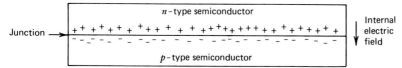

Figure 11.45   A *p–n* junction. On the *n*-type side there is an excess positive charge near the junction. On the *p*-type side it is an excess negative charge. This produces an electric field across the junction.

If the junction is in the dark, the situation becomes stable, and no extra free electrons are produced. However, if light falls upon the junction, the energy of the light "knocks" electrons out of their nice comfortable places in the crystal, and they will be pushed over to the *n*-type side. If a wire is connected from the *n*-type to the *p*-type side, as indicated in Figure 11.46, a current flows. It can then be used to do work in the external circuit. This is then a working solar cell.

Commercially made solar cells can convert solar power directly into electric power with an efficiency of about 10 to 15 percent. This may be increased to about 20 percent if concentrators (mirrors or lenses) are used. Solar cells are very convenient: They can be built onto panels wired in series and parallel combinations to produce whatever voltage and current are desired. An array of solar panels is shown in Figure 11.47. (A 5-cm diameter solar cell will produce about 0.3 W of power at 0.6 V in full sunlight.) Since photovoltaic cells do not depend on thermal energy, they can be cooled sufficiently by naturally circulating air, and large amounts of cooling water are not necessary. They operate at about the same efficiency (but not the same power output) in cloudy and sunny weather, so they will still produce some power when the sun is hidden. They can be mass-produced in quantity and added to an installation, as needed. A power plant that used them could be brought on line gradually as the solar cells were installed, and not have to wait until the entire project was finished to begin power production. They could be just as effective for a single home as for a huge power grid; thus they might be used to eliminate transmission lines. They could be placed on rooftops and thus not use valuable land.

But problems remain. One is the old bugaboo we keep running into: There has not yet been developed a suitable means of storage to provide the needed power at night and in long periods of cloudiness. Perhaps even worse than that is the cost; the manufacturing process for solar cells is very exacting and requires skilled persons, thus the cost is high. Although the price has dropped substantially in recent years, the cost of solar panels is still about $12 or $13 per watt of power output in full sunlight. However, there is a great deal of research underway devoted to finding ways to lower the cost.

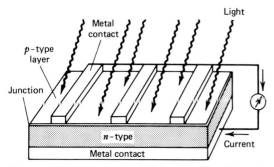

Figure 11.46   A photovoltaic cell (solar cell). Light passes through the very thin layer of *p*-type material and "knocks" electrons free near the junction. The electrons are then moved through the external circuit by the internal electric field.

EXAMPLE

Assume that a photovoltaic cell array large enough to completely power a house and the

Figure 11.47 Operating since December, 1978, this array of solar cells provides power for pumping water, lights, refrigerators, and a community washing machine and sewing machine for the 15 families in the village of Schuchuli in the Papago Indian Reservation in Arizona, 17 miles from the nearest electric power utility. (U.S. DOE photo by NASA.)

associated storage can be installed for a total of $20,000 per kilowatt. Also assume that the system will last 20 years without maintenance. (The solar cells might actually last that long, but presently available storage batteries probably would not. However, let's go with this assumption.) Finally, assume that the array averages a power output equal to 40 percent of its total capacity. With all of these rather optimistic assumptions, how would the cost of the energy produced this way compare to a power company rate of $0.06 per kW h?

Solution

For each kilowatt installed, an average of 400 W (0.4 kW) will be produced. Over a period of 20 years, the total energy produced will be

$$E = Pt = 0.4 \text{ kW} \times 20 \text{ yr}$$
$$= 0.4 \text{ kW} \times 20 \text{ yr} \times 365 \frac{\text{days}}{\text{year}} \times 24 \frac{\text{hr}}{\text{day}}$$
$$= 7 \times 10^4 \text{ kWh}$$

Then, since the cost was $20,000,

$$\text{price} = \frac{\$20,000}{7 \times 10^4 \text{ kWh}}$$
$$\approx \$0.29 \text{ per kWh}$$

So solar electric power is still relatively expensive. Thus, although solar cells are used

MSFC-76-P-A-4000

Figure 11.48   An artist's conception of a Solar Satellite Power Station. (Courtesy of NASA.)

in remote locations (where power lines do not reach), in space, and in other applications where installation cost is not the major factor, they are not yet in common use. Perhaps, as research brings the cost of solar cells lower, and as the cost of energy from the power companies increases, photovoltaic cells will provide a solution to some of our energy needs.

SSPS   An interesting twist on using photovoltaic cells to produce electricity is the "Solar Satellite Power Station" (SSPS). This involves putting a large satellite in a *geosynchronous* orbit; that is, an orbit in which it stays above

the same location on the Earth, as some of our communications satellites do. The satellite would have a very large bank of solar cells to convert solar power to electric power. This has two advantages over having the solar cells on Earth: First the solar intensity is much greater in space, where the sunlight does not have to come through the air. Second, except for brief periods (a maximum of a little over 1 hour a day), the satellite will never be in shadow: The storage problem is solved. The array of solar cells would be very large, indeed, and there are still substantial practical barriers in the way of realizing a working SSPS.

Figure 11.48 is an artist's conception of what such a satellite might look like.

Questions

The solar constant—the amount of solar radiation—is about 1400 W/m^2 or roughly 130 W/ft^2. If the solar cells in an SSPS are 15 percent efficient, how much electric power is produced per square foot of solar cell? How many square feet are required to produce a power output of 1000 MW? If the array of solar cells is square, what is the length of each side of the square in feet? In miles?

As you saw in the questions, the array will be a square well over 1 mile on a side. However, this is not as bad as it might seem: Because of the weightlessness of objects in space, the structure will not have to be as massive or as rigid as a comparable structure on the Earth. An SSPS conceived by the National Aeronautics and Space Administration (NASA) would have two panels, each covering an area of about 11 square miles! The power output would be about 8500 MW. It would have to be constructed in space, and it would require 100 or more heavy-load space vehicles to get the material there.

The question is, how would all of that energy be used? It is conceivable that some automated manufacturing plants might be put into space to use the electric energy right there, but most of the energy would be used on Earth. The energy will probably be transmitted from the SSPS to the Earth by a *microwave beam*. As you may recall from an earlier discussion, microwaves are electromagnetic waves, like radio and light waves, and they carry energy from one place to another. In this case, the microwave frequency will be

chosen so that it passes easily through air and clouds, perhaps with a wavelength of 10 cm. There will be a large receiving antenna on the ground to pick up the energy. For a station of the size suggested, perhaps 5000 MW of microwave power will reach the surface and be converted back into electric power. The receiving antenna might be a circular dish, six or more miles in diameter.

There are still problems to be solved before an SSPS is constructed; primarily they concern the cost of solar cells. Also, we will have to develop means of aiming the solar panels directly at the sun at all times and, more difficult, the microwave beam will have to be aimed precisely at the receiving antenna. If the beam were to wander around the surface of the Earth, it might be sufficiently strong to cause damage to people, animals, and other things. (They would experience something like being cooked in a microwave oven.) The problems are fairly difficult, and they will not be solved overnight, but the potential payoff is large and may possibly be worth the effort.

SUMMARY

In this chapter you have seen the present state of affairs and some future possibilities for generation and transmission of electricity. The section on steam-driven power plants indicates that such plants are all similar, except for their sources of fuel. The current sources of energy to run power plants are primarily fossil fuels and nuclear power. The overall efficiency of a modern fossil-fuel power plant, which is a heat engine, can be greater than 40 percent. Nuclear plants, which run at lower temperatures for reasons of safety, have efficiencies of about 30 percent. A few steam-powered plants run on geothermal energy, but that "free" energy is not sufficiently available to be a major contributor to total electric energy needs. Some progress has been made in recent years in using trash as a source of energy for power plants, and in the cogeneration of energy, where the "waste" steam from electric power generation is used for industrial processes.

Generators in a power plant operate on the simple principle that, when an electric charge moves through a magnetic field, it experiences a force. The negative charges (electrons) in conducting wires moving through a magnetic field can put these forces to use in producing electric currents. The transmission of electricity on wires depends heavily on the use of transformers. The transformers boost voltages to high levels at the power plant, so that losses in the transmission lines will be minimized. When the power reaches the neighborhoods where it is to be used, step-down transformers reduce the voltage to usable levels. In the future, there may be significant changes in power transmission, such as direct current transmission lines and superconducting lines.

Research is under way on a number of alternative methods for producing electric power, some of which hold considerable promise. For example, OTEC—Ocean Thermal Energy Conversion—may tap the enormous thermal reservoirs of the seas, and fuel cells may economically convert chemical energy directly into electricity with no pollution. If some of the problems can be solved, magnetohydrodynamic (MHD) power generation holds promise for producing electric power efficiently and cleanly in large quantities from coal, the only really abundant fossil fuel. Solar electric power generation, using power towers or using photovoltaic (solar) cells also holds promise.

It cannot be emphasized too strongly that none of the possibilities for the future is likely to be an instant cure-all. Some or all of them may contribute significant amounts of energy in the next few decades, provided that we keep the situation under control until then through conservation and the wise use of resources.

ADDITIONAL QUESTIONS AND PROBLEMS

1. An old coal-fired power plant, located on the Ohio River, has boilers that are 80 percent efficient in converting the heating value of the coal to useful heat. At the temperatures at which they run, the turbines have an efficiency of 40 percent, and the efficiency of the generators in converting mechanical energy to electrical energy is 98 percent. What is the overall efficiency of the plant?

2. A 600-MW power plant in New England uses oil as a fuel. The overall efficiency of the plant is 30 percent. Running at full capacity, how much total power must the plant have available from the oil? What is this in Btu/hr (1 W ≈ 3.4 Btu/hr)? If the oil being used has an average heating value of 135,000 Btu per gallon, how much oil is needed each hour? If the plant averages 60 percent of capacity, how much oil is needed for a year's operation?

3. What are the main advantages and disadvantages of geothermal power?

4. The Geysers power plant, in California, can produce about 900 MW of electric power and may one day produce more than 1500 MW. At the higher figure, what is the largest size city in California for which it could provide all of the power? (Assume that the residents of the city consume the national average of about 250 kWh per person per day.)

5. Explain why some sort of energy storage would be likely to be more desirable than peaking units in a power plant.

6. A transformer on a power pole is designed to reduce the voltage from 2400 to 240 V. If there are 2000 turns in the primary, how many turns are in the secondary winding? If none of the homes and businesses served by that transformer have any electric appliances turned on, does any current flow in the secondary? In the primary? Explain. If four 60-W lamps are turned on, how much current flows in the secondary? The primary?

7. Incandescent lamps (light bulbs), being resistors, can be lit either by ac or dc sources of power. For example, a camping trailer may have lights that work on 12-V dc, or that can be used on the equivalent ac, which would also be called 12 V. For the ac to produce the same lighting effect as that produced by the dc from a battery, what would the peak voltage have to be? Since ac is usually provided at 120 V in trailer campsites, how could the proper voltage be obtained?

8. As pointed out in the text, when electric power travels long distances on transmission lines, there are substantial power losses. Where does the lost energy go?

9. Why would the Atlantic Ocean, just off Cape Code, Massachusetts, be a terribly bad place to locate an ocean thermal energy conversion (OTEC) unit? (*Hint:* Think about the maximum possible efficiency.)

10. Hydrogen is a grand fuel, and it can be obtained simply by separating water into hydrogen and oxygen by electrolysis. Further, when it is burned, the separated water is replaced and no pollution occurs. If that is the case, why don't we just solve all of our energy problems by producing huge amounts of hydrogen from ocean water?

11. On a large scale, even solar tower power could upset the delicate energy balance of the Earth. When more of the sun's energy is absorbed, less is reflected back into space. Remember that *all* of that energy eventually appears as heat. In the case of a solar tower, where will that extra heat appear? (*Hint:* It is

not all in the same place.) It has been suggested that, to compensate for the heat gain, extra mirrors could be used to reflect more energy back into the sky from the area near the tower. Desert sand reflects roughly 40 percent of the energy falling upon it into the sky. Ten million square meters of mirrors used for a tower might reflect only 10 percent. If the total insolation is 1000 W/m^2, how much extra power is absorbed? If the extra, compensating mirrors reflect 90 percent of the incident energy (50 percent more than the sand), how many square meters of them would be needed to maintain the energy balance?

12. There is a story that Archimedes, the ancient Greek scientist, once repulsed a Roman fleet attacking Syracuse with solar power in a way that is much like the modern-day power tower. The way the story goes, Archimedes had an army equipped with mirrors. Then with hundreds or thousands of soldiers all aiming their mirrors at a given ship, it was set on fire. Does this story seem plausible? If so, about how many soldiers would you estimate would be required to start a ship burning? (*Hints:* You will have to make some assumptions to answer this. First, what concentration factor would be needed to make the sunlight powerful enough to cause the ship to start burning? In doing Activity 9.2, some students discovered that, on a particular day, they were able to kindle some tissue paper with a concentration factor of about 250. (That is, the area of the lens was about 250 time the area of the focused image of the sun.) I would guess that setting a ship on fire would require much more than this. Then, what might be the size of the mirror each soldier would have? Also, about how big a spot do you think a person could hit and stay within when aiming a mirror? (Assume he had some practice beforehand.) Then, how many mirrors would it take to cover that spot with just one "layer" of sunlight? Finally, how many would be

needed to get the estimated concentration factor?)

13. If the cost of manufacture of photovoltaic (solar) cells were brought down, then what would their major advantages and disadvantages be?

14. What are the major advantages and disadvantages of the proposed solar satellite power stations (SSPS)? What arguments can you think of against putting an SSPS on the moon and aiming its microwave beam at the Earth?

12 *The Energy Crisis*

THE REAL ENERGY CRISIS

Do you remember the Arab oil embargo of 1973? It followed a Mideast war, and it caused long lines at gasoline stations, the shutting down of some factories, and other disruptions. Was that "the energy crisis"? That is when large numbers of people first started talking about the energy crisis—and they have been talking ever since.

How about the very cold winter of 1976–1977? That was the year when there did not seem to be enough natural gas to go around and some industries, particularly in the Northeast, were forced to reduce their activity, resulting in numerous people being laid off from their jobs. Was that the energy crisis?

What happened in the summer of 1979, when there were again long lines at the gas pumps, people were afraid to travel on vacation, and the price of gasoline started to rise at an amazing rate? Perhaps that was the energy crisis?

If those events represent an energy crisis, why from the spring of 1981 through 1983 was there a glut of oil in the United States and elsewhere—more oil than could be immedi-

ately used—and why was there no shortage of natural gas, coal, or electricity?

Why, indeed! It is no wonder that people are confused and some think that the whole energy crisis is a fiction, manufactured by the big oil companies, and perhaps by the government, as well. The confusion arises because the events cited, and other similar events, are *not* the energy crisis, they are merely symptoms of social, political, and economic pressures. The 1973–1974 shortage of oil resulted from pressures put on the United States (and the Netherlands, incidentally) by the Arabs because of our support of Israel. Not incidentally, the price of Mideast oil made a big upward jump immediately following the embargo, and there it stayed until it went *up* again. In the cold winter of 1976–1977, the gas companies preferred to send the gas from Louisiana and Texas to customers within their own states, thus depriving the northeastern states of enough fuel to continue normal activity. The reason was simple: Government pricing regulations held the price of gas piped across state lines to a low level. The gas producers sold as much gas as they could in their own states, where the federal price regula-

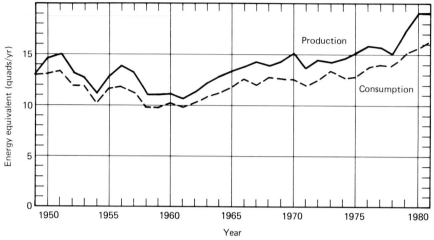

Figure 12.1   Coal production and consumption in the United States.

tions had no effect, thus selling it at a higher price and with a greater profit.

Although these causes are rather over-simplified, those temporary shortages were caused by decisions made for various eco-nomic and political reasons. They were crises, of a sort, and they were good warnings that something is wrong, but they had essentially nothing to do with the total supply of energy available. It is a serious error to confuse them

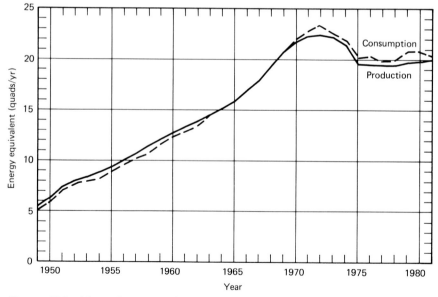

Figure 12.2   Natural gas production and con-sumption in the United States.

with "the energy crisis," and doing so some-times leads people to think there is no prob-lem at all. In fact, as I pointed out earlier, even the word *crisis* probably causes misun-derstanding because it implies that we will be without energy to run our industries, heat our homes, and transport ourselves and our goods next week, or perhaps next month.

A clue to the real nature of the energy problem is revealed by a few graphs. Figure 12.1 is a graph of coal production and con-sumption in the United State for a 30-yr pe-riod, starting in 1949. The units used are *quads;* one quad is 10^{15} Btu, a quadrillion Btus. To

say that 1 quad of coal was consumed means that it was enough coal to produce a quad-rillion Btus when burned. (That would be in the neighborhood of 40 million tons of coal. See Appendix D for a full explanation—if you want a full explanation.) Note that in Figure 12.1 the rate of production of coal very nearly matches the rate of consumption; the small amount of extra coal produced is exported. Note also that the rate of consumption of coal has not changed very much in the past 30 years. Figure 12.2 tells a similar story for nat-ural gas, but with one important difference. The rate of production again closely parallels

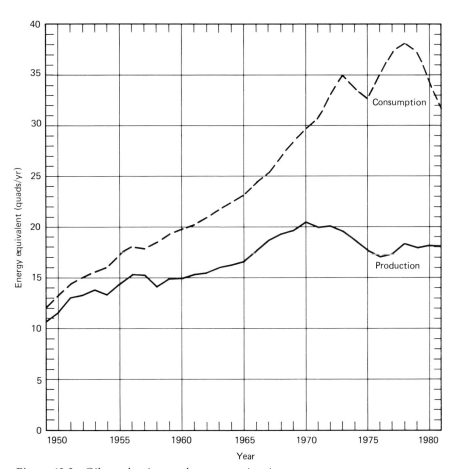

Figure 12.3 Oil production and consumption in the United States.

the rate of consumption. The slight shortage in recent years is made up by importing liquified natural gas (LNG). The difference between coal and gas is that the rate of consumption of gas has increased very substantially in the past 30 years. Note, however, that it peaked in the early 1970s and it now seems to be declining. This is significant and shall be discussed at more length later.

Finally, Figure 12.3 shows the production and consumption of oil in the United States for the same period. This graph should be waving a red flag and shouting "danger!" During the 30-yr period, the rate of consumption of oil tripled. Until the early 1970s the consumption curve looks very much like an exponential curve; the decrease following the oil embargo lasted only two years, then the climb started again. Only in the last three years of the graph does there seem to be any downward motion, and that may or may not continue.

Even more alarming than the fact that the rate of oil consumption has tripled in 30 years is the obvious fact that the rate of U.S. production has not been able to keep pace. Oil wells like the one shown in Figure 12.4 cannot keep up with the rate of U.S. consumption. The difference between consumption and production has to come from those same OPEC nations that cut us off a few years ago. If they were to do so again, and to continue the cut-off for a long period, we would have to find some substitute for nearly one-half of the oil we now use, or cut our energy usage by that much. Neither of these solutions, nor any combination of them, would be easy. Perhaps they are not even possible to implement without severe disruptions to our economy and the society.

However, as important as it is, even the threat of a sudden disruption of the U.S. economy and life-style by an oil embargo is not what I am referring to when I talk about

Figure 12.4 Oil wells in the United States are not able to keep up with the demand for oil, and the difference must be made up by imports. (DOE photo.)

the energy crisis. The graphs showing the *production* of oil and natural gas tell us more about the nature of the real energy crisis. Note that both of those curves peaked in the early 1970s and they have not reached the same levels since then. Yet, in 1973 President Nixon decreed that the United States would be producing all of the energy it needed by 1980, and energy self-suficiency was one of President Carter's top priorities. Why hasn't it happened? We are not yet out of oil or natural gas. Why don't we just produce all we need? Or why don't we use more coal or nuclear power? The answer is that it is not that easy. A complete answer to these questions is complex, depending upon many forces, including those of politics and economics. However, there is one simple underlying reason that we *cannot* simply pump all the oil we need forever. To understand this reason, let's take a look at wine making.

WINE MAKING

Suppose that I would like to make some wine. Being a fairly logical person, I start by planting some grape vines. After two or three seasons of growing, the vines are ready to bear. Early in the spring, the vines start to show buds, which grow into leaves, stems, and blossoms. When the blossoms bloom, bees come to pollinate them, thus assuring that the grapes will develop. Slowly, over a period of many weeks, I check my grapes daily, and find that they continue to grow. At about the first of July, the grapes are a good size and I cannot resist the temptation any longer, so I taste one. It is lip-puckering! During the rest of the summer the grape vines use carbon dioxide from the air and water and, through the chemical magic of photosynthesis, they make sugar in the grapes. The energy for doing this comes from sunlight, and some of that energy is stored in the sugar as chemical en-

ergy. (That is why we say that the sugar "has calories.") By the time they are ripe, the grapes will no longer be sour; the sugar they manufacture will make them very sweet.

After the grapes are picked, if I can resist eating them all or giving them to the neighbors, I am ready to make wine. Doing so is the simplest thing in the world: just crush the grapes and let them sit. In a day or two they will begin to seethe and bubble, giving off carbon dioxide. This is because grapes naturally have the two ingredients necessary for the production of alcohol: sugar in the grapes, and yeast on their skins. When the grapes are crushed, the yeast cells start converting the sugar into alcohol, getting nourishment themselves in the process. I imagine that wine was invented by some careless caveperson who let some crushed grapes sit around too long but, being hungry, ate them anyway. That felt so good, that it was done intentionally the next time.

Actually, a careful wine maker would first kill the wild yeasts on the grapes and add yeast that had been especially developed for wine making. To pick a figure, say that 1 million yeast cells are added to the grapes. (Yeast is a single-celled plant.) Those cells start to convert sugar to alcohol, and after a time—say one hour on the average, just to make it easy—each cell "buds" and divides into two cells. Now there are 2 million yeast cells, and they will use up sugar at twice the rate at which the 1 million cells did so. After another hour passes, the 2 million cells divide and there will be 4 million, using up sugar at four times the original rate. Then it will be eight times the rate, 16 times, and so forth, until the whole mess is bubbling and frothing with the carbon dioxide released.

Does this sound familiar? This pattern of doubling each hour is the pattern of *exponential growth* that was described in Chapter 1. If we make a graph of the rate at which sugar is being used up, it will look like Figure 12.5,

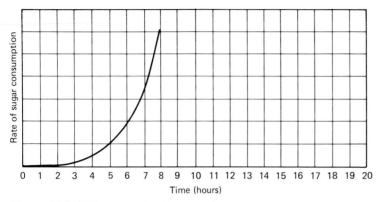

Figure 12.5 When yeast cells grow in a grape-sugar solution, the rate of sugar consumption grows exponentially at first. (The scale of the vertical axis is arbitrary, so no units are shown.)

with a doubling time of one hour. But after a while, something else comes into effect; there will be a shortage of sugar. At some point in the process, yeast cells will start to have trouble finding enough sugar. When that happens the growth of the yeast population slows and, for a short time, it will remain constant. But this will happen only for a short time; as the yeast cells have more and more trouble finding sugar, they will start to die off. Say that

after 1 hr one-half of the yeast cells are unable to find enough sugar to live on, and they die. The other half lives, but uses up more sugar in the process. Then in the next hour, half of *them* will die, and thus it goes. The graph of this is called an *exponential decay* curve. The complete curve of the rate of converting sugar—first an exponential growth, then reaching a maximum, then exponential decay—will look something like Figure 12.6. (In-

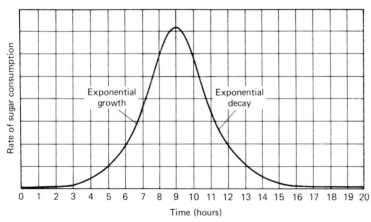

Figure 12.6 As sugar gets harder to find, yeast cells die off and the rate of sugar consumption undergoes exponential decay. (Same vertical scale as Figure 12.5.)

cidentally, I have arbitrarily chosen the time for doubling in the growth period and halving in the decay period both to be one hour. In practice, they would probably be quite different. Also, the size of the vertical axis is arbitrary but the same in Figures 12.5, 12.6, and 12.7.)

Suppose that, being a yeastitarian (something like a humanitarian), I would like to help the yeast to live. I could add a little sugar to the grapes, but I want to do it before any fermentation starts. This will delay the end, but not by much. The curve of sugar consumption will look exactly the same as before, but it will be over a somewhat greater time period and with a higher peak, as shown in Figure 12.7. Finally, suppose I decide to keep the yeast cells alive forever by continually adding as much sugar as they need. Strangely enough, they would die, anyway, for the alcohol they produce is a waste product, a "pollutant." That is, the alcohol is a poison to yeast cells, and after it reaches a certain level it will kill them. The only way I can keep the yeast cells alive is to put some of them into a new batch of grapes. That is, I have to *renew the resource*.

FOSSIL FUELS

The story of the wine is quite similar to the story of fossil fuels. First, the grapevines take a long time—the entire summer—to convert carbon dioxide and water to sugar, and to store the sugar in the grapes. The sugar, a compound of carbon, hydrogen, and oxygen, is a form of stored chemical energy. Likewise, the production of fossil fuels and their storage in the ground took a long time. It started about 600 million years ago, with trees and other plants using sunlight and photosynthesis to store chemical energy in carbon- and hydrogen-containing compounds, called *organic matter*. When the plants died, they decayed. This decay, or rotting, is a process of organic materials combining with oxygen, and the stored chemical energy is released in the form

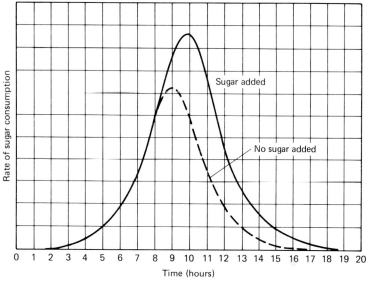

Figure 12.7 If extra sugar is added to the grapes, it takes the yeast cells just a little longer to use up all the sugar.

Figure 12.8 As oil gets harder to find, more of it will come from offshore rigs like this one. Some of the largest offshore platforms require years to build and cost $2 billion or more. (Courtesy Shell Oil Company.)

of heat. It is much like the process of burning, but slower. However, some of the dead plants were covered with mud and other material, and did not get enough oxygen to decay fully. Bacteria worked on them. As they were buried under more and more layers of earth, they were subjected to tremendous pressures and, sometimes, high temperatures. Depending on what happened to it underground, much of that early organic material was converted to coal, which is a burnable rock containing mostly carbon (up to 98 percent). (With greater heat and pressure, a little bit of that material became diamonds, carbon in a crystalline form.) Some of the material became petroleum (*hydrocarbons*, consisting of carbon and hydrogen), and some became natural gas (mostly methane, CH_4). Each of these fossil fuels has carbon as a main constituent. More important, each has some stored chemical energy, which can be released by burning. (Incidentally, the stored energy originally came from the sun, so fossil fuels may be thought of as a very indirect form of solar energy.)

When we use a fossil fuel, nature continues to work on replacing it. Unfortunately, it takes only a few years or centuries to use it and millions of years to replace it. Since none of us is willing to wait that long, we call fossil fuels a *nonrenewable* source of energy. Whatever is in the ground right now is all there is. Like the sugar in the crushed grapes, when the fossil fuels are gone, they will be gone for good. Also like the sugar, the rate of use of fossil fuels has been growing exponentially. New supplies, particularly of oil and natural gas, get harder and harder to find, as illustrated in Figure 12.8. As they become more difficult—and more expensive—to find, they will be used at a lower rate, finally going to zero use when the fuel is "gone." (Here, *gone* means when it is too expensive to recover the remaining fuel.) The graph of the annual rate of oil production will look much like the graph of the rate at which the yeast converted sugar in the wine. Many experts think that the peaks we see for oil and natural gas production in the United States during the early 1970s represent the highest points those graphs will ever reach, and that the future production will continue to decline. What we want to know is how long it will take to decline to zero, or to near zero.

FOSSIL FUEL SUPPLIES

Many experts make estimates of the amount of fossil fuels in the ground, and they use very sophisticated techniques. Unfortunately, it appears that there is still some guessing involved, for the experts do not always agree with each other. This can be seen easily by comparing the estimates from any two independent sources. Coal is the fossil fuel for which the estimates are probably the most accurate. Oil and natural gas are harder to estimate, for they are hidden deep in the earth, often in hard to reach places such as above the Arctic Circle and under the oceans.

However, knowing the exact amounts remaining is not necessary for an understanding of the problem. Even a rather wide range of estimates gives the same results, with just a little difference in the time span. That is, if there is more oil than we now think, the final result would simply be delayed by a few years. Since oil is the most crucial fuel, we'll look at that first.

Oil

Oil is a wonderful fuel. It is relatively clean to burn, easily transported, and it has a high heating value. A gallon of oil will produce approximately 143,000 Btu when burned. A barrel (42 gal) is good for about 6 million Btus. Until recent years, oil was also very easy to find. It is most essential for transportation, although it is still used for many stationary applications such as electric power stations and home heating. It is hard, as yet, to imagine another equally good means of powering automobiles. For example, consider the situations of two friends. One has an electric car. It is very tiny, holding just two persons; it will travel at a maximum speed of 40 mph, and it will go perhaps 100 miles between charges. The batteries, which are about one-half the weight of the car, can be charged overnight, and they must be replaced every two or three years. It cannot be heated in wintertime, and there is no protection for the passengers in the event of a crash. The other friend has a car that will carry five passengers and their luggage 600 miles comfortably at 55 (or more!) mph on a tank of gas. When the tank is empty, it can be refilled in five minutes for another 600 miles.

I suppose one could consider burning coal in automobiles. But, even if a coal-burning engine of the same size and efficiency as a gasoline engine could be built, where would you put the 225 pounds of coal needed for 600 miles? How would you get it to the engine? You see the problem. Imagine how much greater the problem would be for a large jet airplane. Even trains, which once ran on coal, and before that wood, use diesel fuel now, as do large ships.

This marvelous fuel, oil, was once easy to find. When Edwin Drake drilled the first producing oil well in Pennsylvania in 1859, he struck oil at 69.5 ft below the surface. Today oil provides almost one-half of the energy consumed in the United States, but it is getting harder to find, and thus it is becoming more expensive. Productive oil wells are now at an average of 5000 feet deep—a mile into the Earth. Many wells are three times that deep, and the cost of drilling an offshore well can be $1.5 million. There is no guarantee that oil will be found and, for every productive well, several dry ones are drilled. Forty years ago, the yield of oil was about 275 barrels found for every foot of well that was drilled in the United States. Now, it is about 35 barrels per foot, and it will get smaller than that. With oil becoming harder to find, the cost will continue to go up. In the recent past, federal regulations have limited the price of oil produced in the United States. For that and other reasons, many U.S. oil companies found it more profitable to locate new oil fields in the Middle East and elsewhere. Now, with the cost of imported oil rising rapidly, oil companies may again find it profitable to look for new oil in the United States. The most likely place of doing so is offshore in the oceans, where drilling conditions are difficult and oil spills are damaging to the environment.

When oil supplies are estimated, they are often divided into the categories of proven reserves, probable reserves, and future discoveries. *Proven reserves* represent the oil that we *know* is in the ground with a reasonable certainty and that can be recovered using present-day techniques. The *probable reserves* are estimated from reasonably convincing geological evidence of their existence. *Future discoveries* are estimated on the basis of a general knowledge of the geology of the area and the kinds of geological formations that have produced oil in the past.

It is fortunate that we can rely fairly confidently upon more than just the proven reserves. For example, in the United States at the end of 1978 there were about 33 billion barrels of proven reserves. That translates into about 195 quads of energy. If you look back at Figure 12.3, you will see that the 1978 rate of oil production in the United States was over 18 quads. A little division will show you that, if that rate continues, the proven reserve will be exhausted by 1989. Good grief—that is not very far off!

Calm down. But not too much. The situation is serious, but it is not quite as bad as all that. The oil companies *are* continuing to find new oil. The trouble is that in recent years they have not been able to discover new oil as fast as we are using the already-discovered oil. That is at least part of the reason that U.S. oil production has declined since 1970. On a worldwide scale, the estimated recoverable reserve—including oil already used and future discoveries—is about 12,500 quads (2100 billion barrels). (This estimate is about in the middle of a wide range of estimates. It includes oil in Russia, China, and other regions which are likely to use much of it themselves. The estimated worldwide *proven* reserve at the end of 1981 was something over 3500 quads.)

Do you remember the yeast cells changing the sugar in the grapes into alcohol? The graph of oil consumption will look much like the graph of sugar conversion. At this time, on a worldwide scale, the graph still seems to be in the exponential growth stage. Roughly, with the squiggles smoothed out, it looks like Figure 12.9. One important thing to note about this graph is that, from it, we can calculate the total amount of oil used. Figure 12.10 shows how this is done by using a level rate of 150 quads per year for 50 years. In the first year, 150 quads are used. In the second year, another 150 quads are used, and the total is 300 for the two years. At the end of five years, it

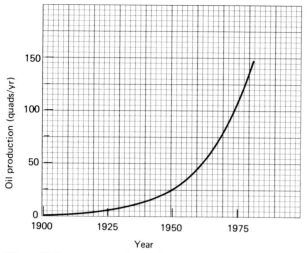

Figure 12.9 An approximate (smoothed-out) graph of world oil production.

is a total of 750 quads. In each case, the total amount of fuel used is the number of years (across the bottom of the graph) times the number of quads per year (the height of the graph). The product of years times quad/year is the area of the rectangle under the line. (The line is often called a curve, even though it is a straight line.) For the entire 50-year period, it is 50 years times 150 quad/year, or

7500 quads. Figure 12.11 shows another kind of "curve," a straight line increasing from zero in 1925 to 150 quads in 1975. The area under this curve is just one-half the area of the previous rectangle, and the total production would be 3750 quads.

Look again at the actual oil production shown in Figure 12.9. You can find the area under the curve by counting squares. The large

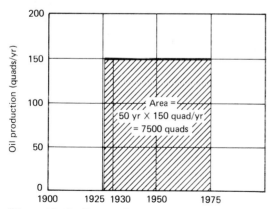

Figure 12.10 For production of 150 quads per year from 1925 to 1975, the total oil production would be the area under the production "curve," or 7500 quads.

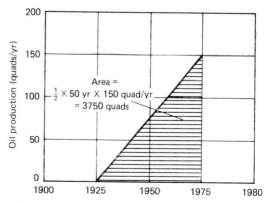

Figure 12.11 If oil production had started at zero in 1925 and risen linearly to 150 quads in 1975, the total production would have been 3750 quads for the 50-year period.

squares have an area representing 10 years × 25 quad/year, or 250 quads. Since there are 25 small squares in each large one, each small square represents 10 quads.

Questions

By counting squares under the production curve of Figure 12.9, how much oil do you estimate has been produced in the world up to 1980?

Now we are in a position to look at the production of oil as it will probably occur in the future. You recall that, in the grape mixture, the yeast cells convert the sugar into alcohol as fast as they can. When a shortage of sugar develops, the rate of conversion slows and finally goes to zero. Unfortunately, we humans have been about as thoughtful and as conservative as the yeast cells. We have

been using oil nearly as fast as we can find it and pump it out of the ground. As oil becomes harder to find, the indications are that the rate of production must soon slow. (In fact, U.S. production seems already to have passed its peak.) In drawing the graph for the future, one must make a guess at when the peak production will occur, and how great that peak will be. However, two things are certain! One is that the graph will peak, then decline; the other is that the area under the curve must not exceed the total amount of oil available. Such a graph was drawn some years ago by M. K. Hubbert, and an approximate version of it is shown in Figure 12.12. As you can see, it predicts that the peak oil production for the world will occur sometime around the year 2000, and most of the oil will be gone before 2050. That is not very far in the future, and we obviously need to start serious work on finding replacements for oil.

Suppose that the estimate of about 12,500

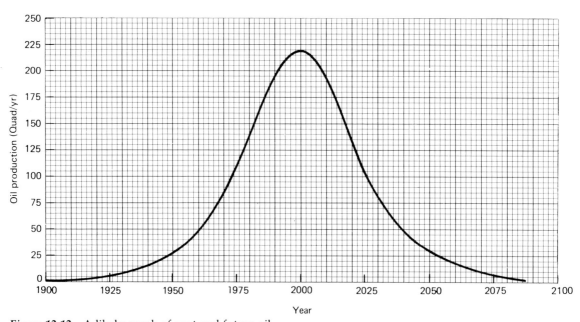

Figure 12.12   A likely graph of past and future oil production, assuming that there are 12,500 quads of total reserve.

quads of recoverable oil is too low, and that quite a bit more oil than that is available. How much longer will the oil then last? The curve of projected oil consumption in Figure 12.13 assumes that there will be enough oil so that the peak production will be 300 quads and the peak will not occur until about 2030. The figure of 300 quads is to allow for development of the developing nations (about which more later), and the deferral of the peak for just 30 years is a modest goal. This results in the oil supply lasting about 50 or 60 years longer, until about 2110, or so. This does not seem to be much to ask, and it does not solve the problem; it only puts it a little farther into the future. Even so, if you count squares under the curve, you will find that over 30,000 quads of oil would be needed. It seems highly unlikely that as much oil as that would be available.

O.K. In that case, suppose we try to do much better. By very severe conservation measures we will hold down the level of oil consumption to 150 quads per year. Also, by being much more active in exploration and in squeezing the last drop of oil out of the wells, let us suppose that production could be maintained at 150 quads per year long enough so that we do not run out of oil until about 2100. The graph of oil production would then look like Figure 12.14, and the area under the graph represents about 18,000 quads. That is still almost 50 percent more than we now believe is available.

Just based on this mechanical method of counting squares under a curve on a graph, it seems obvious that the oil supply is going to run out sooner or later. Probably sooner. This accounting does not take into consideration human needs, the need to save some

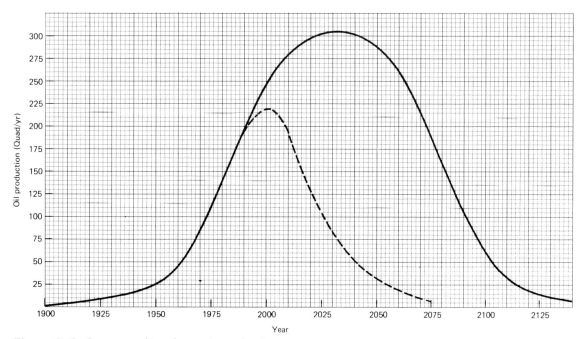

Figure 12.13 Just extending the peak production to 300 quads and the time before running out of oil by 50 years would require a total resource of about 30,000 quads.

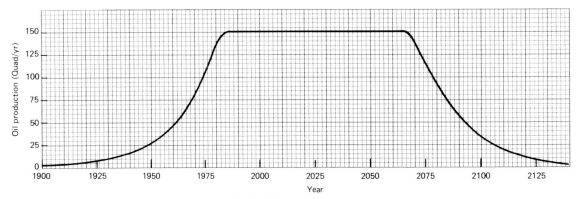

Figure 12.14   If, by severe conservation, the level of oil usage could be held at 150 quads per year for almost 100 years, a total resource of over 18,000 quads would be needed.

oil for use in the chemical industry, or the pollution and other environmental consequences of using oil. It would be foolish to assume that, by the time the worldwide production of oil peaks in 20 years or so, science and technology will have found an easy solution. As you have already seen, the problem is a difficult one and an easy solution may not exist.

Natural Gas

Natural gas is another wonderful fuel. Because it cannot simply be poured into a tank, it is not as useful in transportation as is oil, but it can easily be moved from one place to another far away by means of pipelines. However, no one has yet figured out how to run pipelines across the oceans. Ironically, some of the world's greatest sources of natural gas are at oil wells in remote locations, and there has been no sufficiently economical way to get the gas to potential users. In early satellite photos taken at night, the most outstanding feature on the Earth was light from "flares," where "waste" natural gas was being burned. Much of this flaring continues today for lack of something else to do with the gas, although some installations are now pumping it back into the ground for possible later use.

Nonetheless, natural gas is a fine fuel. It burns very cleanly, is easy to transport where pipelines can be built, and has the fairly high heating value of about 1000 to 1100 Btu per cubic foot (at standard temperature and pressure). Distribution in a city is very easily accomplished by means of pipes. For these reasons, it presently provides about one-fourth of all the energy used in the United States. Natural gas often occurs along with oil, where it is trapped in porous sand or rock below solid rock. The "gushers" one used to see in adventure movies were often caused by natural gas under high pressure forcing the oil out violently through the drilled hole. Even now, when one reads about an oil well burning out of control, the force pushing the oil out is usually provided by natural gas.

As noted earlier, the estimates of the amounts of fossil fuels vary widely. In the case of natural gas, one can find a very wide range of estimates of the reserves available. (Also, the estimates are published in a wild range of units. In a single reference book, I have found natural gas resources reported in quads, trillions of cubic feet, megatonnes, Qs, cubic kilometers, and trillions of Btus. Even worse, when converted into the same units, the different estimates do not agree very well with each other.) However, a reasonable estimate of the proven reserves in the United States would be in the ballpark of 250 quads. At the present rate of use of about 20 quads per year, this will last only a dozen years or so. New reserves are being found, but not fast enough to keep up with consumption. If the total recoverable resource in the United States is about 1100 quads, then the graph of natural gas production might look something like the solid line of Figure 12.15. That graph was drawn by drawing a smooth curve approximately equal to the actual production up to 1975, then letting the curve drop so that the total under the curve is about 1100 quads. Recently, there seems to be a lot more natural gas available than previously believed. This sudden increase was due, at least in part, to the lifting of government regulation of natural gas prices. The greater available amount may extend the consumption curve a little farther but, still, the supply cannot be inexhaustible. Using even the most optimistic estimates, the supply of natural gas in the United States cannot last indefinitely. The dashed line in Figure 12.15 shows an approximate curve for a total natural gas resource of 1600 quads.

On a worldwide scale, the picture might be somewhat better. There are about 2400 quads of proven natural gas reserves in the world, and the estimated amount of total recoverable gas is quite large—something like 10,000 quads. In Figure 12.15, the graph for the United States alone was used because of the problem of transporting the gas across oceans. Much natural gas is in locations far from the users, particularly users in the United States. Presently, when natural gas is transported by ship it is liquified under great pressure and at a temperature of about $-150°C$. However, liquified natural gas (LNG) is dangerous (but then, so is gasoline). In the early days of the use of liquified natural gas, there was a disastrous accident that killed 128 persons. (This was in Cleveland in 1944.) Fear of further accidents prevented the rapid development of the LNG industry for some time. Today, the technology is fairly well developed and giant ships are used to transport LNG across the oceans. In the ships, and land-based storage tanks, sudden losses of LNG are possible as a result of collisions or earthquakes, respectively, or other accidents. The tanks and

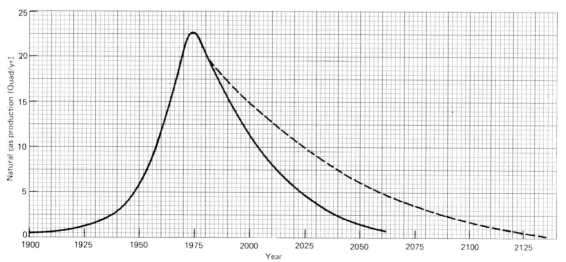

Figure 12.15 Possible future graphs of the U.S. production of natural gas. The solid future line assumes a resource of about 1100 quads, and the dashed line assumes a resource of about 1600 quads.

their surroundings are now designed to minimize the results of any such spillage, and the safety factor is quite high. That is, there is a risk, but it is comparable to the risks involved in other types of fuel storage and production (or to the risks of injury from lightning or automobile accidents). It is likely, then, that the extent of the future shipment of LNG will depend largely upon the economics of the situation. As the price of other fuels increases, LNG may become more attractive. Also, it may become economically feasible to convert the natural gas to methanol (alcohol) and ship it in that form.

The U.S. Government has played a major role in the demand for and supply of natural gas. It has done so through regulations that have held the price of natural gas produced in this country down to much less than its "real" price, the price it could bring in competition with other fuels. Thus, because of the low price, the use of natural gas by both industry and home owners has been encouraged. At the same time, production has been discouraged, also because of the low price. As government price controls are removed, the supply and demand picture seems to be changing and it may change a good deal more. Meanwhile we need to be looking for alternatives to replace natural gas gradually over the next few decades. One possibility is the gasification of coal.

Coal

You have probably heard that there is plenty of coal available. Large coal mines, like the one shown in Figure 12.16, abound in the United States. In fact, many people seem to be fond of suggesting that there is no energy crisis at all because we can simply use more coal. This may sound reasonable, but have you seen any coal-powered automobiles lately?

Coal *is* quite plentiful. On a worldwide basis, there are about 18,000 quads of proven reserves and perhaps as much as 200,000 quads of total reserves, using present methods of mining. This huge amount of coal should be enough to last at least 200 or 300 years. Why don't you see for yourself?

Questions

In 1981, the worldwide consumption of energy amounted to about 290 quads. If consumption were to continue at that rate and coal were used to provide *all* of the world's energy, how long would the estimated reserves of coal last? If the energy consumption were to grow to four times its present size and stay at that level, how long would the coal last?

The huge amount of coal available and the rapidly diminishing supplies of oil and natural gas have led people to think that coal might be substituted for the other two. However, for needs like automobile and truck fuels coal is not very practical, and in power plants it tends to burn "dirty," requiring expensive efforts to control large amounts of pollutants. One partial solution to these problems might be to treat coal, chemically and otherwise, to turn it into fuels that do not have these drawbacks. *Coal gasification* refers to processes by which coal is converted to gaseous fuels, and *coal liquification* refers to processes in which liquid fuels are produced.

Coal was first subjected to changes to make it more useful in about 1709, when coke was first produced in England. This development led to the ability to produce the high-temperature fires needed for the production of steel. In the first third of this century, "town gas" was produced by heating coal, and this relatively low-Btu gas was used for cooking in homes around the United States. The development of natural gas pipelines led to the dropping of localized town gas production. Before and during World War II, Germany produced gasoline and aviation fuel in rela-

Figure 12.16 Surface mining (strip mining) of coal in Gillette, Wyoming. (U.S. DOE photo by Jack Schneider.)

tively large quantities by liquifying coal, but that was a case in which national need overcame cost considerations. Now, with the energy situation changed substantially, there is renewed interest in both gasification and liquification of coal.

The main need in coal gasification processes is to increase the amount of hydrogen in the fuel. A good grade of coal is mostly carbon, whereas natural gas is mostly methane, CH_4. Although there are many different gasification processes, in most of them, the coal is first heated and then reacted with steam, the source of hydrogen. Ideally, coal would be changed directly into methane by the following process:

$$2C + 2H_2O \rightarrow CH_4 + CO_2$$

This reaction is possible, but it does not occur readily with coal. (However, using cellulose rather than coal as the source of carbon, bacteria are able to cause this reaction to happen, thus producing methane out of vegetable matter.) Instead, a variety of other reactions occurs, producing some methane in the process, but with a large component of other gases, like carbon monoxide (CO). The resultant gas produced is called *low-Btu* gas, for it may average considerably less than 200 Btu/ft^3. (Remember that the heating value of natural gas is about 1000 to 1100 Btu/ft^3.) This low-Btu gas is not useful in homes, nor would it be economical to ship it large distances. However, it can be produced on the site of a power plant or large industry and used right there. When the gas is produced, pollutants such as sulfur and particulate matter (small particles) are almost completely left behind in the waste solids. Thus this gas burns much cleaner than coal, producing far less pollution and, in plants that cannot burn coal without violating clean-

Figure 12.17 A coal gasification field model funded by the U.S. Department of Energy and located near Kemmerer, Wyoming. Scientists and engineers from Lawrence Livermore Laboratory have conducted experiments here to determine if underground gasification of deep coal seams is feasible. (Photograph by Lawrence Livermore Laboratory.)

air standards, it can be most useful. Unfortunately, as much as one-fourth of the coal's energy is lost in the gasification process, so the total heating value of the gas is considerably less than that of the coal it came from.

By further processing, medium-Btu gases and substitute natural gas (SNG) may be produced. The latter has very nearly the same qualities as natural gas and it may be used in the same way. But the technology for large-scale production has not been developed and, when the gas is produced, much of the original heating value is lost. However, with advancing technology, coal gasification may become a reasonable substitute for some of our present natural gas use. One of the recent research efforts in economically gasifying coal is shown in Figure 12.17.

Despite the fact that about 90 percent of the aviation fuel used by the Luftwaffe in World War II came from coal, the coal liquifaction technology is also in a relatively undeveloped state. As the need for alternative sources of liquid fuels increases, usable large-scale processes will probably develop. Present technologies include dissolving the fuel components of the coal in liquid solvents and further processing of gasification products to produce liquid fuels. (Some of the "liquid" fuels produced by the former method are liquid only at temperatures above 400°F, and they are sometimes handled and burned in the solid form.) As with fuels from gasification, liquid fuels are more convenient to store and transport than coal and they burn cleanly. Their disadvantages also are similar to those associated with gasification.

Other Fossil Fuels

Peat   Now that we have looked at the supplies of the "big three" fossil fuels, let's take a look at the other fossils. They are peat, tar sands, and shale oil. *Peat* is a soggy, earthlike material, with a high carbon content. It represents an early stage in the development of coal; if the peat were covered with layers of mud and dirt and subjected to great pressures and temperatures, it would eventually become coal. There are estimated to be about 1700 quads of peat in the world, and it is used for energy in some parts of the world. (It is also used for agricultural purposes, being a good material to mix with soil to make it loamier.) However, only about one-half of a quad of peat is used as fuel each year, mostly in the USSR (91 percent) and in Ireland (about 8 percent of world use, but 96 percent of European use, a great deal for a small country). One reason so little is used is that it is a poor fuel. It has only about one-third to one-fourth of the heating value of coal. Thus it gets only local use. In the places where it is used, people dry it out, cut it into logs, and burn it for heating and cooking. About twice as much peat is used for agricultural purposes as for energy. It seems obvious that peat will never be a major source of industrial energy for the world.

Tar Sands   *Tar sands* are presently mined in one place in the world: the Athabasca Tar Sands of Alberta, Canada. The tar is a thick, gooey substance which contains oil. The oil can be removed through a somewhat complex and quite difficult process. The Tar Sands have an area of about 30,000 square miles, and it is estimated that they contain about 300 billion barrels of recoverable oil. This is about 1800 quads of energy and, if it could all be used, it would represent a very substantial addition to the world oil supply. With present mining and processing techniques, only about 35 billion barrels could be eventually recovered. There is presently one plant processing the tar, producing 55,000 to 65,000 barrels of oil per day (a little over one-tenth of a quad per year.) To mine tar sands on a large scale would be very expensive and difficult. If 10 plants, each costing over $1 billion to build, were operating full time, it would take nearly 100 years of operation for them to recover just the 200 or so quads of oil now recoverable. And each plant would have to handle about 400,000 tons of solids every day—no easy matter. In the summertime, with temperatures up to 90°F, the tar is thick and sticky and extremely difficult to handle. Most of the tar sands are covered with muskeg, a swamp containing water and decayed matter, and in the summertime, the machines cannot even operate on the swamp. In the wintertime, temperatures go down to −50°F, and the swampy material is removed while it is frozen. It is then so hard that it has been known to wear 120 100-lb teeth off a digging bucket in eight hours. This is only a beginning of the list of problems that hinder recovery of the oil. I think you can see that, although this may be an important source of oil in the future, it is not going to replace present oil supplies.

Oil Shale   *Oil shale* is rock containing a burnable organic substance called *kerogen*. It has been called "the rock that burns." You may have heard that there are vast reserves of oil shale in the United States and elsewhere. There are, indeed; the worldwide resource may amount to 10 million quads. That would be enough to solve our oil problems for a very long time. But, as you have learned by now, there is always a catch! The catch, of course, is that it is difficult to produce the oil from shale. A good quality shale might yield 20 or 25 gallons of oil per ton of rock. To produce this oil, the rock must be mined, crushed into small pieces, and heated in a vessel called a *retort* to a temperature of about 950°F. In the heating process the kerogen decomposes and is driven from the rock, yielding oil that can be condensed to liquid form. After this has been accomplished, something must be done with the leftover rock, which is called *spent shale*. To produce 100,000 barrels of oil a day (a very small amount) would require processing of about 200,000 tons of shale. Roughly 80 percent of the original material ends up as

spent shale, requiring the disposal of about 160,000 tons of waste. Unfortunately, the spent shale occupys more volume than the original shale, so it cannot just be put back into the hole it came from. Underground mining of shale will probably never be profitable because of the huge mass of material that would have to be moved through the tunnels, so shale will have to be processed in open-pit mines. In the United States, high-kerogen shale is found close enough to the surface for open-pit mining only in some of the western states, where there is not much water. Since the processing of shale requires large amounts of water, this puts another limit on the amount of shale that can be processed. The processing of shale is also potentially damaging to the environment; large amounts of noxious gases and polluted water are produced, and underground water supplies could become polluted by water passing through the spent shale deposits. Some of the environmental hazards may be reduced through the use of *in situ* processing, where the oil is extracted from the shale while leaving the shale in the ground. This will require *fracturing*, or cracking of the rocks in the ground on a large scale and heating them to drive off the kerogen. The *in situ* processing methods are still in the experimental stage.

You can see why the future of shale oil on a large scale is uncertain. The experts seem to be divided in their opinions as to whether shale will contribute large amounts of energy. In the end, much of the answer will depend upon economics. The processing of shale will require the building of multi-billion-dollar plants and the changing of present government regulations. At the present time, the cost of oil from shale is greater than the cost of imported oil. In the future, as the price of oil goes up, it may become economical to produce oil from shale, but we are not likely to produce large amounts of it before the year 2000. By the spring of 1982 three of the four

Figure 12.18 A gas combustion retort for producing oil from shale at the U.S. Department of Energy's Anvil Points Oil Shale Facility near Rifle, Colorado. (DOE photo.)

companies conducting large-scale experimental shale oil recovery operations had given up their efforts as being not promising enough to justify the expense. Figure 12.18 shows part of an experimental government facility for extracting oil from shale.

ENERGY NEEDS

Now that we have had a view of the energy resources available, let's take a look at the projected energy needs. What do you think is meant by the term *energy needs*? To us Americans, it has come to mean a consumption of over 10 kilowatts, day and night, for each man, woman, and child. This comes to over 250 kilowatt-hours per day for each of us. A generation ago, we consumed energy at the rate of about 150 kWh/day per person. In another generation will it be 350? On the other extreme, persons in Yemen average about 1.4 kWh/day per person. To put that in perspective, that amounts to less than 1200 kcal per day per person. Why, we *eat* more energy than that! (Too much more!) Ironically, Yemen borders Saudi Arabia, one of the world's greatest oil producers. Obviously, a citizen of Yemen would (rightly) think that we Americans spend an incredible wealth of energy.

In fact, in today's world the wealth of a nation is fairly closely related to how much energy it uses. One measure of a nation's wealth, and of its overall standard of living, is the *gross national product* (GNP). The GNP is the value of all of the goods and services produced by a nation for a year. Since countries vary in population, a more meaningful figure is the GNP divided by the population—the *per capita* GNP. Figure 12.19 is a chart for several countries showing per capita energy use versus per capita GNP for 1976. The energy is given in millions of Btus and the GNP is in U.S. dollars. The more energy used by each person in the country, the higher on the chart that country appears. The greater the GNP per person, the farther to the right on the chart that country appears. As you ex-

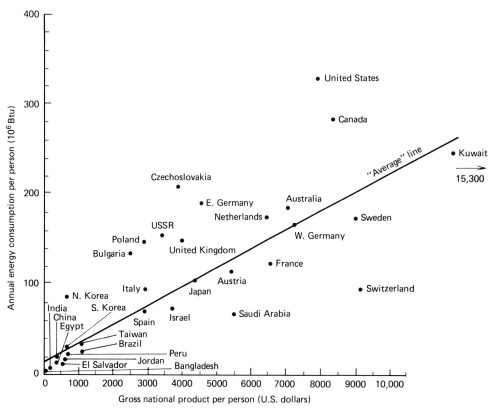

Figure 12.19 Per capita: energy consumption versus gross national product for a few countries.

amine the chart, you can see that, in general, the richer countries use more energy per person.

However, that does not seem to hold true perfectly. Notice, for example, that Canada, Sweden, and Switzerland all have GNPs about equal to ours, but they use less energy. In fact, Sweden, which is a highly industrialized nation, uses about one-half the amount of energy per person the United States uses. Is this a clue that we are using more than our share? One way of looking further at the situation is by drawing a line representing a kind of "average" for all the nations. The line shown on the chart is the straight line that best fits the data for 139 countries, not just those shown on the chart. (The ones on the chart were chosen more or less randomly, but they represent all parts of the world.) Any country that appears above the line uses more than the average amount of energy to produce goods and services. A country that is below the line uses less energy to produce the same goods and services. The United States appears far above the line, indicating a lack of energy efficiency. This has resulted mainly because energy has been so cheap. In recent years the energy efficiency of the United States has improved somewhat. As energy gets more expensive, the efficiency should increase a great deal more.

Let's take a look for a moment at the poor nations, at the lower left corner of the chart. There are many more to put into that part of the chart than space allowed. People in those countries surely have as much right to a comfortable life as anyone else. Yet, they are both energy poor and money poor. Even worse, those are the same countries that will have the greatest population increases in the future. For them just to stay at the same very low level of energy use per person will require large increases in the total energy they use. Where will it come from? An example of a different kind of energy crisis—but one that is very real—is the situation with fuel wood

in some of those countries. Perhaps one-third or more of the population of the world, in Asia, Africa, South America, and other places, depends on wood for most heating and cooking. This might be good, for wood is one fuel that is renewed; new trees grow and can be harvested. The trouble is that the wood is being used up for fuel, building materials, and industry at a far greater rate than it is growing. In the past 25 years, nearly one-half of the world's forests have been lost by being cut down and plowed up. If this continues, by the turn of the century there will be a disastrous lack of firewood. Already, in some countries, people are burning animal dung instead of wood. But that dung was the only source of fertilizer they had before they had to start burning it. As a result the land is poorer and produces less—including less wood.

Well, this still doesn't tell us what the future energy needs of the world will be. Numerous individuals, committees, government agencies, oil companies, and other groups have made predictions. I mentioned earlier that the different sources for estimates of energy resources don't agree with each other very well but, compared to predictions of future needs, those figures are in wonderful shape! The future predictions range all the way from the foretellers of doomsday to those who think there is no problem at all. One hardly knows which set of figures to believe. Often, the same person or group will present several different "scenarios," each one giving different predictions. A scenario is a way of saying "what if?" For example, what if the population continues to grow exponentially? Or what if the world economy stops growing? However difficult, a choice must be made from among the different scenarios. One set of predictions that seems to be reasonable comes from the World Energy Conference, the work of experts from 76 countries. They seem to represent some middle ground, being neither extremely pessimistic nor extremely optimistic.

First take a look at Figure 12.20. It is a

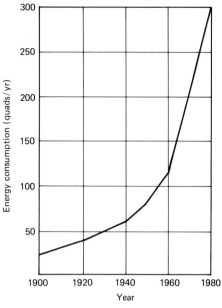

Figure 12.20 Approximate world energy consumption, 1900 to 1980.

rough graph of total world energy use since 1900. Naturally, it has grown exponentially, approximately. Figure 12.21 represents future world energy demand according to two scenarios of the World Energy Conference. The boxed-in part of the graph is the same graph as Figure 12.20. The only difference in the two future projections is that for one, low growth of the world economy was assumed, and for the other high growth was assumed. (The difference between the two growth assumptions is not great. The low-growth scenario assumed an economic growth rate of 3 percent per year, and the high-growth was 4.2 percent per year.) Both scenarios assume that the developed nations, such as the United States, will practice considerable energy conservation in the future. Both assume that the developing nations will have population growths and higher energy demands in the future. As you can see, even the lower demand line predicts that the demand for energy will grow from the present 300 quads per year to almost

800 quads per year by 2020. That is a great deal of energy!

How about the supply of energy? The World Energy Conference has been fairly optimistic in presenting the projections shown in Figure 12.22. Again, the only difference between the high and low lines is in the assumption about the growth of the economy. Both of the projections assume that coal and nuclear power will provide much of the future energy. As you can see, even the very large resources predicted do not meet the predicted demands. Where will the necessary energy come from?

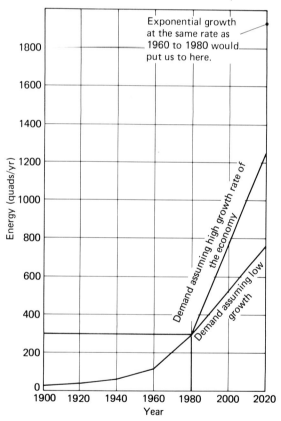

Figure 12.21 Two projections of the world energy demand to the year 2020. Slightly different assumptions are made about the rate of the growth of the world economy.

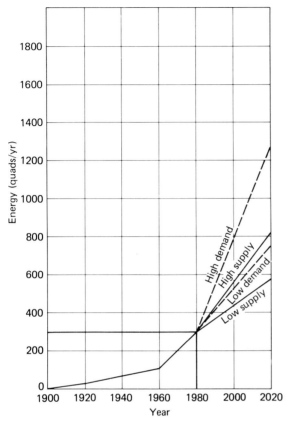

Figure 12.22 Estimated world energy supplies to the year 2020. The huge increase of the *high* supply projection is barely enough to meet the needs of the *low* demand projection.

WHY NOT COAL?

Obviously, during the coming century the world is going to "need" more energy than it now consumes. As you have seen, the supplies of oil and natural gas are not going to last terribly long, and either some substitute must be found or we will have to do without sufficient energy. Of course, we could just shut our eyes and hope that science and technology will somehow bail us out. I have heard intelligent and thoughtful people argue that no planning at all for the future is a better course than any planning we could do. Their argument rests on the idea that government and corporate bureaucracies will *always* make the wrong choices and thus make things worse. However, I still believe that intelligent planning can and *must* be pursued. Otherwise, the world will just blunder ahead into an intolerable situation.

One course of action which seems tempting is to use coal in a big way to buy time. Even if coal were used to provide all of the world's energy needs, it would last for perhaps a couple of hundred years. By then, surely technology will have advanced to the point where *renewable* resources, such as solar energy, could be used. Besides, you and I will be dead by the time the coal runs out. So, why not use coal? Why not run all of our power plants on coal? Why not convert some of the coal to liquid fuels for transportation? Why not convert still more into substitute natural gas?

Whoa! By now you must be aware that the theme that there is no easy solution keeps popping up. Or, as has been said before, you can't get something for nothing. The United States is very fortunate in having vast reserves of coal. (Four countries, the United States, the Union of Soviet Socialist Republics, the Peoples Republic of China, and Australia have, between them, almost 90 percent of all the known coal reserves in the world.) However, as with all of the other possibilities we have discussed, there are serious difficulties associated with using coal. Some of them are economic in nature. It takes about 10 years and an investment of as much of $100 million to get a large underground coal mine into full production. Such a mine might produce 2 million tons of coal per year for 20 to 30 years; then it would have to be replaced. Underground mines presently produce about half of the U.S. coal. The present consumption of energy in the United States is almost 80 quads per year. Suppose this is allowed to grow a great deal, and that exports of coal increase to the extent that 300 quads per year of coal will be needed by the year 2020. If half of this comes from underground mines, about 6000 million tons of coal per year will be required from underground. In other words, there would have to be about 3000 of those underground mines! Further, if all of them have to be replaced every 30 years on the average, in any given year one-thirtieth of them would have to be replaced. That is 100 new mines each year, at $100 million per mine for startup costs. You figure out the total cost. That is only half the required supply.

The other half of the coal supply will have to come from open-pit strip mines. The figures are similar, but the problems are somewhat different. Strip mining is disruptive to the surrounding area. (Although coal companies have made more of an effort in recent years to fill in the land and reclaim it.) Coal from east of the Mississippi River has a high sulfur content and coal from west of the Mississippi has a low heating value. Also, much of the coal is far from where it is needed. In

a few locations, pipelines are used to transport coal. The coal is made into a *slurry*, by grinding it into small pieces and mixing it with water, and it is pumped through pipelines. One such pipeline is pictured in Figure 12.23. This technique may work, but it will be more expensive than pumping oil or natural gas. (Also, the railroads, which now move most of the coal, will undoubtedly object, and politics will enter into the decisions.) Coal mining can also be dangerous. In addition to accidents, underground coal miners suffer from several occupational diseases, usually lumped under the term *black lung disease*.

However, it is possible that all of these difficulties will be overcome and enough coal will be produced in the United States to provide the energy we "need." Even if that happens, the problems are not solved. For one thing, what will happen to the large fraction of the rest of the world that has little coal? Are we going to be able to produce enough coal to satisfy our huge demands and theirs,

Figure 12.23 This is an operating coal slurry pipeline which carries coal from the mine in Kentucky to barges in the Ohio River. (Photo courtesy of R.L. and J.L. Guernsey.)

as well? This is not likely. What about future generations? Even though we may all be gone by the time the most serious problems arise, I believe we owe our children's children a reasonable world to live in. If, somehow, we could produce all of the coal that seems to be demanded, the problems would not cease. Let us return one final time to the yeast cells working on the grapes to produce wine. You recall that, if the yeast were fed as much sugar as they needed, they died anyway, because they produced "pollution" in the form of alcohol, a poison to themselves. The same sort of thing can happen to humankind. If we had all the "sugar" (coal or other fossil fuels) that we wanted, we could end up "poisoning" ourselves. I do not mean this necessarily in the usual sense of air pollution. It is true that burning coal—especially that from east of the Mississippi—puts large amounts of sulfur dioxide (SO_2) into the air, and that sulfur dioxide combines with water vapor in the air to form sulfuric acid, which is harmful to humans, animals, plants, and even buildings. However, it will probably someday be possible to remove most of the sulfur, before or after burning, and thus to reduce that problem. The same may be true for other pollutants such as carbon monoxide (CO) and particulates (tiny particles in the smoke).

On the other hand, there are problems associated with the burning of coal or any fossil fuel that cannot be solved. For example, if coal were pure carbon and it were burned completely, the only chemical reaction would be

$$C + O_2 \rightarrow CO_2$$

The result is pure carbon dioxide, and it is produced in large quantities whenever a fossil fuel is burned. (In fact, burning a ton of coal produces around *three* tons of carbon dioxide.) Carbon dioxide is not a pollutant. Humans and other animals could stand many times more of it than is now in the air. Plants use carbon dioxide and water to produce their leaves, stems, and fruit, and they also produce pure oxygen in the process. But carbon dioxide can still be dangerous to us. Remember the "greenhouse effect" discussed earlier? Carbon dioxide and water vapor in the atmosphere have the effect of "trapping" the sun's radiation and thus warming the Earth. If it were not for this effect, the Earth would be much colder than it is. Well, the more carbon dioxide there is in the air, the greater will be the trapping effect. The carbon dioxide content of the atmosphere has increased about 10 percent in the past century, due to its production in an industrial society and also due to the decrease in the world's forest area. Although we cannot be certain, this increase in carbon dioxide has probably contributed to a slight increase in the average temperature of the world. If the carbon dioxide content of the air increases greatly in the century to come, that average temperature could increase by two or three degrees.

Well, that doesn't sound too bad. The trouble is that the climate of the world seems to be in a very delicate balance. This is particularly true in the Arctic Ocean, near the North Pole. In the wintertime, the ocean freezes to a thickness of several meters. In the summer, some melting of the ice occurs, and in some areas there will be open water. Still, the ice freezes again the next winter. It is thought that if the average temperature were just a little higher, the melting might be greater. If that happens, two things will occur to further affect the temperature. One is that the greater amount of open water will absorb more energy from the sun than would be absorbed by the ice and snow, and thus it will tend to raise the temperature even more. Also, evaporation from the open water will occur much faster than the rate of evaporation from the ice. Since the water vapor in the air contributes to the greenhouse effect, once again this will tend to raise the temperature. Thus, everything will work together to melt more ice. An increase of carbon dioxide in the atmosphere could then trigger a general and permanent melting of much of the polar ice caps. This would mean that many major cities of the world—those near sea level—would be flooded and great land areas would be lost. That would not mean the end of humankind, but it would work considerable hardship. (For instance, much of the farming area of Holland would disappear under water.)

On the other hand, the particles added to the atmosphere by burning coal have the opposite effect. They reflect and absorb radiation from the sun, causing less to reach the ground, and thereby causing some cooling. It is not known how long these particles stay in the atmosphere, thus how much they contribute to cooling. Nor is it known whether this effect or the heating from the increased carbon dioxide will be greater. One might imagine that, instead of melted polar ice caps, the future might hold another ice age, with ice creeping down into the middle of the United States and into Europe.

One thing is clear: The works of humans are now on a scale large enough to affect the worldwide climate. Already we see local effects near large population centers, with higher average temperatures, smog, and other effects. What the total effect on the worldwide climate will be is not certain, but most of the things we do tend to warm the atmosphere. In addition to increasing the concentration of carbon dioxide, we do such things as change the reflectivity of the Earth's surface. Cutting down forests, plowing, and building cities, generally result in more heat being absorbed and less reflected. Even such activities as the widespread irrigation of farmland has an effect. When the extra water used for irrigation evaporates, it may cool the surface a bit, but the end result is warming. This is because the extra water vapor in the atmosphere adds to the greenhouse effect.

Finally, all of the energy we use, in whatever form, ends up as heat in the atmosphere.

I have just done a little calculation which indicates that the amount of heat we now produce in the United States amounts to something like one-half of one percent (0.5 percent) of the average solar energy absorbed on the land surface. This may not sound like much, but suppose we multiply it by a factor of five or ten in the next century? No one knows just what effect that would have on the worldwide climate, but it could be a serious problem. Note well that this problem exists for *whatever* form of energy is used, even solar. If solar energy is used on a large scale, whether it is in the form of photovoltaic cells, power towers, or another form, the net result will be to increase the amount of solar energy absorbed by the Earth (and decrease the amount reflected back into space). Perhaps this effect will prove to be unimportant. We do not yet know enough about climate to predict all the subtle future effects of our present activities, but, obviously, we need to be careful in planning for the future.

SUMMARY

The message of this chapter is that there is a real and serious energy problem for the world in the not too far distant future. The result could be worldwide disaster but, with careful and thoughtful planning, such a fate can be avoided. However, we must take into consideration every aspect of energy production and use. Simply finding more energy resources and using them as fast as possible will not solve the long-term problems.

Of the fossil fuels, only coal is still very abundant and relatively easy to get. The most immediate shortages of fossil fuels are likely to be of oil, and the United States is already relying heavily on foreign sources of that fuel. Diminishing supplies of natural gas are also likely to occur in the next generation, or so. Even if heretofore unsuspected supplies of fuel are discovered, the world will eventually "run out" of fossil fuels, and alternatives must be sought for future generations. Coal may provide some time to prepare, but neither it nor other fossil fuels, like shale oil, is likely to provide easy or complete solutions to the problems. This is particularly true in light of the needs of underdeveloped nations for much more energy than they presently have access to.

You may have noticed that there has been hardly a mention of nuclear power so far. Some people believe that nuclear power can provide the answer to all of the questions raised in this chapter. Others think that it is the way of disaster. Although we have discussed numerous controversial and emotional subjects thus far, nuclear power tends to be the most controversial and to cause the strongest emotional reactions of all. In the next two chapters we shall take a look at nuclear power and attempt to assess the situation in a rational manner. Those chapters are not intended to persuade you to be either "antinuke" or "pronuke." Rather, it is hoped that, with what you will learn, together with what you have already learned, you will be able to come to some intelligent, unemotional conclusion of your own.

ADDITIONAL QUESTIONS AND PROBLEMS

1. Looking at Figure 12.3, you can see that the consumption of oil in the United States doubled in the period from 1952 to 1970, starting at about 15 quads/year and going to 30. If the growth were to continue this way, growing exponentially with a doubling time of 18 years, what would be the U.S. consumption of oil by the year 2060? Make a graph of this. By counting the squares under the curve, estimate the total amount of oil that will have then been used by the United States. How does this compare to the estimated worldwide reserves of 12,500 quads?

2. Suppose that the worldwide rate of producing coal were to increase roughly exponentially for the next century or so, then to start leveling off, with the peak coming in about 2150. The present rate of producing coal is about 80 quads per year, and that has gradually grown from nearly nothing in 1850. Draw a rough graph of the rate at which coal might be produced for the next few centuries. Assume roughly exponential growth at the beginning. Remember that, because coal will gradually become more difficult to find, the rate of production will go down after it peaks. Assume roughly exponential decay after the peak. Remember, also, that the area under the curve should represent about 200,000 quads, the estimated total reserve. Using this graph, when would you judge that the world will essentially be out of coal?

3. In your own words, briefly summarize the true nature of the energy problem.

4. Why is the loss of about half of the world's forests in the past 25 years a serious loss? What problems might that, and further cutting down of forests, have on the atmosphere? On the world's climate?

5. Using Figure 12.19, name several nations that use more energy than the average to produce a given amount of goods. Name several that are more efficient than average.

6. Figure 12.22 tells us that the future energy "needs" of the world are expected to be greater than the available supply. What are some steps we can start taking now to see that this does not happen?

7. What effects would a new ice age have on your life?

13 Atoms and Nuclei

Less than 100 years ago, near the turn of the twentieth century, many scientists thought that all of the "fundamental" physics was already known. All that was left, it was thought, was to learn additional details. Then suddenly, with the discovery of x-rays in 1895, whole new vistas were opened up. The 50 years following brought a tremendous expansion of knowledge about the physical universe, coming to a climax with the explosion of the first nuclear devices ("atomic bombs") in 1945. Those explosions had for us today two very important messages. First, science and technology may work wonders of which we have not yet even dreamed. Second, those wonders may be for the *benefit* or for the *harm* of society, depending on how we use them. We need to pay attention to those messages, especially the second one. In the case of nuclear power, decisions concerning how and if it is used will depend primarily not on science but on politics, economics, and public opinion. In order to help make decisions that are in the best interests of our society, every citizen needs an understanding of nuclear power. This chapter and the next will help you to get that understanding at some reasonable level. (It is

not necessary to become a nuclear physicist to have enough understanding to make intelligent decisions.) This chapter will introduce you to some of the basic physics involved in nuclear power, and Chapter 14 will deal with nuclear power and its consequences directly.

ATOMIC STRUCTURE

The ancient Greeks understood the idea of the atom and, in fact, the word comes from the Greek *atomos* (meaning indivisible). The idea is that if you have a chunk of a pure substance, with no other substance mixed in, you can divide it in half. Then you can divide one of those pieces into halves, and one of the resulting pieces into halves, and so on. If you could keep this up indefinitely, sooner or later you would get to the smallest possible piece that is still the same substance. If you divide it any further, it will be changed into something else. That smallest piece is the *atom*. We have since refined the idea to include *molecules*. An atom is the smallest piece into which a pure *element*, such as carbon, gold, or uranium, can be broken. However, many substances are composed of chemical combinations of two or more elements. These are called *compounds;* the smallest piece of a compound is a molecule. For example, water is a compound of hydrogen and oxygen. Each molecule of water contains two hydrogen atoms and one oxygen atom.

In this chapter we are mainly interested in atoms rather than molecules. As mentioned earlier, every atom has only three components; they are *protons, neutrons,* and *electrons.* The protons and neutrons are in a very small core, the *nucleus,* and the electrons whiz around the nucleus at very high speeds. An atom is very tiny, with a diameter of about 10^{-10} m. However, small as it is, it is mostly empty space, as the nucleus is only about 10^{-15} m in diameter and the electrons are much smaller than that. In other words, the tiny electrons are running around the nucleus at a distance that is 100,000 times the size of the nucleus itself. If the whole atom was enlarged until the nucleus was the size of a basketball, the nearest electrons would be 13 or 14 miles away.

Since atoms are so tiny, we cannot see them, even with the most powerful microscopes. Thus we have to imagine what they look like. It is useful to use a "model" to help form a mental image of an atom. One model with which you are familiar is like a miniature solar system, with the nucleus at the center and electrons in orbits around the nucleus. Figure 13.1 shows hydrogen, the simplest of all atoms, with one proton in the nucleus and one electron in an orbit. Since the nucleus has a positive charge and the electron has a negative charge, they are attracted to one another, much like a planet and the sun are attracted, but with an electric rather than a gravitational force. (There is a gravitational force between them, also, but it is much, much smaller than the electric force and can be ignored.) What keeps the electron from falling

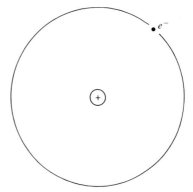

Figure 13.1 A simple model of the hydrogen atom. The single electron is in an "orbit" about the nucleus, which consists of a single proton.

into the nucleus? It is just the fact that it is moving so rapidly; the force pulling it toward the nucleus is the only thing that prevents it from flying off into space.

Figure 13.2 shows three possible orbits for this model of the hydrogen atom. The interesting thing about them is that the farther the electron is from the nucleus, the greater is the energy of the atom. (Remember from an earlier chapter that we must examine the energy of a *system.* In this case, the system is the entire atom, consisting of the nucleus and

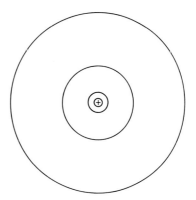

Figure 13.2 Three possible orbits for the hydrogen atom. (The electron can only be in one of these orbits at a particular time, of course.) The larger the orbit is, the greater is the energy of the atom.

the electron.) It is something like putting a satellite into orbit around the Earth. If we want the orbit to be 1000 miles above the surface of the Earth, we must give a particular amount of energy to the satellite to lift it off from the Earth and put it into the orbit. If, however, we want the satellite in an orbit 2000 miles above the surface we must be prepared to give it more energy.

In order to test this model of the atom, it is necessary to observe atoms. Observing a single atom is very difficult, so usually one observes the behavior of a collection of similar atoms. One way to do so is to examine the light given off by collections of similar atoms under different conditions. You may do this in the following activity.

Activity 13.1

For this activity you will need a *diffraction grating* and a shoe box. A diffraction grating is a piece of glass or plastic material with very closely spaced parallel grooves scratched or stamped on it.

With the diffraction grating close to your eye, look through it at a source of light. What do you see? Still looking through it, rotate the grating slowly and observe the result. Try this with a bare incandescent lamp (light bulb), a fluorescent lamp, and any other sources of light you might have handy. What differences do you see when looking at different light sources?

What you are seeing when you look through the grating is several *images* of the light source, each in a different color and each in a slightly different position. There are two sets of images, one to either side or above and below the light source, depending on how you have the grating turned. Mark on the grating frame the directions into which the light is spread. The grating has the property of making the light which comes to your eye seem to be coming from a place different than the source of the light, and each color comes

from a slightly different angle. This may not be too obvious from what you have done thus far, for the different-colored images are not completely separated, and they are smeared out as a result. The effect is shown in an exaggerated way in Figure 13.3.

You can convince yourself that this description is correct by looking through the grating at a moving object, perhaps your hand, under a strong light. The images will be dim and fuzzy with the different colors running into each other, but you will definitely see a moving hand. One way to separate the images is to make the object you are looking at very narrow. You could do this with a thin strip of white paper on a background of a large piece of black construction paper under a strong light. An easier and better way is to use the shoe box to make a thin slit of light to view.

On each end of the shoe box, and near one side, cut a rectangular hole, a centimeter or so on each side, as in Figure 13.4. On the inside of the box, tape the diffraction grating over one of the holes, in such a way that the images will be spread out to the sides, rather than to the top and bottom. Over the other hole, make a narrow, vertical slit for light to enter. You can do this by taping a double thickness of index cards over the hole, leaving

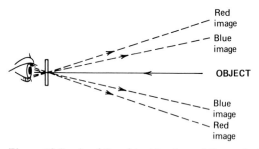

Figure 13.3 A white object is viewed through the diffraction grating. The light passing through the grating is "bent," with the red light being bent more than the blue. The colored images then appear to be in locations to the sides of the object, with the red image being farther away.

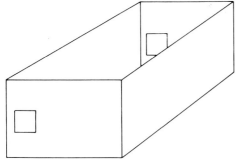

Figure 13.4 Placement of the holes in the shoe box.

a narrow strip of space between them. Be sure that the edges of the slit are straight and parallel to one another. The finished job should be like Figure 13.5. The slit width should not be more than about the thickness of one index card.

Put the cover on the box and look into it, through the diffraction grating and toward the slit. The box provides a dark space for easier viewing of the images of the slit. Look at several sources of light and describe carefully what you see. Try looking, especially, at some neon signs like the one in Figure 13.6, at mercury vapor lamps (the bright bluish lamps which light some streets), and at so-

dium lamps (the bright yellow street lights). You might also try looking at the sky *near* the sun. (Under *no* circumstances should you look *at* the sun, even through a narrow slit. The ultraviolet radiation from it can permanently damage your eyes. Stand in the shadow of a house and with the sun just below the roof line, look at the sky just above the roof line.)

The device you have made and used is called a *spectrometer,* and it is much like those used by research scientists to investigate properties of matter. When you looked at the light from an incandescent lamp or the sky near the sun, you saw a spread of color into a light *spectrum.* The light coming from either of these sources is close to that for a blackbody radiator, as we discussed earlier. However, when you observe the light from a fluorescent lamp, a neon sign, and the mercury or sodium lights, each looks distinctly different from the others. With the fluorescent light, for example, you see a spread of color, but with a few particularly bright bands. These are called *spectral lines.* When you look at a neon sign, this effect is particularly noticeable, with most of the light being distributed into just a few

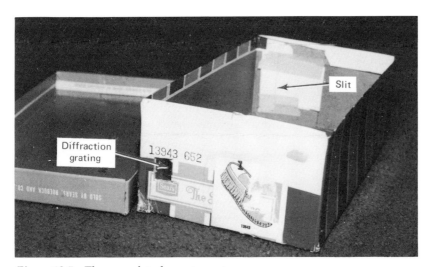

Figure 13.5 The completed spectrometer.

Figure 13.6 Las Vegas has lots of neon signs (and other kinds) to lure the customers. (Owen Franken/Stock, Boston.)

spectral lines, with practically no light between. Were you surprised at some of the colors that appear in that light that looks so red to the eye?

If you could look at the light emitted by pure samples of the various elements, such as neon, sodium, and mercury, you would see that the spectral lines of each are quite different. In fact, no two elements have exactly the same spectrum, and an element may easily be identified by its spectrum. It is sort of a "fingerprint" for elements.

These *spectra* (plural of spectrum) of elements were most puzzling to scientists for many years. For one thing, according to the theory then known, a charged particle, such as an electron, moving in an orbit about the nucleus should lose all of its energy in the form of electromagnetic radiation in a fraction of a second, thus causing the electron to spiral rapidly into the nucleus. Also, the spectral lines could not be explained at all by the theory of orbiting electrons without some additional puzzling assumptions. It required the

development of *quantum mechanics*, an entirely new model of atoms, to explain the experimental results. According to quantum mechanics, which was developed rapidly over a four- or five-year period in the 1920s, the model of a hydrogen atom still consists of a nucleus and an electron. However, the electron, instead of being in an orbit like a satellite, is whizzing about the nucleus in an undefined way. Even with perfect measuring instruments it is impossible to locate the electron and measure its velocity exactly. The best that can be done is to specify a probability that the electron will be a given distance from the nucleus at any specific time. This is *not* because the experiments are not precise enough; it is a fundamental property of matter on the atomic (small) scale. That is, according to quantum mechanics no experiment will ever measure the location and velocity of an electron exactly at a given instant of time without disrupting the atom itself. This may be a bit hard to swallow but, as far as we now know, that is the way things are.

Figure 13.7 shows the probability of finding the electron at any given distance from the nucleus for the lowest energy state of the hydrogen atom. The most likely place for the electron to be is at the peak of the curve, where the probability is the highest. (This is at a distance of about 5×10^{-11} m from the nucleus.) It is less likely to be at other distances, but the probability (except at the nucleus) is greater than zero, so the electron will spend some of its time at other distances. The probability decreases farther from the nucleus, but it goes to zero only infinitely far away. That means that a small fraction of the time the electron could be quite far away from the nucleus to which it is attached.

Have you ever seen a cartoon in which one of the characters moved in a particular space, faster and faster, until he was nothing but a blur? The quantum mechanical picture of the electron in the hydrogen atom is something like this. The electron moves around the nucleus at very high speeds, giving us a blurred, fuzzy picture. This is often called an *electron cloud,* and it is thickest where the probability of finding the electron is the greatest. Figure 13.8 shows a rough representation of the electron cloud surrounding a hydrogen

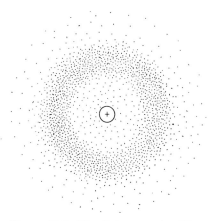

Figure 13.8 The "electron cloud" surrounding a hydrogen nucleus. The electron could be anywhere in the cloud. The probability that it will be at a particular distance from the nucleus is given by Figure 13.7.

nucleus in the lowest energy state. The heavier the shading, the greater the probability of finding the electron at that distance.

This electron cloud is not the only one possible for the hydrogen atom. The atom can be given a greater total energy, perhaps by being heated or by being hit by another electron which transfers some energy in the collision. This new cloud will be larger than the old one, and it may be either *spherical*

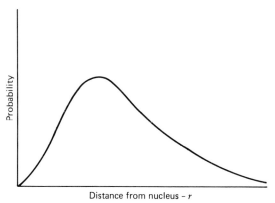

Figure 13.7 The lowest energy state of a hydrogen atom. The curve represents the probability that the electron is a given distance from the nucleus at any instant of time.

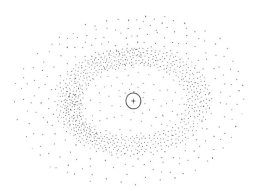

Figure 13.9 An ellipsoidal electron cloud surrounding the hydrogen nucleus. The energy of the atom in this cloud is different than that of the same atom in the circular cloud.

(circular) or *ellipsoidal* (like a football) in shape. The ellipsoidal shape is indicated in Figure 13.9.

ENERGY LEVELS

Let us continue with the investigation of the hydrogen atom. The total energy of the atom is greater for larger electron clouds and less for smaller electron clouds. The energy is also slightly different for electron clouds that are about the same size but different shapes (i.e., spherical or ellipsoidal). In quantum mechanics, the energy and several other properties are specified by a set of numbers called *quantum numbers*. We shall be interested in only one of these quantum numbers. It is given the symbol "*n*" and called the *principal quantum number*. It specifies the size of the electron cloud (or the size of the circular orbit, using the old model of the atom). When $n = 1$ the cloud is smallest, and the atom has the least possible energy. This is called the *ground state* of the atom. For the spherical electron clouds we can write down the distance of maximum probability. That is, what the most likely distance of the electron from the nucleus is. That distance is given by

$$r_n = kn^2$$

where $k \approx 5.3 \times 10^{-11}$ m. Thus, for $n = 1$, $r_1 \approx 5.3 \times 10^{-11}$ m, for $n = 2$, $r_2 \approx 2.1 \times 10^{-10}$ m, $r_3 \approx 4.8 \times 10^{-10}$ m, and so on. The quantum number n can only be a whole number: 1, 2, 3, 4, ... It cannot be, for example, 3.5. This, in fact, is a fundamental result of quantum mechanics: Only certain states are allowed. States in between cannot exist at all. The ones that exist are those for which n is a whole number. These are called *discrete states*.

What we are really interested in, however, is the energy of the atom. The energy also is given by the principal quantum number, but before writing down the equation,

let's define carefully what is meant by the energy of the atom. The energy will be the total energy, both kinetic and potential. It is possible to remove the electron completely from the atom (to *ionize* the atom). In order to do so, work must be done *on* the atom from the outside to pull it apart. That means that the complete atom has less energy than the separated, or ionized, atom. However, it is customary to call the energy of the ionized atom zero. Therefore, the energy of the complete atom is always less than zero, or negative. Also, the joule, or any other unit of energy we have discussed thus far, is much too large to conveniently describe the energy of atoms. We need a smaller unit. The unit usually used is the *electron volt* (eV). Its relation to the joule is

$$1 \text{ eV} = 1.6 \times 10^{-19} \text{ J}$$

or

$$1 \text{ J} = 6.2 \times 10^{18} \text{ eV}$$

So this is a very small unit indeed.

EXAMPLE

If an electron is placed in an electric field and allowed to move through a potential difference of 1 V, how much work is done on it by the field? What will be its change in potential energy in the process? What will be its change in kinetic energy?

Solution

An electron is a charged body, with a negative charge of about 1.6×10^{-19} C. If an electron is placed in an electric field, it will experience a force. If the electron is then allowed to move, the electric force will do work on it, and it will gain kinetic energy.

Suppose that the electric field strength is $E = 1$ volt/meter. Then the electron would have to move 1 m along the lines of force (in a direction opposite that of the electric field) to be accelerated through a potential difference of 1 V. If the electric field strength were 10 V/m, then it would have to move only 0.1 m to experience the same potential difference. As you saw in Chapter 10, the work done on the charge is the potential difference through which it moves times the electric charge:

$$W = qV = 1.6 \times 10^{-19} \text{ C} \times 1 \text{ V}$$

Since a volt is a joule per coulomb (J/C), the units come out to be joules, as they must:

$$W = 1.6 \times 10^{-19} \text{ J}$$

Another way of looking at this is that the potential energy of the *system* (charge and electric field) decreases as the charge moves through a potential difference. That is:

$$\Delta PE = -1.6 \times 10^{-19} \text{ J}$$

where the change in potential energy is just the negative of the work that would have to be done on the charge to move it back to its original position. Finally, since this is a conservative system, the change (increase) in kinetic energy is just the negative of the change (decrease) in potential energy:

$$\Delta KE = -\Delta PE = 1.6 \times 10^{-19} \text{ J}$$

As you can see by comparing the numbers, an electron volt is simply the work done on a charge of one electron moving through a potential difference of 1 V.

Now we are in a position to write down the energy of the hydrogen atom. In electron volts, it is

$$E_n = -\frac{13.6 \text{ eV}}{n^2} \tag{13.1}$$

That is, the *only* energy states allowed are those given by this equation, where n is a whole number. These are called *discrete energy states* or *energy levels*. Figure 13.10 is an *energy-level diagram* for hydrogen, showing the various possible energies. All of the energies are negative, with the lowest energy state (the *ground state*) being at -13.6 eV. The higher energy levels are called *excited states*. If we were to look at the other quantum numbers, specifying the ellipticity of the electron cloud, the behavior of the atom in a magnetic field, and other characteristics, each of the energy levels shown would be split into several very closely spaced levels. This will not be necessary for our purposes, so we shall use just the energy levels shown.

Now that you know about discrete energy levels for the hydrogen atom (or any atom, for that matter), there is just one more important fact to be pursued. When an atom is in an excited state—with an energy higher than the ground state energy—it can sud-

Figure 13.10 The principal energy levels of hydrogen. Each level has "fine structure," which causes it to have several closely spaced levels, but those are not shown here.

denly jump to a lower energy state. That is, the total energy of the atom suddenly goes from a higher to a lower value; the atom loses energy. Where does the lost energy go? If we believe in the law of conservation of total energy (and we do, we do!), then the "lost" energy must appear in another form. In this case, it appears as a small bit of light energy, called a *photon*. The frequency of the photon depends on its energy according to the equation

$$E = hf$$

Where E is the energy of the photon, f is its frequency, and h is a constant called the *Planck constant*. It is equal to about 4.1×10^{-15} eV-seconds.

EXAMPLE

If a hydrogen atom changes suddenly from the $n = 3$ energy state to the $n = 2$ state, what is the frequency of the photon emitted as a result?

Solution

The energy of the photon is just the difference in the energies of the two states of the atom. That is:

$$E = E_3 - E_2$$

$$= -13.6 \text{ eV} \times \left(\frac{1}{3^2} - \frac{1}{2^2} \right)$$

$$= -13.6 \text{ eV} \times \left(\frac{1}{9} - \frac{1}{4} \right) = 1.9 \text{ eV}$$

Thus the photon has an energy of about 1.9 eV. Its frequency is given by

$$f = \frac{1.9 \text{ eV}}{4.1 \times 10^{-15} \text{ eV-s}}$$
$$= 4.6 \times 10^{14} \text{ Hz}$$

From its frequency we know that this photon is a tiny bit of red light.

Figure 13.11 shows some of the energy jumps that can occur for hydrogen and the frequency of the light emitted as a result of each.

Now, we finally have an explanation of what is seen with the spectrometer. Remember that when looking at a large collection of similar atoms, such as neon, the light emitted is not a continuous spectrum. Most of the colors are missing, and only a few bright lines appear. They are called *discrete lines*. Those lines correspond exactly to the light produced by jumps in energy levels. In the case of a neon sign or a fluorescent tube, the needed

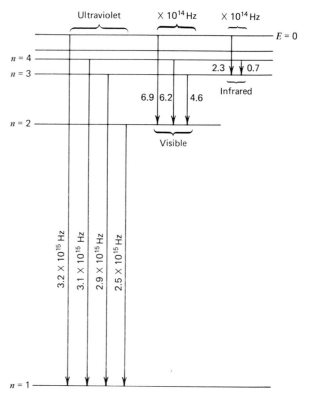

Figure 13.11 Some possible energy jumps of the hydrogen atom, showing the frequencies of the resulting photons.

energy is supplied continuously to the atoms by an electrical discharge in the tube. Since there are huge numbers of atoms in the tube, many photons are produced, giving a continuous stream of light with the several different allowed frequencies.

As you may deduce from Figure 13.11, hydrogen also has a spectrum. If you can, take a look at a sample of hydrogen energized by an electrical discharge using your shoe-box spectrometer. The spectrum you will see looks something like Figure 13.12. (Naturally, you can see only the visible part of the spectrum.) For the sake of comparison, the spectra of sodium, mercury, hydrogen, and helium are shown in Color Plate 2. You can see that each has its own distinctive "signature."

Here is one final word on this topic, to clear up a question that may have come to your mind. When you observe the spectrum of a fluorescent lamp, there seems to be a continuous spectrum, with one or two discrete spectral lines overlaying it. In this case, you are observing light that originates from mercury atoms excited by an electric discharge. The mercury vapor gives off light, much of it in the ultraviolet range, in a discrete spectrum. Most of that light is absorbed by the coating on the inside of the glass, a *fluorescent* material. The fluorescent material has the property that it is able to absorb an ultraviolet (high-energy) photon and release the energy in several jumps, thus releasing several lower energy photons. Those lower energy photons are in the visible range. The

fluorescent material is a complex compound, with numerous energy levels. Thus, the light emitted has a wide range of wavelengths, and it appears to be a continuous spectrum. The bright lines come from the visible part of the mercury spectrum, some of which passes through the fluorescent material and the glass.

THE PERIODIC TABLE

Thus far we have examined only the hydrogen atom because it is the simplest and the only one easily analyzed. The single electron accounts for the chemical and optical properties of hydrogen. The next simplest atom is helium, having two protons in the nucleus and two electrons surrounding it. There are also usually two neutrons in the nucleus. This is indicated in Figure 13.13. The chemical symbol for helium is "He". (For hydrogen, it is "H".)

When examining helium or more complex atoms, the theoretical analysis of the atom using quantum mechanics suddenly becomes *much* more difficult and the calculations become less precise. The electrons still account for the chemical, thermal, and optical properties of the atom, but those properties are

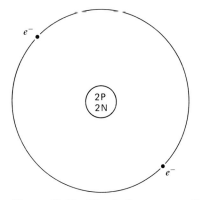

Figure 13.13 The helium atom. (In all of these drawings, the nucleus is drawn much too large, to enable you to see it.)

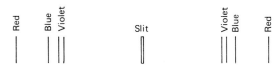

Figure 13.12 Part of the visible spectrum of hydrogen, as seen with the shoe-box spectrometer. There is a complete spectrum on each side of the slit.

very different for hydrogen and for helium. Hydrogen is a very active element. For example, if mixed with oxygen and ignited, it burns or explodes. Helium, on the other hand, is very stable. It is almost impossible to make it react chemically with any other element. For this reason, and because it is a gas at room temperature, helium is called an *inert gas* (also known as a *noble gas*).

The next element, lithium (Li), which is shown in Figure 13.14, is very active. It has some properties similar to those of hydrogen. An interesting thing has happened here: The third electron is not with the first two. The first two both have a principal quantum number, $n = 1$. The third electron has $n = 2$. (Incidentally, we are talking only about *neutral* atoms in the *ground state*. That is, the number of electrons is the same as the number of protons, and the atom has its lowest possible total energy.) If you were to examine all of the elements, you would find that all of them except hydrogen have two electrons in the $n = 1$ "shell." The $n = 1$ shell seems to be filled with as many electrons as it can accept when it has two. This helps to explain why helium is so stable: The two electrons are attracted by two protons and so they are firmly attached (*bound*) to the atom. It is difficult to remove one of the electrons from the atom (ionize it), and it is impossible to add an elec-

tron to the shell occupied by those two. Lithium also has a full $n = 1$ shell, and its third electron is in the $n = 2$ shell. The two negative electrons in the $n = 1$ shell partially "shield" the outer electron from the influence of the positive nucleus, so it is not tightly bound. The third electron can be rather easily removed, or an additional electron can move into the $n = 2$ shell, so the atom is very active chemically.

Moving on, elements 4, 5, and 6 are beryllium (Be), boron (B), and carbon (C). It is not until we get to number 10, neon (Ne), that some familiar properties are found. Neon, like helium, is an inert gas. It is very stable chemically, and has its electrons arranged as shown in Figure 13.15. Once again, an examination of all of the elements would reveal a filled shell: the $n = 2$ shell can accept only eight electrons.

The next element, sodium, is shown in Figure 13.16. It has just one electron in the outermost shell, and it has properties remarkably similar to those of lithium. In fact, it was noticed long ago that when going through the elements in order, some chemical and physical properties seemed to repeat *periodically*. All of the ele-

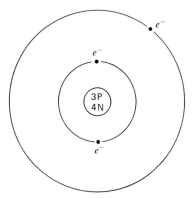

Figure 13.14 The lithium atom.

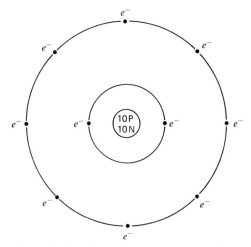

Figure 13.15 The neon atom.

ments were then arranged into a *periodic table* like the one in Figure 13.17. In each block, the large print in the center is the chemical symbol of the element and the number at the top is the number of protons in the nucleus. (This is called the *atomic number*.) We will talk about the number at the bottom of the block later.

In the periodic table, the elements are arranged so that elements in any column have similar properties. For example, the elements below hydrogen in the leftmost column are *alkali metals,* and they are very similar chemically and physically. They are all very active elements. The elements in the rightmost column are all inert gases.

If you were to examine the electron structure for each element in the periodic table, you would see a pattern in the filling of the electron shells. Although the periodic table was invented over 100 years ago, it remained for quantum mechanics to explain it. Without going into details, we can understand the explanation in terms of two principles. One is called the *exclusion principle,* and it states that in the same atom, no two electrons can be identical. Saying this in a different way, no

two electrons in the atom can have all of the same quantum numbers. This turns out to place a limit on how many electrons can have the same principal quantum number, n. As indicated already, only two electrons may have $n = 1$, so there are two electrons, at most, in the $n = 1$ shell. For the $n = 2$, $n = 3$, $n = 4$, and $n = 5$ shells, the corresponding numbers are 8, 18, 32, and 50 electrons, maximum.

However, a close examination of the periodic table will show that the $n = 5$ shell never gets as many as 50 electrons (32 is the maximum), and that the $n = 6$ and $n = 7$ shells do get some electrons in them in the heavier elements. This is a result of the second principle: A neutral atom will assume the lowest energy possible. For detailed reasons we need not go into, the $n = 6$ and $n = 7$ shells have some energy levels that are at lower energies than the higher levels of the $n = 5$ shell. In fact, as we move through the periodic table, the $n = 3$ and $n = 4$ shells are the first to exhibit this behavior. The $n = 3$ shell starts filling at sodium (11 electrons), and continues through argon (18). Then the $n = 4$ shell accepts the next two electrons for potassium (19) and calcium (20) before the $n = 3$ shell finishes filling up with scandium (21) through zinc (30).

Applying these two principles in detail has led to a complete explanation the structure of the periodic table. In fact, by seeking elements to fit in places where gaps existed in the periodic table, elements previously unknown in nature have been discovered.

ISOTOPES

The number at the bottom of each block in the periodic table is also of interest. It relates to the number of *nucleons* (protons plus neutrons) in the nucleus and it is called the *atomic mass*. Note that this number is not generally a whole number. This results, in part, because

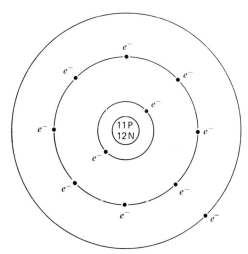

Figure 13.16 The sodium atom.

PERIODIC TABLE OF THE ELEMENTS

Transition Elements

Period	Group I	II												III	IV	V	VI	VII	0
1	1 H 1.00797																		2 He 4.0026
2	3 Li 6.939	4 Be 9.0122												5 B 10.811	6 C 12.01115	7 N 14.0067	8 O 15.9994	9 F 18.9984	10 Ne 20.183
3	11 Na 22.9898	12 Mg 24.312												13 Al 26.9815	14 Si 28.086	15 P 30.9738	16 S 32.064	17 Cl 35.453	18 Ar 39.948
4	19 K 39.102	20 Ca 40.08	21 Sc 44.956	22 Ti 47.90	23 V 50.942	24 Cr 51.996	25 Mn 54.9380	26 Fe 55.847	27 Co 58.9332	28 Ni 58.71	29 Cu 63.54	30 Zn 65.37		31 Ga 69.72	32 Ge 72.59	33 As 74.9216	34 Se 78.96	35 Br 79.909	36 Kr 83.80
5	37 Rb 85.47	38 Sr 87.62	39 Y 88.905	40 Zr 91.22	41 Nb 92.906	42 Mo 95.94	43 Tc (99)	44 Ru 101.07	45 Rh 102.905	46 Pd 106.4	47 Ag 107.870	48 Cd 112.40		49 In 114.82	50 Sn 118.69	51 Sb 121.75	52 Te 127.60	53 I 126.9044	54 Xe 131.30
6	55 Cs 132.905	56 Ba 137.34	57–71 *	72 Hf 178.49	73 Ta 180.948	74 W 183.85	75 Re 186.2	76 Os 190.2	77 Ir 192.2	78 Pt 195.09	79 Au 196.967	80 Hg 200.59		81 Tl 204.37	82 Pb 207.19	83 Bi 208.980	84 Po (210)	85 At (210)	86 Rn (222)
7	87 Fr (223)	88 Ra (227)	(89–103) †	(104)	(105)	(106)													

*Lanthanide rare-earth elements

57 La 138.91	58 Ce 140.12	59 Pr 140.907	60 Nd 144.24	61 Pm (145)	62 Sm 150.35	63 Eu 151.96	64 Gd 157.25	65 Tb 158.924	66 Dy 162.50	67 Ho 164.930	68 Er 167.26	69 Tm 168.934	70 Yb 173.04	71 Lu 174.97

† Actinide rare-earth elements

89 Ac (227)	90 Th 232.038	91 Pa (231)	92 U 238.03	93 Np (237)	94 Pu (242)	95 Am (243)	96 Cm (245)	97 Bk (249)	98 Cf (249)	99 Es (254)	100 Fm (252)	101 Md (256)	102 No (254)	103 Lw (257)

Key:

26 — Atomic number (Z)
Fe — Element symbol
55.847 — Atomic mass of the naturally occurring isotopic mixture; for the elements that are naturally radioactive, the numbers in parentheses are mass numbers of the most stable isotopes of those elements

Figure 13.17 The periodic table of the elements. (From *General Physics With Bioscience Essays*, J.B. Marion, John Wiley and Sons, 1979.)

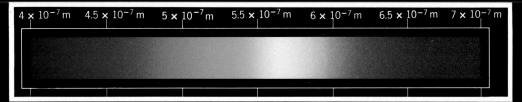

Plate 1. The spectrum of visible light. Light at a wavelength of 4×10^{-7}m is violet, and at 7×10^{-7}m it is red.

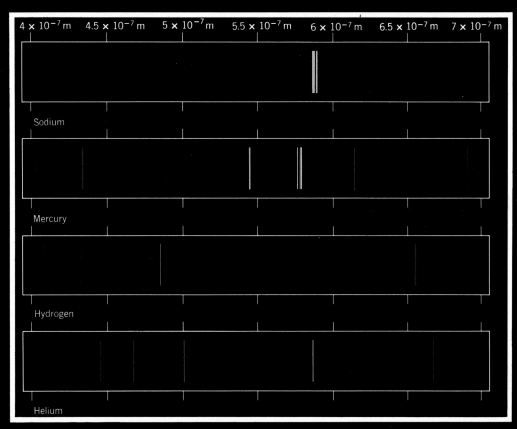

Plate 2. Comparison of the spectra of four different elements. Each element has its own distinctive "signature."

Plate 3. A Landsat 4 Thematic Mapper image of the Washington, D.C. area. Taken from a satellite, this "false-color" image portrays information in different wavelength bands as different colors. The red portions of the image result from the reflection of infrared radiation from vegetation, such as trees and grass. The blue portions are mostly reflections from buildings and other structures. By comparing this image with a map of the Washington area, many well-known features, such as the Washington Monument and the Pentagon can be picked out. (Image courtesy NOAA/NASA/EROS Data Center.)

two atoms of the same element may have different numbers of neutrons. When this occurs, the two are called *isotopes*. For example, naturally occurring lithium (Li) has three protons and either three or four neutrons in the nucleus. That is, there are two natural isotopes of lithium. (There are also three artificially made isotopes of lithium.) Lithium atoms with four neutrons comprise about 92.6 percent of the lithium found in nature, and lithium atoms with three neutrons are the other 7.4 percent. The two isotopes are called lithium-6 and lithium-7, indicating the total number of nucleons in each nucleus.

The symbols used for these two isotopes are

$$_3Li^6 \text{ and } _3Li^7 \quad \text{or} \quad Li^6 \text{ and } Li^7$$

The number at the upper right of the chemical symbol is the *atomic mass number,* the total number of nucleons. The number at the lower left is the *atomic number,* the number of protons in the nucleus. It isn't absolutely essential to write the lower number in, for the chemical symbol itself gives the same information. (That is, Li always has three protons.) To find the number of neutrons, just subtract the protons from the nucleons: There are four neutrons in the nucleus of Li^7 and three in Li^6.

It is customary to give letter symbols to the number of nucleons as follows:

Z = number of protons
N = number of neutrons
A = number of nucleons $(Z + N)$

Li^7 has $Z = 3$, $N = 4$, and $A = 7$, whereas Li^6 has $Z = 3$, $N = 3$, and $A = 6$.

Li^6 and Li^7 are identical atoms chemically. That is, they behave exactly the same way in chemical reactions. This means, among other things, that they cannot be separated chemically. When natural lithium is purified, it re-

mains a mixture of 92.6 percent Li^7 and 7.4 percent Li^6. Separating them requires a *physical* (rather than chemical) process which takes advantage of their difference in mass. One method is to shoot ionized atoms through a magnetic field. Since the ions are moving charged particles, the magnetic field causes the paths they take to curve. However, because Li^6 has less mass, its path will curve more than that of Li^7, and the two can be collected in different locations. The machine for doing this, shown schematically in Figure 13.18, is called a *mass spectrometer*. It is an important tool for separating the isotopes of uranium to make fuel for nuclear reactors. It is also useful for making precise measurements of nuclear masses.

Perhaps you noticed that I have used the term *nuclear mass* as well as *atomic mass*. This is not done in a deliberate effort to confuse you. The two are slightly different, and it is important to know which you are discussing. The nuclear mass is just the mass of the nucleus. The atomic mass (also called *atomic weight*), however, is the mass of the nucleus

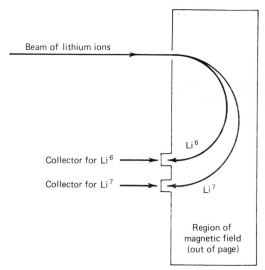

Figure 13.18   In a mass spectrometer, the heavier ionized atom moves in a circular path larger than the path of the lighter ionized atom.

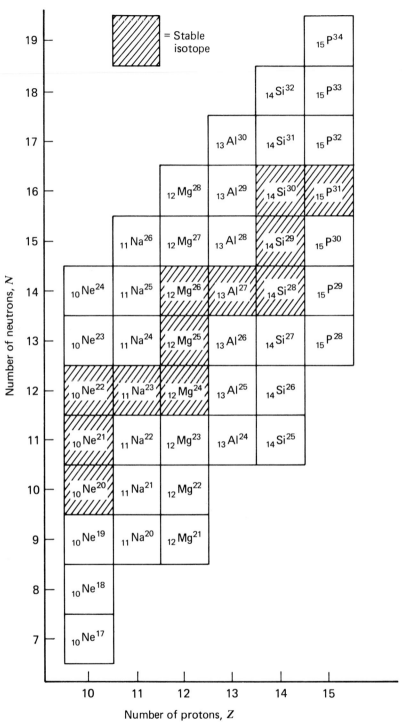

Figure 13.19 A partial chart of the nuclei. Each column contains the isotopes of a given element. The shaded boxes are stable (nonradioactive) isotopes.

plus the mass of the electrons in a neutral atom, and it will thus be slightly larger than the nuclear mass. The atomic mass is nearly equal to the number of nucleons in the nucleus but, except for carbon-12 (C^{12}), not exactly equal. That is because C^{12} is arbitrarily defined to have an atomic mass of exactly 12 *atomic mass units* (amu). The masses of all of the other atoms are measured by comparing them to C^{12}. Using this definition, 1 amu = 1.66057×10^{-27} kg, a very small unit of mass, indeed.

However, if you will notice, even the atomic mass of carbon is not given as 12.00 amu in the periodic table. That is because the number in the periodic table is an *average* atomic mass of the naturally occurring isotopes. Carbon occurs in nature as about 98.89 percent C^{12} and 1.11 percent C^{13}. Thus, the average atomic mass is just slightly over 12 amu.

Figure 13.19 is a chart of a few *nuclei* (plural of nucleus). It is something like a graph, with the atomic number, Z, running from left to right, and the number of neutrons, N, running from bottom to top. All of the nuclei in a single column are isotopes of the same element. For example, magnesium (Mg) has eight isotopes. Of the eight, the three that are in shaded blocks occur in nature. The others are artificially made. Later, we will look at some isotopes of the heavy elements that are important in nuclear power.

RADIOACTIVITY

Activity 13.2

For this activity you will need an old wristwatch or alarm clock, a pack of instantly developing black and white film (such as Polaroid type 667 film), and a camera for processing the film.

The clock or watch should be one of the old-fashioned kind that glows in the dark. It should have been made 20 or more years ago, because radium was then used to paint the

dials, and this has been illegal for some years. (One could also use a sample of a slightly radioactive mineral. This can be purchased from a scientific supply house or found around almost any science department.)

Working in dim light, remove the film pack from its wrapper, but be sure that the film is still protected from exposure to light. Place the film pack on a level surface, film side up. (That is, with the side that would normally face the front of the camera up.) Gently place the watch, face down, on the film pack and tape it there. If you wish, you could position a metal object, such as a key, so that it partly shields the film pack from the watch. If you have more than one radioactive object, you may tape several on the same film pack. If you have any suspicion that light from the watch dial could reach the film, wrap the watch in paper before laying it on the film. Leave this setup in a dark place (e.g., in a closet, or a shoe box) where it will not be disturbed for about two days.

At the end of two days, run the film through the camera, following the directions, to develop it. Do not expose the film to the light while doing so. If the only way to develop the film is to press the camera shutter release each time, hold a folded handkerchief firmly over the lens so that no light enters the camera when the shutter opens. Develop all of the film, numbering each sheet as it comes out of the camera.

What you have observed here is *radioactivity*. The exposure of your film was very similar to the accidental discovery of radioactivity near the end of the last century. Obviously, something is penetrating through the protective covering of the film pack, and through the layers of film and paper, as well. Figure 13.20 shows three photos resulting from exposure of a film pack. Your results probably look quite similar except that you may not have had as many sources available. It seems

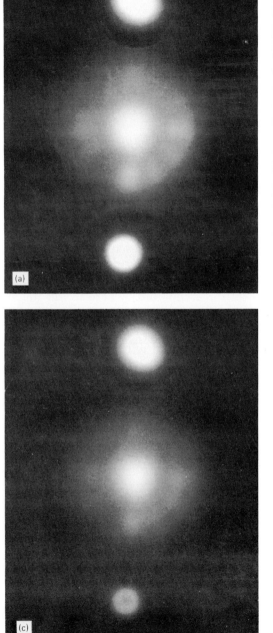

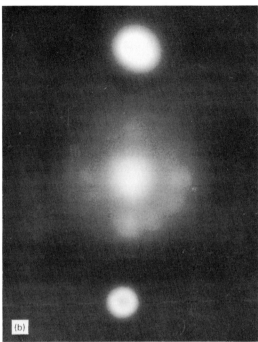

Figure 13.20 A film pack exposed to some radioactive sources: strontium-90 at the top, a watch with a radium dial at the center, and cesium-137 at the bottom. (*a*) Top sheet in the pack. (*b*) Middle sheet. (*c*) Bottom sheet. Note that the beta particles from the strontium source are more penetrating (more energetic) than those from the cesium source.

that whatever is penetrating the paper cover also penetrates the layers of film and paper. However, it does not penetrate completely. The bottom layer is noticeably less exposed than the top layer, so some of the radiation must have been stopped in the layers of film and paper. (Incidentally, the main reason that watches ceased to be manufactured with radium dials was to protect the workers who painted the dials with a radium-containing paint. The finished watch releases only a very small amount of radiation, so the government has made no effort to recall watches already in existence. There will be a discussion of the health hazards of radiation in the next chapter.)

Experiments which followed the discovery of radioactivity soon showed that these radiations came from a number of naturally occurring minerals. Further, it was discovered that there were three different kinds of radiation. This can be determined by passing the radiation through a magnetic field. When this is done, it is found that the path of one type of radiation is curved in one direction, the second is curved in the opposite direction, and the third is not curved at all. Thus, one must have a positive charge, one a negative charge, and the third no charge. It could also be determined that the positive ones had much more mass than the negative ones, and the neutral ones had no detectable mass. Since their nature was otherwise unknown at first, they were called by the first three letters of the Greek alphabet: *alpha rays(α), beta rays(β),* and *gamma rays(γ).* These names are still in use.

We now know that radioactivity comes from the disintegration (break-down) of a nucleus. The alpha rays turned out to be particles with two protons and two neutrons ($Z = 2$, $A = 4$). Thus, an alpha particle is just a helium nucleus. (That is, it is a "doubly ionized" helium atom—an atom with both electrons stripped away.) The beta rays are nothing more than electrons, and the gamma rays are the same electromagnetic radiations we have already called by that name. The alpha particles do not penetrate materials very well, being large and massive. The beta particles penetrate far more readily. The gamma rays penetrate easiest of all, and it may take a lead shield many inches thick to stop them. Thus, the radiation you observed in the activity must have been beta particles: alpha particles would not penetrate the protective paper cover; gamma rays, if present, would pass through the film and paper with little effect.

Questions

Suppose you would like to determine for certain what radiation is coming from your radioactive watch. One thing you could do would be to "collimate the beam." That is, you could put the watch into a thick lead box with just one hole in it, as indicated in Figure 13.21. Using the collimated beam, some film, and the strong magnet you used earlier, describe how you would determine the type of radiation emerging. Suppose that both beta and gamma radiation are present, how would you find out if both exposed the film? How would you make some measure of the penetrating ability of the radiation?

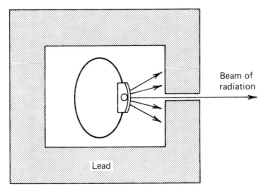

Figure 13.21 "Collimating" radiation. Only a thin beam can escape the lead box.

RADIOACTIVE DECAY

When a nucleus disintegrates (*decays*), what is left? This will depend, of course, on the radioactive particle that is emitted. For example, radium-226 ($_{88}Ra^{226}$) decays by emitting an alpha particle. That leaves a nucleus with two fewer protons ($Z = 86$) and two fewer neutrons ($N = 136$) than the radium. A look at the periodic table will tell you that this new nucleus is radon, and since this "daughter" nucleus has four fewer nucleons than the original "parent" nucleus, it will be radon-222 ($_{86}Rn^{222}$). All of this information can be written in a simple equation:

$$_{88}Ra^{226} \rightarrow \, _{86}Rn^{222} + \, _2He^4 + energy$$

There are two important things to note in this equation. First, there is a *conservation of nucleons.* That is, there are 226 nucleons on the left side of the equation and 226 nucleons on the right side. On the left side, they are all in the radium nucleus; on the right side they are divided between the radon and the helium. There is also *conservation of charge*; in this case there are 88 positively charged protons on each side of the equation.

The second thing to note is that energy is produced as a result of the radioactive decay. Most of the energy appears as kinetic energy of the alpha particle, with a little going into kinetic energy of the daughter nucleus. But, in this decay as in many others, a gamma-ray photon often is also emitted, carrying energy with it. This is explained by the fact that nuclei, like atoms, have discrete energy levels. After the alpha particle is emitted from the radium nucleus, the daughter radon nucleus may be left in an excited energy state. After a very short time, the protons and neutrons in the nucleus rearrange themselves, dropping to the ground state of energy and emitting a photon in the process. The reason this is a gamma-ray photon instead of visible or ultraviolet light is that its energy is much

greater. The energies involved in nuclear decays are typically millions of times greater than those involved in electron energy-level jumps so they are expressed in terms of *millions of electrons volts* (MeV). The two ways in which $_{88}Ra^{226}$ can decay are shown in Figure 13.22. In one case, the radon nucleus is created in the ground state and the emitted alpha particle has a kinetic energy of 4.78 MeV. In the other case, the radon nucleus is in an excited energy state and the alpha particle has an energy of 4.59 MeV. The additional 0.19 MeV is very soon carried away by the gamma ray.

The other type of radioactive decay is the emission of a beta particle—an electron. This is shown in the following equation:

$$_{82}Pb^{214} \rightarrow \, _{83}Po^{214} + e^- + 0.7 \text{ MeV}$$

In this equation nucleons are conserved—214 on each side of the equation—but the number of protons has increased from 82 to 83! How-

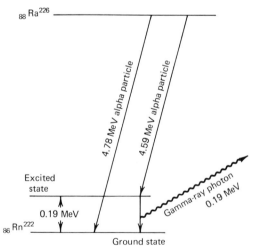

Figure 13.22   A radium-226 nucleus may decay directly to the ground state of radon-222, in which case the emitted alpha particle will have a kinetic energy of about 4.78 MeV. Or it may decay to the excited state, giving about 4.59 MeV to the alpha particle and 0.19 MeV to a gamma ray which is emitted.

ever, charge is also conserved: There are 82 protons on the left side of the equation and there are 83 protons and 1 electron, for a net charge of $+ 82$, on the right side. A neutron has decayed into a proton and an electron, as in the equation:

$$_0N^1 \rightarrow \,_1P^1 + \,_{-1}e^0$$

or, using the more common symbol for the electron:

$$_0N^1 \rightarrow \,_1P^1 + e^-$$

Note that there is one nucleon and zero net charge on each side of the equation. A free neutron is a radioactive particle, and it decays as shown by the equation. In a nucleus the neutrons are usually stable and do not decay but, under the right conditions, the decay occurs with the emission of an electron (beta particle) from the nucleus.

The radioactive decay equation shown above for $_{88}Ra^{226}$ is part of a *radioactive decay chain*. In such a chain, one radioactive element decays into another one, which then decays into another, and so on. There are four chains of this sort, but only one is of interest to us. It starts with uranium-238 and ends with lead-206, which is stable and does not decay. The entire chain is shown in the chart of Figure 13.23. This chart is just like that of Figure 13.19 except that it shows only the nuclei in the chain. Starting with uranium-238, which decays to thorium-234, each nucleus decays in turn. Some of the decays—those jumping two spaces down and two spaces to the left on the chart—are alpha decays. The ones that jump one space down and one space to the right are beta decays. The numbers in MeVs on the chart are the energies of the particles thrown off in the decay processes. For example, when radon-222 decays it emits an alpha particle with a kinetic energy of 5.49 MeV. Since this whole chain starts with uranium-238, a pure sample of that element would

soon have atoms of elements in the entire chain mixed in with the uranium. How soon? We shall soon see.

Questions

What are the equations for the jumps in the chain of Figure 13.23 from polonium-218 and from bismuth-210?

EXAMPLE

What are the equations showing the decays leading from bismuth-214 to lead-210?

Solution

Note that, at bismuth-214 there is a *branch* in the chain. That is, the $_{83}Bi^{214}$ nucleus can decay either by alpha or beta emission. Therefore, there are two paths for transforming bismuth-214 to lead-210. For the path passing though polonium-214, the first decay is a beta decay. The equation is

$$_{83}Bi^{214} \rightarrow \,_{84}Po^{214} + e^- + 1.51 \text{ MeV}$$

The next decay is by means of an alpha particle, giving the equation

$$_{84}Po^{214} \rightarrow \,_{82}Pb^{210} + \,_2He^4 + 7.69 \text{ MeV}$$

or, putting the two together:

$$_{83}Bi^{214} \rightarrow \,_{82}Pb^{210} + e^- + \,_2He^4 + 9.20 \text{ MeV}$$

Questions

What are the equivalent equations for the other path from bismuth-214 to lead-210? Note that the total energy is different from that of the first path. If this were accounted for by the fact that a gamma ray is given off in the sec-

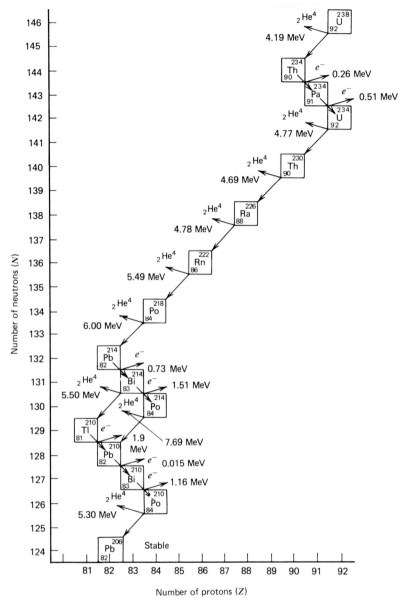

Figure 13.23   A radioactive decay chain. The particle emitted and its approximate maximum energy
are indicated for each decay.

ond path and not in the first, about what would be the energy of the gamma ray?

EXAMPLE

There is a kind of radioactive decay that has not yet been mentioned. The particle emitted in that process is a *positron*. A positron is identical to an electron—same mass, same charge, and so forth—except that its charge is positive rather than negative. It has the symbol e^+. Mercury-192 ($_{80}Hg^{192}$) decays by emitting a positron, and so does the daughter element formed, to produce a stable element. What is the final element in this small chain? What are the equations showing these two decays?

Solution

Since there will be two fewer positive charges—protons—in the nucleus, the final element must have $Z = 78$. A peek at the periodic table shows this to be platinum. The two equations will be

$$_{80}Hg^{192} \rightarrow {}_{79}Au^{192} + e^+$$

and

$$_{79}Au^{192} \rightarrow {}_{78}Pt^{192} + e^+$$

Note that the total number of nucleons remains 192.

Before leaving this section, I would like to mention one more item, lest you be hoodwinked. The energy balance for beta decay (and positron decay), does not come out quite as shown in the decay-chain chart. First, as is the custom, I have listed in the chart just the kinetic energies of the alpha and beta particles, and not the much smaller kinetic energies of the daughter nuclei following a decay.

The latter must also be included to get total energy exactly.

However, that is not the only problem. For a very long time physicists could not get the energy equations to come out quite right for any electron or positron decay process. Try as they might, there always seemed to be some energy lost in the process, and it seemed the electron or positron could come out with a whole range of energies. Experiments were done repeatedly and in different ways to ensure that the loss was not just due to experimental error. No, the measurements were too exact to account for the loss. Still, the energy just did not add up. It was enough (almost) to shake one's faith in the law of conservation of energy. Finally a new particle was invented to take away the missing energy. It was called a *neutrino*, which means "little neutral one," for it could have no charge nor any mass to speak of, and it would have to be very difficult to detect. Because of other supporting evidence, the neutrino gradually gained acceptance and, finally, after several years it was detected in an experiment. Thank goodness! The law of conservation of energy survived again and, in fact, faith in it led to the discovery of a new particle. We will not need to show the neutrinos in the nuclear equations that follow, but I thought you would like to know they are there.

HALF-LIFE

Suppose that you are playing a roulette wheel, like the one in Figure 13.24, with 100 numbers on it. Your number is 27 and, for you to win, the ball has to land on 27. Say that a complete spin of the wheel takes a minute, on the average. Here are two possible scenarios:

1. On the first spin 27 comes up, and you win.
2. On the first spin 26 comes up, and you lose. On the second spin it is 63, then 45,

Figure 13.24 Each time the roulette wheel is spun the chance of a particular number coming up is 1 in 100. (Courtesy of Las Vegas News Bureau.)

14, 91, 2, 7, 28, 36, 11, ... You stay with it, day and night, and a week later your number still has not come up. Equipped with a cot to sleep on and someone to bring in pizza, and with shifts of operators for the wheel, you stay on. Nineteen years later, after 10 million spins (and 7000 pizzas), your number finally comes up.

How likely is either of these scenarios? Not very. (Unless the wheel is "funny.") Your number has one chance in a hundred of coming up on each spin. In other words, there will be one chance in a hundred of getting your number on the first spin. These are not very good odds. On the other hand, if the wheel spins 100 times, there is a pretty good chance of getting your number at least once— about a 63 percent probability. (Incidentally,

some people think that if your number does not come up for 99 spins it is "due" on the hundredth spin. That is nonsense; the chance of your number coming up on *each and every spin* is one in a hundred.) If the wheel is spun 1000 times, your number will *probably* have come up about 10 times. If it spins 10,000 times, you should see your number about 100 times. And for the 10 million spins in 19 years, your number should be in there about 100,000 times. In fact, the greater the number of trials, the closer (in percentages) the result will be to what you would expect.

Suppose that, instead of watching a roulette wheel, you are watching a single atom of uranium-238, waiting for it to decay into thorium-234. Could it do so in the first second? Possibly. Might you have to wait for 10^{20} years for the decay to occur? Also possible.

The most likely event is somewhere between those two extremes. All that can be stated with any certainty is a *probability* that the decay will occur in a certain time.

When dealing with such probabilities, it is always easier to consider large numbers of events. As another example of this, consider tossing a coin. Will it come up heads or tails? You do not know, but you know that there is a 50 percent chance for either. Could it come up heads for 10 tosses in a row? That is quite possible, though not too likely and, if you will try tossing a coin repeatedly, you may get 10 heads in a row (but not necessarily the first 10). On the other hand, it would be most unusual if you got 100 heads in a row (unless the coin has heads on both sides!). To illustrate what it would be like with radioactive decay, I have tried tossing 500 coins. (I hasten to add that I used pennies. It's fun—just shake up a bag full of coins and dump them on the floor. Then count heads and tails as you pick them up.) I then considered those coins that came up heads to have "decayed." That is, I put them to one side and tossed the remaining coins—those that came up tails. Then, I put aside those that came up heads this time and tossed the remainder, repeating this until all of the coins were gone. The results of this fascinating experiment are shown in Table 13.1 and the graph of Figure 13.25.

The curve of Figure 13.25 should look quite familiar to you: It is an exponential decay curve.

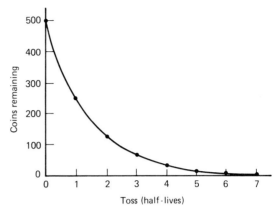

Figure 13.25 The results of tossing coins.

This coin tossing fits the definition of exponential decay exactly: After one toss (about) half of the coins are gone. After the next toss, half of those are gone, and so forth.

A sample of radioactive atoms does much the same thing. As with a coin toss, there are two possibilities: Either a given nucleus decays or it does not. Now, a sample of radioactive material will contain billions of nuclei,

TABLE 13.1 *Coin Tossing*

Toss number	Coins remaining
0	500
1	259
2	127
3	65
4	35
5	18
6	6
7	1
8	0

TABLE 13.2 *Half-Lives of Elements in the Radioactive Chain*

Nucleus	Half-life
U^{238}	4.5×10^9 yr
Th^{234}	24.1 days
Pa^{234}	1.18 min
U^{234}	2.48×10^8 yr
Th^{230}	76,000 yr
Ra^{226}	1620 yr
Rn^{222}	2.82 days
Po^{218}	3.05 min
Pb^{214}	26.8 min
Bi^{214}	19.7 min
Po^{214}	1.64×10^{-4} s
Tl^{210}	1.3 min
Pb^{210}	22 yr
Bi^{210}	5.0 days
Po^{210}	138.4 days
Pb^{206}	Stable

so the use of a probability should work very well indeed. At the end of a given time, half of the original atoms will have decayed and become something else. Say it is one hour. Then, in the next hour half of the remaining atoms will decay, leaving one-fourth of the original kind. So it goes, with half of the remaining atoms decaying in any given hour.

In any sample of radioactive material, the time required for half of the nuclei to decay is called the *half-life*. For purposes of comparison, I list in Table 13.2 the half-lives of the nuclei in the radioactive decay chain we have already examined. You will see that half-lives can vary enormously, from very short to very long. Figure 13.26 shows the radioactive decay of radium-226, which has a half-life of 1620 years. In 1620 years, half of the original N_0 radium nuclei remain and in 3240 years one-fourth remains.

Questions

Uranium-235 is the fuel used in nuclear reactors. Its half-life is about 7×10^8 years.

The age of the Earth is in the ballpark of 4.5×10^9 years. If there were N_0 (a very large number) atoms of U^{235} when the Earth was new, about how many would remain today? (*Hint*: About how many half-lives have elapsed?) Now try the same thing for U^{238}, which has a half-life of 4.5×10^9 years. Suppose the Earth started out with approximately equal amounts of U^{235} and U^{238}. Do you see why natural uranium is over 99 percent U^{238}? Suppose the half-life of U^{235} had been only 7×10^7 years. How would this have affected the present debates about nuclear power?

BINDING ENERGY

Have you wondered at all yet about *why* some nuclei are unstable and thus decay radioactively? It all goes back to force and energy. First consider what would happen if one tried to place two protons close together. As you know, they would repel each other, because

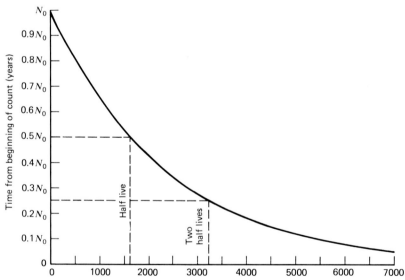

Figure 13.26 A radioactive decay curve for radium-226. After one half-life, half of the original radium nuclei remain.

each has a positive charge. As they were forced closer and closer the repelling force would become greater, going to very large values as they get very close. The system, consisting of two protons near each other, has a potential energy because of the work we have done in moving them that close. If released, they would fly apart.

However, if we can throw in a couple of neutrons and push the four nucleons *very* close to one another the situation suddenly changes. When the nucleons get very close they attract one another and form a stable nucleus—$_2He^4$. To pull the nucleons apart again would require doing work *on* the nucleus. As with the atom, it is customary to consider the state of zero energy as the condition in which the nucleons are completely separated. Then, the energy of the nucleus is negative, since work must be done on it to pull it apart. This is called the *binding energy*, and it will be different for different nuclei. Later we will examine the binding energy divided by the number of nucleons, the energy per nucleon.

Since the nucleus stays together, and work must be done on it to rip it apart, a force in addition to the electric force must be acting, because the electric force causes the protons to repel each other. There is, of course, the attractive gravitational force between nucleons, but this is much, much smaller than the electric force. An additional force must be involved. Also, it must be a powerful but short-range force because the nucleons have to be very close together before it shows up. But at close range it is large enough to overcome the huge electric force of repulsion.

It is now known that there are not one, but two additional forces acting in the nucleus. They are called the *strong nuclear force* and the *weak nuclear force*. Each is responsible for a different set of experimental results, but we shall not need to distinguish between them in what follows. The four forces that have now been mentioned—gravitational, electric, strong nuclear, and weak nuclear—are the

only known forces in the universe. Everything that we know about the interactions of matter can be explained with these four forces.

Figure 13.27 shows a chart of the known nuclei. It is much like the charts we have already examined (Figures 13.19 and 13.23), but it is on a smaller scale so that all of the nuclei can fit. The filled-in blocks represent stable nuclei. The blocks containing a cross are radioactive nuclei that are found in nature, and the empty blocks are artificially made radioactive nuclei. Obviously, there are now many more artificial than natural nuclei.

The filled-in blocks form a path, called the *valley of stability*, more or less in the center of the groups of nuclei. Nuclei above the valley of stability have an excess of neutrons. They will tend to decay by emitting an electron, thus changing a neutron to a proton in the nucleus. (Occasionally one of them will emit a neutron directly, and heavier nuclei can emit alpha particles.) Nuclei below the valley have an excess of protons, and they tend to decay by emitting a positron, thereby losing a proton and gaining a neutron. The line on the chart shows the region where the number of protons and neutrons would be equal ($Z = N$). Up to about $Z = 20$, the stable nuclei have approximately equal numbers of protons and neutrons. Above $Z = 20$ the stable nuclei have more neutrons than protons. This is because, as nucleons are added and the nucleus gets larger, the nucleons are farther apart, on the average. But the nuclear force is a very short-range force, whereas the electric force is long range. Therefore, as nucleons are added to a larger and larger nucleus, the average nuclear force gets smaller. On the other hand, the electric force on newly added protons, being a long-range force, stays about the same. As a result, there comes a time, with a sufficiently big nucleus, when the electric repulsive force is greater than the nuclear binding force, and the protons will not "stick." Thus, it is easier to put another neutron into a nucleus than another proton.

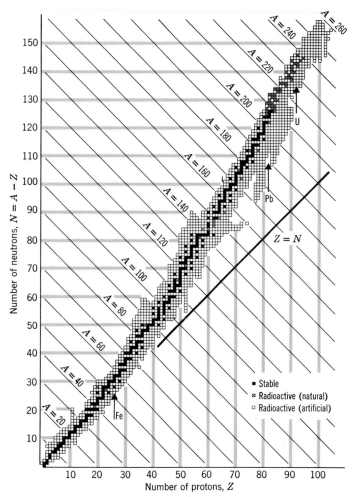

Figure 13.27 A chart of the known nuclei. The solid blocks are stable nuclei. Blocks containing a cross represent naturally occurring radioactive nuclei and empty blocks are artificially made radioactive nuclei.

The situation might be likened to an inept handyman plastering a ceiling. As he puts the first layer on, the plaster is fresh, and it sticks to the ceiling very well. However, he is slow and clumsy and he mixed too much plaster at once. It is drying out before he can get it applied. With each successive layer of plaster, the force holding it to the ceiling becomes weaker. The gravitational force on the plaster, however, is the same for each layer. Being a long-range force it does not change much. Finally, a point comes when the gravitational force is greater than the binding force, and the last layer of plaster will not stick. In fact, trying to make it stick may bring down some previously stuck layers of plaster. So it goes with the nucleus; there comes a point where the nuclear "glue" is not strong enough to hold the enlarged nucleus together, and pieces of it may "fall off."

However, that cannot be the whole story, or it would be possible to add as many neutrons as desired without making the nucleus unstable. The answer is that the exclusion principle discussed earlier in relation to electrons also applies to neutrons and protons. That is, no two neutrons (and no two protons) can have exactly the same energy state in the same nucleus. Thus, as neutrons (or protons) are added into a nucleus, each one goes into a higher (less negative) energy level. When the energy of the nucleus gets sufficiently high, the nucleus becomes unstable. When a radioactive decay occurs, the new nucleus has less total energy than the original nucleus. The excess energy goes into kinetic energy of the emitted particles and the new nucleus and into any gamma rays that are produced.

MASS DEFECT

Another way to understand the energy release in radioactive decay is to examine the mass of the particles. This requires only simple addition and subtraction, but it must be done with precision. (This is one of the few times in this book that we need to carry out the calculations to several decimal places.)

Why in the world are we talking about the mass, when energy is the topic? To get a clue, let's take a look at the masses of the particles involved in the alpha decay of radium-226 into radon-222. Precise measurements of those masses yield the following:

$m (_{88}Ra^{226})$ 226.0254 amu
$m (_{86}Rn^{222})$ 222.0175 amu
$m (_2He^4)$ 4.0026 amu

Now, add together the masses of the radon and the alpha particle, which are left after the decay:

$m (_{86}Rn^{222})$ 222.0175 amu
$m (_2He^4)$ 4.0026 amu
—————————————————
Total   226.0201 amu

But this is less than the mass of the radium nucleus by 0.0053 amu! You might think that such a small mass difference could be simply due to an experimental error in the measurements. That is not the case; the experiments have been carefully done and thoroughly checked, and the numbers are good to the fourth decimal place.

Where, then, has the missing mass gone? This was explained by Einstein early in this century in his famous Theory of Relativity. In it, Einstein concluded that there is a relationship between mass and energy, and that it can be expressed by the simple equation:

$$E = mc^2 \tag{13.2}$$

where E is the *total* energy of the particle, m is its mass, and c is the speed of light. Does this mean that, if an object is moving, and thus has kinetic energy, its mass is greater than the mass of the same object when it is stationary? It does indeed, and experiments bear this out. Why then, when you run a hundred-yard dash, do you not weigh 300 lb? It is because, unless the speed gets very large, the effect is very small. For example, when the 70,000 kg space shuttle is travelling at 25,000 mph, its mass is only about .00005 kg greater than its mass at rest—not very much. Effects this small are difficult, if not impossible, to measure. To produce a large effect, the speed of the object has to be very large, somewhere in the neighborhood of the speed of light. But we cannot give ordinary objects, such as a human body or a space shuttle, such high speeds, for it would require enormous amounts of energy to do so. We can do it, however, for very tiny objects, such as an electron or a proton. When such a particle is moving at about 90 percent the speed of light, its mass is about 2.3 times its rest mass. One way this can be experimentally measured is by injecting a fast-moving, charged particle into a magnetic field, where it moves in a circular path. If the speed is known, the diameter of the circle is a measure of the mass. (This is the mass spectrometer described earlier.) Careful experiments show that Einstein's prediction about mass and energy is correct. The law of conservation of energy has therefore been extended to include mass. *Mass–energy must be conserved.*

Since the E in $E = mc^2$ is the *total* energy, it must include potential energy, also. When neutrons and protons are collected together into a nucleus, there exists a negative binding energy. In order to pull the nucleus apart, work must be done on those nucleons. Thus, the separated nucleons have more total energy than the nucleus they came from. An exactly equivalent statement is that the separated nucleons have more *total mass* than the nucleus they came from.

We shall not go into this in greater detail here, but the result is that large amounts of energy are available from the conversion of a small amount of mass. This is because of the very large magnitude of the speed of light (3×10^8 m/s). For example, science fiction stories sometimes have magical nuclear machines that convert garbage, or any other material, completely to energy. If one kilogram of matter *could* be converted completely into energy (it cannot), the amount of energy produced would be

$$E = mc^2 = 1 \text{ kg} \times (3 \times 10^8 \text{ m/s})^2$$
$$= 9 \times 10^{16} \text{ J} \approx 2.5 \times 10^{10} \text{ kWh}$$
$$\approx 0.09 \text{ quad}$$

This would be enough energy to supply the energy needs of the entire United States for about half a day! However, the science fiction example is impossible for mass is never converted entirely into energy. When a nuclear decay, or other nuclear process, takes place,

the total number of nucleons remains the same. They are merely rearranged, so that they have a slightly different total mass. Any mass lost in such a process appears as energy.

The Einstein mass–energy relationship can be written in terms of the mass loss (Δm) in a radioactive decay. One simple way to write it is

$$\Delta E = k\Delta m$$

where the units of mass are amus and energy is in MeV. The constant k is just c^2 expressed in strange units. The conversion to the needed units gives the following result for that constant:

$$k = 931.5 \ \frac{\text{MeV}}{\text{amu}}$$

That is, for every amu of matter converted, 931.5 MeV of energy are produced. Thus, for the radioactive decay of radium, the energy produced is

$$\Delta E = k\Delta m = 931.5 \text{ MeV/amu} \times 0.0053 \text{ amu}$$
$$= 4.94 \text{ MeV}$$

If the decay is to the ground state of radon-222, the alpha particle has a kinetic energy of about 4.78 MeV, and the rest of the energy (0.16 MeV) goes into the kinetic energy of the radon nucleus. If the decay is to the excited state, the alpha particle has a kinetic energy of only 4.59 MeV, and about 0.19 MeV is emitted in the form of a gamma ray.

EXAMPLE

If you are a thoughtful person, you may yet be wondering what this mc^2 business has to do with the binding energy, anyway. This example will help you to understand. Take the case of the $_2\text{He}^4$ nucleus. It has a binding energy which is measured to be about -28.3 MeV. That is, work in the amount of 28.3 MeV must be done on the nucleus to separate it into four nucleons. Also, the mass of the $_2\text{He}^4$ nucleus is somewhat less than the total mass of the four separate nucleons. This is called the *mass defect*. The question is, how are the binding energy and the mass defect related?

Solution

From a table, I find that the atomic mass of $_2\text{He}^4$ is 4.0026 amu. Although atomic mass is the quantity usually given in the tables, we need nuclear mass for this computation. Thus, the mass of the electrons—0.00055 amu each—must be subtracted from this figure. Subtracting the mass of two electrons gives a nuclear mass of 4.0015 amu.

Now, let us calculate the mass of the separate neutrons and protons. The mass of a proton is about 1.00728 amu and for a neutron it is 1.00867 amu. The arithmetic follows:

$$
\begin{aligned}
\text{2 proton masses} &= 2 \times 1.00728 = 2.01456 \text{ amu} \\
\text{2 neutron masses} &= 2 \times 1.00867 = 2.01734 \text{ amu} \\
\hline
\text{total mass} &= 4.0319 \text{ amu} \\
\text{mass of helium nucleus} &= 4.0015 \text{ amu} \\
\hline
\text{mass difference} &= 0.0304 \text{ amu}
\end{aligned}
$$

So the mass of the helium nucleus is less than the mass of the four separate nucleons. A mass defect of this sort is found in all stable nuclei. Now let us see how much energy that mass represents:

$$
\begin{aligned}
\Delta E &= k\Delta m \\
&= 931.5 \text{ MeV/amu} \times 0.0304 \text{ amu} \\
&= 28.3 \text{ MeV}
\end{aligned}
$$

This is just equal to the negative of the binding energy of the nucleus.

Repeating the result of this example in words: The mass defect of a stable nucleus is equivalent to its binding energy. Since the total energy of the nucleus is less than that of the separate nucleons, the total mass is also less.

NUCLEAR REACTIONS

In the olden days, the *alchemists* searched night and day for a means of changing metals such as lead into gold. They were unsuccessful in that, but some of their work laid the foundations for modern chemistry.

Now, using nuclear reactions, scientists are able to change one element into another, almost at will. (They can even change other materials into gold, but the cost of doing so is much more than the value of the gold produced.) A *nuclear reaction* may take place when nuclei are bombarded by other particles or gamma rays. The bombarding particles may be protons, neutrons, electrons, alpha particles, or others. The first successful such reaction involved bombarding nitrogen with alpha particles (helium nuclei). The result is shown in the equation

$$_7\text{N}^{14} + {}_2\text{He}^4 \rightarrow {}_8\text{O}^{17} + {}_1\text{H}^1$$

As with all nuclear equations, there is a balance of nucleons and a balance of charge on the two sides of the equation. On the left side of the equation there are nine $(7 + 2)$ positive charges, and on the right side there are also nine $(8 + 1)$. There are 18 nucleons on each side. We can also examine the mass–energy balance, as we did earlier:

$$
\begin{aligned}
m(\text{N}^{14}) &= 14.003074 \text{ amu} \\
m(\text{He}^4) &= 4.002603 \text{ amu} \\
\hline
\text{total left side} &= 18.005677 \text{ amu}
\end{aligned}
$$

$$
\begin{aligned}
m(\text{O}^{17}) &= 16.999133 \text{ amu} \\
m(\text{H}^1) &= 1.007825 \text{ amu} \\
\hline
\text{total right side} &= 18.006958 \text{ amu}
\end{aligned}
$$

The total mass after the reaction is 0.001281 amu *greater* than the mass before the reaction. Put in terms of energy, this is about 1.19 MeV (931.5 × 0.001281). This is called the *Q-value* and is expressed as $Q = -1.19$ MeV.

The negative sign of Q for this nuclear reaction shows that energy had to be *added* in order for the reaction to occur. In other words, a bombarding alpha particle must have an energy of at least 1.19 MeV in order for this reaction to be possible. There are also reactions for which the Q-value is positive. For example, in the case of lithium-7 being bombarded by neutrons:

$$_3Li^7 + n \rightarrow {_3}Li^8 + \text{gamma ray}$$

In this case, $Q = +2.03$ MeV and the excess energy produces a gamma ray.

Questions

By using conservation of charge and conservation of nucleons, fill in the blanks in the following nuclear equations.

$$_6C^{12} + n \rightarrow \underline{\hspace{1.5cm}} + \text{gamma ray}$$
$$_5B^{10} + n \rightarrow \underline{\hspace{1.5cm}} + {_2}He^4$$

FISSION

In about 1939 experiments done by bombarding various nuclei with neutrons led to the discovery of the following reaction:

$$_{92}U^{235} + n \rightarrow {_{56}}Ba^{139} + {_{36}}Kr^{95} + 2n$$

Here was quite an amazing result! What might be expected is for the U^{235} to absorb a neutron and become U^{236}, which might then decay radioactively. What happened, instead, is that the U^{235} split (*fissioned*) into two smaller nuclei, giving off two neutrons in the process. Actually, a U^{236} nucleus is formed for a short time, but it is in a very unstable energy state and undergoes fission almost immediately. We shall follow the common practice of ignoring that intermediate step and referring to the fissioning of the U^{235} nucleus.

The resulting nuclei shown in the equation—called the *fission fragments*—are by no means the only ones possible. There are many different possible combinations. For example, the fission may yield the following:

$$_{92}U^{235} + n \rightarrow {_{57}}La^{143} + {_{35}}Br^{90} + 3n$$

or

$$_{92}U^{235} + n \rightarrow {_{55}}Cs^{143} + {_{37}}Rb^{90} + 3n$$

In each of these equations there is, of course, conservation of protons and of total nucleons. There are several features that we need to explore further.

One of them is the mass–energy balance for each equation. It would be very easy to do the same kind of arithmetic shown earlier to find the mass difference between the left and right sides of the equation, and thus to find the Q-value. Unfortunately, it is not easy to get precise mass values for many of the fission fragments. So we will use a somewhat different technique which will give a general, ball-park figure for all fission events. This can be done by looking at a curve of binding energy, as shown in Figure 13.28. The binding energies of nuclei, found by measuring their mass defects, is graphed as a function of the number of nucleons. To make the graph more meaningful it shows the binding energy per nucleon. Remembering that the binding energy is negative, an interesting feature shows up: There is a region where the binding energy per nucleon is greatest (most negative), occurring at the lowest spot of the graph. That minimum occurs at iron-56; therefore it is a very stable nucleus. As more nucleons are added, the larger nuclei become less stable, with less binding energy per nucleon.

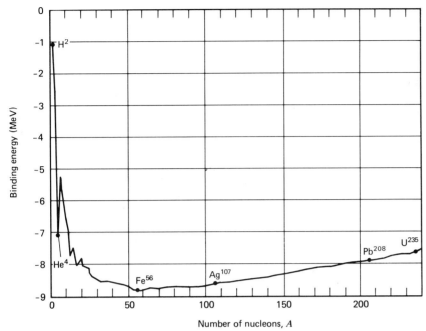

Figure 13.28 The binding energy per nucleon as a function of the number of nucleons in the nucleus.

We can now use the graph to get an estimate of the energy balance for a nuclear fission equation. In order to find the total binding energy of any nucleus from the graph, find the binding energy per nucleon and multiply it by the number of nucleons.

EXAMPLE

What is the difference in binding energies before and after the fission shown in the last equation?

Solution

From Figure 13.28, the binding energy per nucleon of U^{235} is about -7.6 MeV. Since there are 235 nucleons, the total binding energy is about 235×-7.6 MeV $= -1786$ MeV. In the last of the fission equations the two fission fragments are cesium-143 and ru-

bidium-90. For Cs^{143} the binding energy per nucleon is about -8.3 MeV for a total of -1187 MeV, and for Rb^{90} it is about -8.7 MeV and -783 MeV total. The neutrons are free and have no binding energy. Thus, the total binding energy on the left side is about -1786 MeV and on the right side it is -1970 MeV, about 184 MeV more negative. This means that the nuclei have lost about 184 MeV during the fission.

The energy "lost" during a fission goes into the kinetic energies of the fission fragments and the neutrons and into gamma rays. In addition, the fission fragments formed tend to have too many neutrons to be stable, so they decay radioactively, usually by beta decay. The half-life of Cs^{143} is about 2 seconds, and for Rb^{90} it is 2.9 minutes. As you will see in Chapter 14, this radioactive decay of the fission fragments is very important when considering the safety of nuclear reactors, for it yields an additional 10 MeV or more of energy and more radioactive isotopes.

Questions

After absorbing a neutron, one way in which $_{92}U^{235}$ can fission produces $_{53}I^{137}$ and two neutrons. What is the other fission fragment? Write the equation of this fission. By using Figure 13.28, estimate the energy released in this process.

In the fission process, whatever fragments are produced, the average energy output is roughly 200 MeV per fission. That is a truly huge amount of energy from one atom. To get an idea of how huge it is, take a look at the energy available from 1 kg of U^{235}. There are approximately 2.5×10^{24} atoms in 1 kg of U^{235}. Since each atom can produce about 200 MeV in a fission, the kilogram has the ability to produce about 5×10^{26} MeV if all of the nuclei fission. Since 1 MeV = 4.45×10^{-20} kWh, this is over 22 million kWh—from a single kilogram! By comparison, burning 1 kg of coal could produce approximately 8 kWh of energy. (Neither of these figures takes into account power plant inefficiencies.) The energy potentially available from uranium is simply enormous. This is why nuclear power has great appeal as a source of energy.

Chain Reactions

In order for a fission of U^{235} to occur, the nucleus must capture a neutron. In the early experiments the neutrons were provided from an external source and the sample of U^{235} was bombarded with them. It was soon realized, however, that the neutrons produced by a fission event could be used to trigger more fission events. This can lead to a *chain reaction*. For example, suppose every fission yielded exactly two neutrons and that both of them

could be captured by other U^{235} nuclei. Those two would then fission, producing four neutrons which would trigger four fissions, and so forth. The process is illustrated in Figure 13.29. You probably recognize the signs of an exponential growth pattern: Each fission produces two more. Since the fission process is so rapid, a chain reaction like that shown would be an *uncontrolled* reaction–a nuclear bomb. (Variously called an atomic bomb, an atom bomb, and even a "newculer" bomb by people who should know better.) In fact, making a nuclear bomb is ridiculously simple, in principle. All that is necessary is to bring together a large enough mass of U^{235} (called the *critical mass*) suddenly. The reason it has to be sudden is that the reactions occur so fast that, unless the critical mass is brought together suddenly, too many of the neutrons will escape, and the chain reaction will not occur fast enough to cause an explosion. Some bombs are made by surrounding two chunks of uranium with TNT or some other conven-

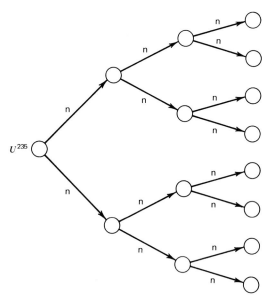

Figure 13.29 A chain reaction. Each fission produces two excess neutrons which then cause two additional fissions. The fission fragments are not shown.

tional explosive. Then the TNT is exploded, cramming the two pieces of uranium into one another to make a single piece of greater than critical mass. Then, boom!

However, we are interested in *controlled* chain reactions. In a controlled chain reaction, exactly one neutron per fission produces another fission. The other neutrons escape. This is indicated in Figure 13.30. Actually, this does not mean that each fission event produces one effective neutron; rather, the average for a large number of events is exactly one. For this to occur, two things must be done. First, it must be arranged that just the right number of neutrons are allowed to escape or are absorbed by something other than U²³⁵. In a nuclear reactor, this is accomplished by using the right arrangement of the fuel elements and using *control rods*. I will discuss this more fully in Chapter 14, but will say here that the control rods are arranged so that they can be pushed between the *fuel elements*, and they can be moved in and out to allow just the right amount of activity.

Also, the neutrons that are to be captured by the U²³⁵ nuclei must be slowed down. When neutrons are emitted by a fission event, they typically have kinetic energies of several MeV. They are going so fast that they go right past the U²³⁵ nuclei without being captured. In order to be captured, they must hang around the neighborhood a while. To slow them down

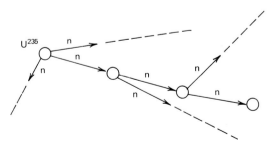

Figure 13.30 In a controlled chain reaction, an average of exactly one neutron per fission produces another fission. The other neutrons escape or are absorbed by other materials in the core.

(*moderate* them), a material called a *moderator* is placed between the fuel rods. Several different materials can be used, but often it is just very pure water. Neutrons that are slowed down to the proper speeds are called *thermal neutrons,* because they are moving at about the speeds given to them by their thermal energy.

Uranium Enrichment

As you no doubt have noticed, we have been talking all along about the fission of uranium-235. But, as uranium appears in nature, it is 99.27 percent U²³⁸ and only 0.72 percent U²³⁵, with the very small remainder being U²³⁴. So why do we not use the more abundant U²³⁸? It is because U²³⁸ does not fission so readily. In fact, a chain reaction can be sustained in only three isotopes: uranium-235, uranium-233, and plutonium-239. Of the three, only U²³⁵ occurs naturally.

However, there is a price to pay in order to use U²³⁵. The 0.7 percent of it in natural uranium is not enough for a chain reaction. For some types of nuclear bomb, the material needs to be almost 100 percent U²³⁵. A reactor will work, however, if the U²³⁵ is 3 to 4 percent of the total uranium. Thus, the uranium has to be "enriched." That is, it is brought to the point where there is 3 percent or so of U²³⁵ and 97 percent U²³⁸. Since the two isotopes are identical chemically, no ordinary chemical process will separate them. One way to separate the isotopes is by using a mass spectrometer, as discussed earlier. Another is by combining the uranium chemically with fluorine to produce a gas, uranium hexafluoride. The gas is then forced through a sort of "strainer," a material with very tiny pores. The slightly larger U²³⁸ molecules do not pass through quite as readily as the U²³⁵ molecules. Thus, a slightly enriched gas appears on the other side. A thousand or more passes are needed to get the appropriate enrichment. Both processes are time consuming and very ex-

(a)

Figure 13.31 Part of the interior of the Gaseous Diffusion Plant at the Oak Ridge National Laboratories in Tennessee. (*a*) Thousands of stages like the ones pictured enrich uranium to a higher than the natural percentage of U²³⁵. (*b*) A cutaway drawing showing three stages in the same gaseous diffusion cascade. (Courtesy Union Carbide Corporation. Photo by Frank Hoffman.)

pensive. An isotope separation facility is expensive to own and operate. Figure 13.31 shows part of a facility operated for the United States government.

Breeding

The more abundant isotope of uranium, U^{238}, can be made to fission, but only with high energy neutrons. The probability of fissioning is so small that a chain reaction cannot occur in U^{238}. However, another interesting reaction can occur when U^{238} is bombarded with neutrons. First, the uranium captures a neutron to become U^{239}. Then, the U^{239} decays radioactively by emitting a beta particle, with a half-life of about 23.5 minutes. The reactions are given by the equations:

$$n + {}_{92}U^{238} \rightarrow {}_{92}U^{239}$$
$${}_{92}U^{239} \rightarrow {}_{93}Np^{239} + e^-$$

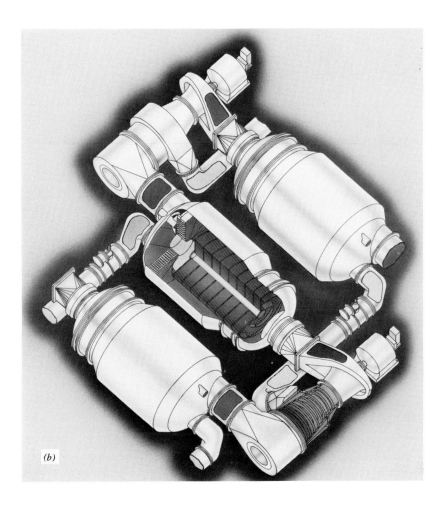

(b)

The element produced, neptunium, does not occur naturally on Earth. (No element with Z greater than 92 occurs naturally on Earth.) Of itself, neptunium is not terribly important. However, it is unstable and decays with a half-life of about 2.4 days, as follows:

$$_{93}Np^{239} \rightarrow {}_{94}Pu^{239} + e^-$$

This new element is the famous *plutonium*. It, too, is radioactive, but its half-life is 24,360 years. Thus, a sample of it will last a long time. It is important because it is a *fissile* material. That is, like U^{235}, it will sustain a chain reaction with thermal neutrons. It is used to make nuclear bombs, and it can be used in nuclear power reactors. Plutonium can be made in large quantities in *breeder reactors,* but there is yet considerable debate over the wisdom of doing so. There are also unsolved technological problems. The advantage to breeding plutonium for use in power reactors is that much more fuel would then be available, since the original fuel is U^{238}, which is about 140 times more abundant than U^{235}.

However, because of its radioactivity, plutonium presents a serious danger to the populace should any sizable quantity ever escape. It has been called "one of the most toxic substances known." Plutonium is not highly

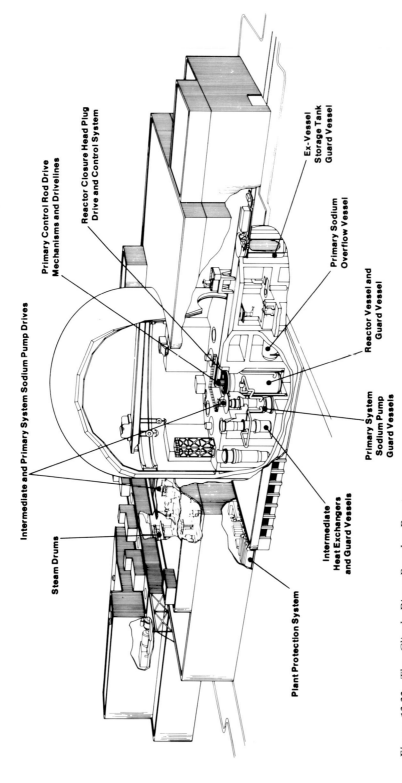

Intermediate and Primary System Sodium Pump Drives

Primary Control Rod Drive
Mechanisms and Drivelines

Reactor Closure Head Plug
Drive and Control System

Ex-Vessel
Storage Tank
Guard Vessel

Primary Sodium
Overflow Vessel

Reactor Vessel and
Guard Vessel

Primary System
Sodium Pump
Guard Vessels

Steam Drums

Intermediate
Heat Exchangers
and Guard Vessels

Plant Protection System

Figure 13.32 The Clinch River Breeder Reactor Project. (Courtesy Project Management Corporation, Oak Ridge, Tenn.)

radioactive, unlike some elements with shorter half-lives. Also, it emits alpha particles, which cannot even penetrate a sheet of paper; they penetrate only a few hundredths of a millimeter in human tissue. Just the layer of dead skin normally on your body would be enough to protect you from harm from plutonium's alpha particles, as long as it is kept outside the body. However, plutonium combines readily with oxygen to form plutonium oxide and it does so in *aerosol* form. An aerosol is a suspension in air of very fine particles. The uranium oxide aerosol has particles just the right size to get past all of the body's defenses and into the lungs. There the 5-MeV alpha particles can do serious damage to lung tissue. It has been said that a millionth of a gram of plutonium in the lungs would be almost certain to cause lung cancer.

An additional danger from plutonium arises from the fact that it is chemically different from the uranium in which it is bred. It is, therefore, relatively easy to separate, and plutonium bombs are not terribly difficult to make. If plutonium becomes quite readily available, it is feared that a presently nonnuclear nation or even a terrorist group might get enough to make a nuclear bomb.

Nonetheless, breeder reactors may be used commercially in the fairly near future, and we shall take a closer look at them in Chapter 14. A diagram of the experimental Clinch River breeder facility is shown in Figure 13.32.

FUSION

Whenever a group of nucleons can be rearranged so that their binding energy becomes greater (more negative), energy is released. You saw this with both radioactive decay and nuclear fission. Now, look once more at the binding energy curve of Figure 13.28. This time, look at the left side of the graph, where the very light nuclei are. The binding energies per nucleon of those nuclei are much smaller than those for the nuclei nearer the middle of the graph. This suggests that if two light nuclei can be joined together (*fused*) into a heavier one, then energy would be released.

For example, suppose we could fuse together two oxygen-16 nuclei to produce a sulfur-32. The binding energy of O^{16} is about -8 MeV per nucleon, or a total of -256 MeV for two oxygen nuclei. The binding energy of S^{32} is about -8.5 MeV and -272 MeV total. The change in binding energy that would occur is then about 16 MeV, and this "lost" energy would appear as kinetic energy of the resulting nucleus.

The problem with doing this fusion is that each nucleus of O^{16} has eight protons. When two such nuclei get close together, the repelling forces are very great indeed. Getting them close enough together for the nuclear forces to take over is far beyond present technology. To get around this, we could use nuclei with fewer protons. The simplest nucleus of all is that of hydrogen, with a single proton. However, two protons alone cannot be made to fuse because the repelling force is great and there are not enough nucleons to overcome it. With the addition of two neutrons, the two protons will be a little farther apart in the fused nucleus, and the very stable He^4 nucleus can be formed. One way to do this, then, is to use the "heavy" isotope of hydrogen, H^2. Because of its importance, this isotope is given a name of its own, *deuterium*, but it is chemically identical to ordinary hydrogen. Its nucleus, with a single proton and a single neutron, is called a *deuteron*. The fusing of two deuterons is represented by the equation

$$_1H^2 + {_1}H^2 \rightarrow {_2}He^4 + 23.8 \text{ MeV}$$

The energy produced goes into kinetic energy of the He^4 nucleus and gamma radiation. Actually, this reaction is not very likely. When two deuterons are combined, one of the following two reactions is more probable, with either a neutron or a proton being emitted:

$$_1H^2 + {_1H^2} \rightarrow {_2He^3} + n + 3.3 \text{ MeV}$$
$$_1H^2 + {_1H^2} \rightarrow {_1H^3} + {_1H^1} + 4.0 \text{ MeV}$$

The fusion process does not release as much energy per nucleus as does fission. However, the nuclei involved are much lighter, and the energy released *per nucleon* is roughly the same—about 1 MeV. This means that the total energy release from the same amount of mass is about the same. For example, suppose a kilogram of deuterium could be fused completely according to the last equation. There are about 3×10^{26} deuterium atoms in 1 kg. Since it takes two of them to produce a fusion reaction, there are about 1.5×10^{26} fusions possible, with a total energy release of about 6×10^{26} MeV. Using 1 MeV = 4.45×10^{-20} kWh, this converts to a total energy of 27 million kWh. This enormous amount of energy is in the same "ballpark" as that potentially available from 1 kg of U^{235}.

Questions

Deuterium occurs as about 0.015 percent of all of the hydrogen found in nature. This means that about one part in 6500 of all the water found in nature has deuterium in place of hydrogen. (Water which has more deuterium than the natural amount is called *heavy water.*) The chemical formula for water is H_2O, with two atoms of hydrogen and one of oxygen in each molecule. Since each atom of hydrogen contributes about 1 amu and the oxygen contributes about 16 amu, for a total of 18 amu, water is roughly $\frac{1}{9}$ hydrogen. If you had 500 gallons of sea water, at about 4 kg/gal, what would the mass of the hydrogen be? Of that hydrogen, how much would be deuterium?

Show that the total energy potentially available from the fusion of the deuterium from 500 gal of sea water would be nearly 1 million kWh.

Considering the huge amount of sea water in the world, the energy available if deuter-

ium fusion (often called D–D fusion) can be made practical would be vast indeed. In fact, it would be so enormous that this is often considered a *renewable* source of energy. It is hard to imagine ever running out, whereas the supply of U^{235} is limited.

Another advantage of using deuterium rather than uranium is that it is much easier to separate the two isotopes that occur in nature. Deuterium has twice the mass of ordinary hydrogen, whereas U^{235} and U^{238} have masses that differ by a little over 1 percent. Since the separation is done on the basis of the difference in mass or size, the latter separation is much more difficult. Deuterium fusion is also much "cleaner" than fission, with far fewer radioactive products resulting.

But alas, there always seems to be a catch. We shall look at some of the practical problems of fusion power in Chapter 14, but one disadvantage of the D–D reaction is that it is difficult to make happen. There is still that very large force of electric repulsion to be overcome, and that is not easy. One way in which it can be done is to "shoot" deuterons at one another at very high speeds, as indicated in Figure 13.33. This has been done by directing a very high energy beam of deuterons at a target containing deuterium in a *particle accelerator.* (Generally, these are very large machines which will accelerate relatively few particles.) While this method can cause fusion to occur, it is not suitable for the large-scale production of energy.

Another way to cause fusion to occur is to heat the particles to very high tempera-

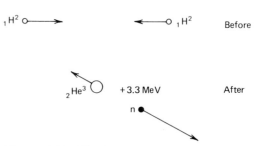

Figure 13.33 The fusion of two deuterons.

tures. As you will recall, the hotter the material is, the faster the particles will move. If there are enough particles moving fast enough, then some of them will collide at high speeds and fusion will occur. Unfortunately, in order for deuterium fusion to occur, the temperature reached must be about 4×10^8 K, hotter than the interior of the sun!

Perhaps it would be best to look for some other material to fuse. A material worth trying is H^3, the third isotope of hydrogen. Also called *tritium*, this isotope does not appear in nature but it can be made in nuclear reactions. It is radioactive, emitting beta particles (electrons) with maximum energies of only about 0.018 MeV, and it has a half-life of about 12.26 years. A fusion that appears to have promise is between deuterium and tritium:

$$_1H^2 + _1H^3 = _2He^4 + n + 17.6 \text{ MeV}$$

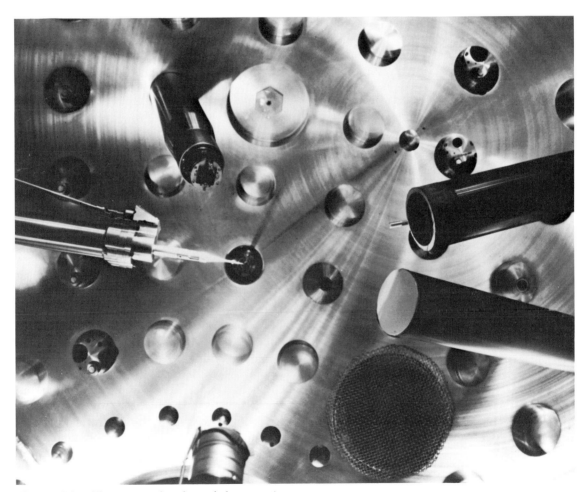

Figure 13.34 The target chamber of the experimental Shiva laser fusion facility at the Lawrence Livermore Laboratory. The needlelike device, coming from the left side, holds a tiny sample exactly in the center of the 4-ft chamber. (DOE photo by Lawrence Livermore Laboratories.)

Figure 13.35 Part of the Shiva laser banks. The lasers can deliver 3×10^4 joules of energy to the target in less than 10^{-9} seconds, for a power rating of 3×10^{13} watts. (DOE photo by Lawrence Livermore Laboratories.)

This reaction—called a D–T reaction—produces a great deal of energy. For it to occur, the material must be heated "only" to about 4×10^7 K. This is still a very high temperature, but it is only $\frac{1}{10}$ of that needed for a D–D reaction. In hydrogen bombs the high temperature needed is realized by exploding a fission (U^{235} or plutonium) bomb surrounded by deuterium and tritium. Obviously, an H-bomb explosion is not "controlled" power production. To produce usable power the fusion must occur in some controllable way. Figures 13.34 and 13.35 show part of an experimental effort to produce controlled fusion by blasting a tiny pellet of fusion "fuel" with lasers from several directions simultaneously. The practical difficulties of producing controlled fusion will be discussed in the next chapter.

SUMMARY

In this chapter you have investigated the characteristics of atoms and their nuclei. Atoms consist of a positively charged nucleus surrounded by fast-moving electrons in electron clouds. Each element has its own spectrographic "fingerprint," consisting of a spectrum not duplicated by any other element. The spectrum of an element results from the possible electron energy states of its atoms. The total energy of a neutral atom is always negative, meaning that energy must be added to it to remove one or more electrons. The electrons, grouped together in "shells," are also responsible for the chemical properties of the element. All of the known elements have specific places in the periodic table; they are located according to their total number of electrons and the number of electrons in the outermost shell. Elements in a given column of the periodic table have similar chemical properties.

The nuclei of atoms consist of protons and neutrons. The atoms of a given element have a fixed number of protons, equal to the number of electrons in a neutral atom. The number of neutrons may vary, however. This produces isotopes, atoms with identical chemical properties but with different masses. Many elements appear in nature as mixtures of several isotopes. The elements of particular interest are those with radioactive isotopes. A radioactive nucleus is unstable and, after a time, it may decay, producing a new nucleus. The decay products may include an alpha particle, which we now know to be a He^4 nucleus; a beta particle, which is just an electron; and a gamma ray, which is electromagnetic radiation of very high frequency. When radioactive decay occurs, there is always conservation of nucleons and conservation of charge: Neither nucleons nor charge will be created or destroyed. The time at which a particular radioactive nucleus will decay cannot be predicted but the half-life of a sample containing large numbers of identical nuclei can be specified. The graph of radioactive decay versus time is an exponential decay curve.

The only known forces in the universe are gravitational, electric, and two types of nuclear force. The nuclear forces are very short range and provide the binding energy to hold nucleons together against the repulsive electric forces between protons. The binding energy of a stable nucleus is always negative, since it requires work to separate the nucleons. When a nuclear reaction results in the release of kinetic energy, the new nuclei have less total energy (more negative binding energy per nucleon) than the original nuclei. Another way of regarding this is to examine the mass defect: The resulting nuclei and particles have less total mass than the original. The relationship between energy and mass is given by Einstein's famous equation: $E = mc^2$.

Nuclear fission is the process of a heavy nucleus, such as U^{235}, splitting into two nuclei. It usually requires the capture of an extra neutron, and free neutrons are found along with the daughter nuclei. In the process, energy is released, amounting to roughly 1 MeV per nucleon. A controlled chain reaction occurs when, on the average, exactly one neutron from each fission produces another fission. When a controlled chain reaction is sustained, fission frees large amounts of energy which can heat water to produce steam to run electric generator turbines. To fuel a nuclear reactor, uranium found in nature must be enriched to contain about 3 percent of U^{235}, for the most abundant isotope, U^{238}, will not sustain a chain reaction. Breeder reactors convert U^{238} to Pu^{239}, which is a fissionable material. The plutonium can be used in reactors (or bombs), but it is highly toxic and, thus, dangerous to handle. If breeder reactors are successful on a large scale, they will increase greatly the amount of available usable nuclear fuel.

Nuclear fusion is the fusing of two light nuclei, producing a heavier nucleus and extra energy. The amount of energy produced in the reaction is similar to that produced by fission, about 1 MeV per nucleon. Controlled fusion is not yet practical on a scale that would be suitable for electric power generation. However, if the problems can be solved, fusion holds great promise, since the supplies of available fuel could be enormous.

ADDITIONAL QUESTIONS AND PROBLEMS

1. You have seen that atoms in an excited energy state can undergo transitions, or jumps, downward in energy, giving off photons in the process. In addition, atoms can receive energy from photons and undergo transitions upward in energy. What is the frequency of a photon which can raise the hydrogen atom from the ground state ($n = 1$) to the first excited state ($n = 2$)? What is the wavelength of that photon? In which part of the electromagnetic spectrum does it fit?

2. What is the minimum energy required to ionize—remove an electron completely—a hydrogen atom in the ground state?

3. Using the periodic table and what you know about the filling of electron shells, sketch the shell structure for a neutral rubidium atom. How many electrons are there in the outermost shell? What does this tell you about the chemical activeness of rubidium?

4. Using the periodic table and what you know about the filling of electron shells, sketch the shell structure for a neutral krypton atom. How many electrons are there in the outermost shell? What does this tell you about the chemical activeness of krypton?

5. Describe what is meant by the word *isotopes.* Give examples.

6. Using the periodic table, write the complete chemical symbol for the following radioactive isotopes: carbon-14, strontium-90, iodine-131, cesium-133, bismuth-210, uranium-233, and plutonium-240.

7. Insert the correct symbols in the blanks for the following nuclear equations:

$$_{90}Th^{230} \rightarrow \underline{\hspace{1.5cm}} + \, _{2}He^{4}$$
$$_{55}Cs^{137} \rightarrow \underline{\hspace{1.5cm}} + \, e^{-}$$
$$\underline{\hspace{1.5cm}} \rightarrow \, _{7}N^{14} + \, e^{-}$$
$$_{92}U^{235} + n \rightarrow \, _{53}I^{137} + \underline{\hspace{1.5cm}} + 3n$$

8. The isotope $_{90}Th^{232}$ decays radioactively by emitting an alpha particle. The daughter nucleus then decays with a beta emission, and the next generation also emits a beta. What is the final isotope in this small chain?

9. Lanthanum-142 ($_{57}La^{142}$) decays by beta emission to cerium-142 ($_{58}Ce^{142}$), with a half-life of 1.4 hr. Cerium-142 has a half-life of 5×10^{16} yr. Starting with a pure sample of lanthanum-142, about how long would be required for the sample to become 90 percent or more cerium-142? In that period, would a great deal of the newly formed cerium have decayed? Explain.

10. In the radioactive decay chain we have studied, one step is from $_{90}Th^{230}$ to $_{88}Ra^{226}$ by means of an alpha decay, followed by the emission of a gamma ray. The masses of the nuclei involved are:

Th^{230}	230.0331 amu
Ra^{226}	226.0254 amu
He^{4}	4.0026 amu

How much energy, in MeV, is produced by the conversion of mass? Figure 13.23 gives the kinetic energy of the alpha particle as 4.69 MeV. If all of the excess energy were to go into a single gamma ray, approximately what would be the energy of the gamma ray? What would be the frequency of that gamma ray?

11. In the radioactive decay chain we have studied, one step is from $_{86}Rn^{222}$ to $_{84}Po^{218}$ by means of an alpha decay, followed by the emission of a gamma ray. The masses of the nuclei involved are:

Rn^{222}	222.0175 amu
Po^{218}	218.0089 amu
He^{4}	4.0026 amu

How much energy, in MeV, is produced by the conversion of mass? Figure 13.23 gives the kinetic energy of the alpha particle as 5.49 MeV. If all of the excess energy were to go into a single gamma ray, approximately what would be the energy of the gamma ray? What would be the frequency of that gamma ray?

12.   The following equation shows one possible fission of U^{235}:

$$_{92}U^{235} + n \rightarrow {}_{57}La^{144} + {}_{35}Br^{90} + 2n$$

Using Figure 13.28 estimate the energy released in the fission.

13.   One way in which U^{235} can fission, after absorbing a thermal neutron, results in the production of barium-141 and three neutrons. What is the other fission product? Write the equation of this fission process. Using Figure 13.28 estimate the amount of energy produced by the fission.

14.   If $_{94}Pu^{239}$ absorbs a thermal neutron and fissions, producing $_{58}Ce^{145}$ and three neu-

trons, what is the other fission product? Write the equation for this fission. Using Figure 13.28 estimate the amount of energy produced by the fission.

15.   Describe a chain reaction. What is an uncontrolled chain reaction? What conditions are necessary for a controlled and sustained chain reaction?

16.   Why is it desirable to perfect breeder reactors?

17.   Using Figure 13.28 and using the difference in mass before and after the reaction, estimate the energy released in the following fusion reaction:

$$_1H^2 + {}_1H^2 \rightarrow {}_2He^4$$

Do the two results agree? Do they agree with the handbook figure listed in the text? The masses needed are:

m $(_1H^2)$	2.01410 amu
m $(_2He^4)$	4.00260 amu

18.   What are some of the possible advantages of fusion power? Why aren't we using it now?

14 *Nuclear Power*

Before reading any further, write the answer to the following question down somewhere and save the answer. Are you "antinuke" or "pronuke"? That is, are you for or against the use of nuclear power to provide electricity?

Nuclear power plants, like the one pictured in Figure 14.1, arouse strong emotions on both sides of the nuclear controversy. That controversy is one of the most bitter and emotional political issues in this country and in the world. It is also an extremely important issue, and it should be decided on the facts and not on emotion. Unfortunately, the facts are a bit complex and few people bother to learn what they are. Further, there are some very real unknowns and gaps in the facts. These are such that two groups of equally intelligent and equally sincere people can look at the same information and come to completely opposite conclusions. Sadly, even the most intelligent and sincere people seem to become "polarized" in their thinking. That is, once they are convinced in one direction or the other, then every fact and every event tends to be interpreted according to what they already believe. I had an excellent example of this just the other day, during a trip from Dallas to Chicago. At the Dallas/Fort Worth

airport I was literally grabbed by an articulate young woman sporting a sign which screamed, "Don't be a pea brain like ____ _____!" (In the blanks was the name of a prominent U.S. senator.) She and her colleagues were totally for nuclear power and determined to convince every passerby. She was well-informed, persuasive, and almost impossible to get away from. After conceding that she was very knowledgeable, I pointed out that she greatly oversimplified a very complex subject. I had to be downright rude to get away from her. Having made my escape, I had an uneventful flight to Chicago. There, in the main terminal, I was grabbed by an articulate young man whose sign shouted, "Don't be a child murderer!" This one was strongly antinuclear, picturing a prominent scientist as the villain. Now some of the same facts were thrown at me, but with a completely different interpretation. Once again, a terribly complex subject was greatly oversimplified and, once again, I had to be rude to escape.

Another example of the mind set that intelligent people can develop was provided by the accident at Three Mile Island. The committed antinukes used this incident as evidence that nuclear power is unsafe. But the

Figure 14.1 The Calvert Cliffs nuclear power plant on the Chesapeake Bay in Maryland. The two units pictured use pressurized water reactors. (Photo courtesy Baltimore Gas and Electric.)

pronukes use the same event—and the fact that no one was seriously injured by the accident—as proof that nuclear power is very safe, even when things go wrong. What can we believe? Sometimes it seems that we might just as well toss a coin or, much the same thing, simply pick one group of experts to support over another.

Whatever your present inclination, I hope you will approach this chapter with an open mind. Most of the issues are easily understandable by any reasonably intelligent person who is willing to make the effort. I shall try to present the facts to the best of my own understanding and without bias. In many instances, the true facts are unknown and I shall point this out, along with the *opinions* of both sides of the controversy. There is not enough room here to exhaustively investigate the situation—after all, dozens of books have been written on the subject. However, after completing study of this chapter, you should have a much better understanding than the average person, and at least the beginning of a basis for making your own judgments. You

should also be able to read further material on the subject with understanding. We shall start with a description of present-day nuclear reactors.

LIGHT-WATER REACTORS

There are about 80 commercial power reactors presently operating in this country, with another 80 or so under construction. All but one of them are of the "light-water" type, called LWRs, *light-water reactors. Light water* is ordinary water, containing the common isotope of hydrogen. *Heavy water* is water that contains more than the normal amount of deuterium; that is, $_1H^2$. The water in the reactor serves two purposes: cooling of the reactor core, with transfer of power to the turbines, and moderation. *Moderation* refers to the slowing down of the neutrons to thermal speeds. Remember that the slower neutrons have a greater probability of being captured in a U^{325} nucleus and causing fission. The following activity will give you some insight into why water is a good moderator.

Activity 14.1

For this activity you will need a pool table or billiard table, a couple of billiard balls, and one or two Ping-Pong balls. The aim is to find out what happens when two objects of different masses or of the same mass collide.

First, try some collisions between one of the Ping-Pong balls and one of the billiard balls. A Ping-Pong ball can be rolled at a high speed directly at a billiard ball. This can be done either by pushing it by hand or hitting it with the cue stick, whichever you can do more easily. Try a few of these head-on collisions and describe what happens. (That is, what are the motions of the two balls before and after the collision?) Now try some "glancing" collisions, with the Ping-Pong ball hitting nearer the edge of the billiard ball than the center.

Now try the same thing, but with a rolling billiard ball hitting a stationary Ping-Pong ball. Finally, do the experiment with two billiard balls. Before you do this, see if you can use your experience with the two preceding cases to predict what will happen. Summarize your observations.

The results of this activity are summarized in Figure 14.2, and they apply to moving neutrons. If a neutron hits a much larger object, such as a large nucleus, it bounces off, still having much of its original kinetic energy. If a neutron hits a much smaller object, say a free electron, the smaller object is pushed ahead at high speed, with the neutron following behind, again with much of its original kinetic energy. However, when a neutron hits an object of equal mass head-on, then it stops completely, losing all of its kinetic energy. In a glancing collision with an object of equal mass, the neutron loses some of its energy. The amount it loses depends on how close to head-on the collision is.

In a reactor, there will generally not be

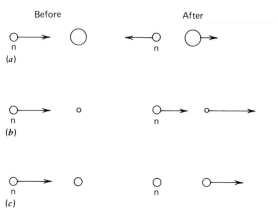

Figure 14.2 If a fast-moving neutron should hit a particle of equal mass head-on, it would be stopped. Neutron hits (*a*) larger mass; (*b*) smaller mass; (*c*) equal mass.

stationary neutrons hanging around to slow down the neutrons from fissions. However, protons have very nearly the same mass as neutrons, and there are lots of protons (hydrogen nuclei) in water. Thus, water acts as a very good moderator, slowing down neutrons very effectively.

Now we are ready to examine the workings of a typical reactor. Although a working reactor is a very complex maze of pipes, wires, valves, sensors, motors, pumps, and so on, the main parts of it are quite simple and easy to understand. They are the fuel, the moderator, the coolant, and the control rods. In the case of light-water reactors, the moderator and the coolant are the same, consisting of water circulating through the fuel.

The Fuel

In the United States at present, the fuel used in light-water reactors is uranium in the form of uranium dioxide (UO_2). As you may recall from Chapter 13, the uranium is somewhat enriched, with about 3 percent being U^{235} and the remainder U^{238}. The uranium dioxide is compressed into pellets, usually cylinders about 1 cm long and 1 cm in diameter. These pellets are then enclosed in long tubes (about 3.7 m) made of stainless steel or an alloy of zirconium (*Zircaloy*). Zirconium is usually chosen for this *cladding* because it is strong and does not absorb neutrons very well. The neutrons pass right through the cladding, which is generally only about 0.025 in. thick. Figure 14.3 shows fuel pellets being stacked for insertion into zircaloy tubes. The finished *fuel rod* results when a zircaloy tube is loaded with fuel pellets, filled with helium, and sealed off. (It is also called a *fuel pin* in much of the literature.) The finished fuel rods are then assembled into *bundles*, with perhaps 200 rods per bundle, shown in Figure 14.4. The rods are attached to spacers at both ends and along their lengths; this keeps them separated so that water can flow through the assembly and between the rods. Then these bundles are made into the *core* of the reactor. A 1000 MW reactor would have over 200 bundles with 40,000 to 50,000 fuel rods in the core. The core might be 4 to 5 m in diameter and have a mass of 125 to 250 metric tons. (In what follows, I shall use as a "standard reactor" a 1000 MW reactor, approximately the size of the large reactors now being built. This reactor is capable of producing 1000 MW of electric power. To make it clear, this will be referred to as 1000 MWe, "megawatts–electric." Since the efficiency of light-water reactors is roughly 30 percent, this means that the heat generated is over 3000 MW, or 3000 MWt for "megawatts–thermal.")

In a LWR the fuel enrichment will typically be about 3 percent U^{235}. When the core is fabricated, the fuel is of varying degrees of enrichment, with the richest fuel being near the outside and the least enriched being near the center. That is because the center fuel rods will tend to have more neutrons passing through them than pass through the outer

Figure 14.3 Uranium fuel pellets being stacked for insertion into zircaloy tubing to form fuel rods. (Photo courtesy of the Atomic Industrial Forum, Inc.)

Figure 14.4   A fuel-rod assembly being loaded into the reactor core of the San Onofre Station in California. (Photo courtesy DOE.)

ones. As the fuel is used up it must be replaced. In a typical replacement, the outer rods are moved to the center of the core and new rods are placed in the outer positions. The rods that were in the center are then discarded or reprocessed.

In operation the U^{235} in the fuel fissions, producing fission fragments and neutrons. Most of the 200 MeV produced with each fission goes into kinetic energy of the fission fragments. Since they are relatively large, the fission fragments stay within the fuel rods, giving up their kinetic energy in the form of heat. Many of the neutrons produced in fissions, on the other hand, pass through the fuel and zirconium cladding into the water.

The Moderator

When the neutrons get into the water, they strike protons and are slowed down. That is, they are "thermalized." Of course, the neutrons also collide with other particles, but they slow down much less in each of those other collisions. (Remember that the neutron is slowed the most when it hits a particle with a mass equal to its own.) Some of the *thermal neutrons* then enter other fuel rods, where they can be absorbed by U^{235} nuclei and cause fissions. The trick is to have the right number of fuel rods spaced so that, on the average, each fission produces *exactly* one neutron that causes another fission. The excess neutrons produced are absorbed by other materials in the reactor or they escape. Thus, the chain reaction is sustained, neither increasing nor decreasing as time goes by.

There is a built-in safety feature resulting from the use of water as a moderator. If the chain reaction rate rises abnormally, the resulting higher temperature will expand the water slightly. The slightly less dense water then is less effective at moderating the neutrons, and the chain reaction therefore slows down. In fact, if the water in the core is lost for any reason the chain reaction will stop completely. (However, this does not mean that all activity in the core ceases—more about this later.)

The Coolant

The water that moderates the neutrons also removes heat from the core. There are two types of light-water reactors. In the United States about one-third of the commercial reactors are boiling-water reactors (BWRs) and the other two-thirds are pressurized-water reactors (PWRs). In a *boiling-water reactor* the water is allowed to boil at high pressure. That is, there is steam above the core. Around and in the core the water is liquid, so that moderation can continue. The steam is used to power turbines directly and, from the turbines on, the power plant is essentially the same as a fossil-fuel power plant. In a typical BWR the steam pressure is over 1000 pounds per square inch (nearly 70 times atmospheric pressure) and the temperature is about 545°F (285°C). Generally, the fuel-rod assemblies of boiling-water reactors are made with fewer fuel pins per bundle (50 or 60) and more bundles (about 750) than those of pressurized-water reactors, with the total pins being roughly the same.

In a *pressurized-water reactor* (PWR) the water is kept under high pressure and not allowed to boil. Typically, the pressure might be about 2250 pounds per square inch (over 150 times atmospheric pressure), with a temperature of about 620°F (330°C). The *primary coolant* water is piped to a heat exchanger to produce steam in a *secondary coolant*, also water. The steam in the secondary loop runs the turbines. This has the advantage that the water in the turbines does not pass through the reactor core, so it does not become radioactive. On the other hand, the extremely high pressure in the core adds to the danger of an accident occurring.

The Control Rods

Remember that, in order for a chain reaction to be sustained *and* controlled, an average of exactly one neutron per fission is used to produce another fission. There must, then, be some way of controlling the number of thermal neutrons in the reactor at any time. If the reaction starts to increase it must be slowed down, and if it starts to slow down it must be increased. *Control rods* are used as the means of doing this. The control rods are made of a material which is a good absorber of neutrons, commonly an alloy of silver, indium, and cadmium. The control rods can be inserted in place of some of the fuel rods in the fuel bundles. They are generally inserted about halfway in from the top, as shown in Figure 14.5. There they absorb neutrons that

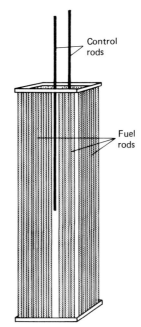

Figure 14.5 A simplified reactor core.

would otherwise cause fissions. If the activity slows, some control rods are pulled out a little, if the activity becomes too great, the control rods are pushed in a little. An additional control mechanism is to put some boric acid in the cooling water. Boron is a good absorber of neutrons.

The control rods are able to provide control because of *delayed neutrons*. When a fission occurs, the neutrons emitted right away are called *prompt neutrons*. These come out about 10^{-17} seconds after the fission. It takes these neutrons about 10^{-4} s to slow down and cause additional fissions. Thus, if only prompt neutrons were involved, the control rods would have to react in less than a ten-thousandth of a second. This they could not do. Fortunately, not all of the neutrons emitted are prompt neutrons. When the fission fragments are first formed, they often contain an excess of neutrons and are in excited energy states. This means they are very unstable and soon disintegrate radioactively by emitting neutrons. These delayed neutrons are emitted on the order of seconds after the fission, and they provide enough of a delay time for the control rods to react.

About half of the control rods are used for routine control of the reaction rate. The other half are reserved for emergencies. If, for any reason, the reactor starts to go out of control, all of the control rods are automatically shoved all the way into the core. This is called a *scram*, and it stops the chain reaction quickly.

Figure 14.6 is a schematic drawing of a nuclear reactor. The core contains fuel rods, some empty spaces, and the control rods, which are inserted from above. Water circulates in and around the core. The *pressure ves-*

sel, which surrounds the core and contains the coolant, is made of steel with walls from 8 to about 11 in. thick. A typical pressure vessel is pictured in Figure 14.7. Its inner surface is lined with stainless steel. It is made to withstand operating pressures of 2500 lb/in.^{2}, with a considerable safety factor built in. In a pressurized-water reactor, the heated water goes to a heat exchanger, where it generates steam in a secondary loop. In addition to the major components shown in Figure 14.6, there are many auxiliary components. These include, for example, sensors to detect numbers of neutrons present in various regions and others to detect leaks, temperatures, flow rates, and the like; pumps, valves, and many other

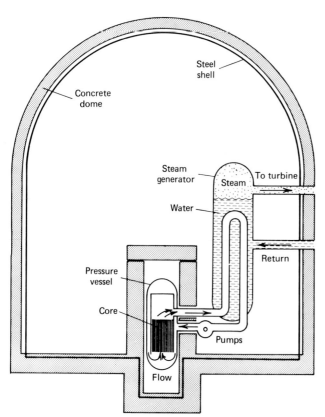

Figure 14.6 A nuclear reactor (pressurized-water type).

Figure 14.7 A pressure vessel, made to contain the core of a nuclear reactor. (Courtesy Project Management Corporation.)

parts. A very important component of every reactor is an *Emergency Core Cooling System* (ECCS). Its function is to provide emergency cooling water to the core in the event that there is a loss of coolant accident (LOCA). This is discussed further in the section on reactor safety.

EXAMPLE

Just to give you some idea of the size of the undertaking, let's calculate the rate at which water has to be pumped through a nuclear plant.

Solution

In our "standard" 1000 MWe plant the efficiency is about 33 percent. Thus, about 2000 MW of heat must be removed. (One-third of the thermal power goes into electric power and the other two-thirds go into wasted heat.) Remember that a megawatt is 1 million watts, and that a watt is a joule per second. Then, the amount of heat (power) discarded is

$$
\begin{aligned}
P_H &= 2000 \text{ MW} \times 10^6 \text{ W/MW} = 2 \times 10^9 \text{ W} \\
&= \frac{2 \times 10^9 \text{ J/s}}{4200 \text{ J/kcal}} \\
&= 4.8 \times 10^5 \text{ kcal/s} \times 60 \text{ s/min} \\
&= 2.9 \times 10^7 \text{ kcal/min}
\end{aligned}
$$

The rate of getting rid of heat was put in units of kilocalories per minute, since most pumps are rated in terms of the amount of water they can pump per minute. For most power plants, it is planned that the temperature of the river water used to cool the plant should not rise more than about 8°C (15°F) in the process. Now we use the familiar equation $\Delta H = mC\Delta T$. I will take the liberty of letting ΔH be kilocalories per minute, in which case it is the same thing as P_H. This will make the mass come out in kilograms per minute to be pumped. Solving for m, gives

$$
m = \frac{\Delta H}{C\Delta T}
$$

If the maximum temperature change is to be 8°C, then

$$
\begin{aligned}
m &= \frac{2.9 \times 10^7 \text{ kcal/min}}{1\dfrac{\text{kcal}}{\text{kg °C}} \times 8°C} \\
&= 3.6 \times 10^6 \text{ kg/min}
\end{aligned}
$$

We can then use the fact that 1 gal of water has a mass of about 3.8 kg to change this to gallons per minute:

$$
\begin{aligned}
\text{flow rate} &= \frac{3.6 \times 10^6 \text{ kg/min}}{3.8 \text{ kg/gal}} \\
&\approx 9 \times 10^5 \text{ gal/min}
\end{aligned}
$$

The answer of nearly 1 million gallons per minute is roughly the amount used in some actual plants, This is a rather sizable flow rate!

OTHER TYPES OF REACTORS

Although U.S. reactor technology has been dominated by light-water reactors, there are alternatives which have some interesting features. It is certainly conceivable that in the future, some of the other types might provide safer and more efficient nuclear power. However, because of the commitment to light-water reactors by the entire industry in the United States, a change in the near future seems somewhat unlikely. Let's take a brief look at a couple of alternatives.

Heavy-Water Reactors

As the name implies, *heavy-water reactors* use heavy water as the moderator and coolant. A good example of this type is the CANDU (Canadian Deuterium–Uranium) reactor produced in Canada. The heavy water contains much more deuterium than ordinary water. The deuterium nucleus, with a proton and a neutron, is about twice as massive as a neutron. Therefore, it does not slow down the neutrons as effectively as ordinary water does, and more collisons are needed to thermalize the neutrons. This means that, on average, a neutron travels farther in the moderator before being thermalized and thus it has a high probability of causing a fission.

One advantage of the CANDU is that it does not need enriched fuel; it runs on natural uranium, with only 0.7 percent U^{235}. Its ability to do so depends on subtle differences in design. For one thing, the neutrons moderated by heavy water are slightly more energetic when they cause fissions. This increases slightly the average number of neutrons released per fission. Also, both ordinary water and heavy water absorb some neutrons. The light water does so by having H^1 converted to H^2, deuterium. In heavy water some of the absorbed neutrons convert H^2 to H^3, tritium. The difference in the two is that there are slightly fewer average absorptions in the heavy water. The net result is that there are more neutrons in the core able to produce fissions. Further, the CANDU is designed so that it has smaller fuel-rod assemblies, with shorter and fewer fuel pins in each bundle. A 600 MWe CANDU has about 4500 such bundles. The bundles are enclosed in pipes in such a way that they can be removed and replaced individually, while the reactor is operating. This on-line refueling capability means that refueling occurs every day, with the replacement of about 15 bundles. Thus, the fuel is always "fresh," and it can be replaced as needed. In present-day light-water reactors, the fuel is replaced once or twice a year on a schedule. If the rate of "burn" has not been as great as expected, the fuel may well be removed before maximum benefit has been gained from it. (Nuclear reactors are designed as base-load units running at near full capacity. However, the experience in the United States in recent years is that demand has not been as high as predicted and reactors are averaging about 60 percent of capacity. Thus, the decision often has to be made about whether to refuel on schedule or to burn up more of the fuel. Refueling is such a large undertaking that violating the schedule is not easy.) The burned-up fuel, just before refueling, has a much lower percentage of U^{235} than does the fresh fuel, so light-water reactors must be designed to operate in a wide range of fuel conditions. This is part of the reason for the necessity to start with enriched fuel.

Because in the CANDU the fuel is in hundreds of individual pressurized pipes in the core, a gross failure of the pressure system is not possible. However, the pipes are connected to *headers*, large pipes collecting all of the flow. Thus, if a header were to break, a large LOCA (loss of coolant accident) could still occur. The efficiency in converting heat to electricity by CANDUs is about 29 percent. Thus far, no CANDUs have been sold to U.S. utilities. One reason may be that the heavy water used costs more than $100 per kilogram.

High-Temperature Gas-Cooled Reactors

The lone commercial reactor in the United States which is not of the light-water type is a *high-temperature gas-cooled reactor* (HTGR). (*Aside*: The use of so many initials in this chapter seems like falling into a giant bowl of alphabet soup and it may strike you as ridiculous. However, if you do any reading on the subject elsewhere, you are bound to run into them. Therefore, I include them here.) The design of this reactor is quite different from the design of the water cooled types.

For one thing, the fuel must be very highly enriched (over 90 percent U^{235}) in facilities like the one shown in Figure 14.8. Then the fuel pellets are embedded in carbon blocks. The carbon serves to give structure to the core and it also acts as the moderator. The whole thing is buried in a huge (about 100 million pounds) concrete structure. The need for the high fuel enrichment results from the fact that carbon is not as good a moderator as water. (Although it is thought that fuel with a considerably lower enrichment would serve nearly as well.) The primary coolant used is helium under pressure. It produces steam for the turbines in a heat exchanger. The high temperature attained means that the thermal efficiency can be about 40 percent. The single HTGR in the United States is at Fort St. Vrain, Colorado, and it has a capacity of 330 MW.

Figure 14.8 A truck prepares to leave the Oak Ridge Gaseous Diffusion Plant in Tennessee. There are five 2.5 ton cylinders, in protective outer drums, containing uranium hexafluoride (UF_6) that has been enriched to contain a higher proportion of U^{235}. (DOE photo by F.W. Hoffmann, OROO.)

The company that made it has left the reactor business.

One advantage of the HTGR is that the carbon provides structural strength, and it gets stronger as it gets hotter, rather than the other way around. Also, if there were to be a LOCA, the huge mass of concrete would act as a heat sink to protect the fuel for an hour or more without coolant. A disadvantage is that the highly enriched fuel is of weapon grade. That is, nuclear bombs could be made from it. Thus, it is vulnerable to being stolen and used by terrorists.

NUCLEAR FUEL RESOURCES

Estimating the amount of uranium in the ground seems to be an even riskier business than estimating fossil-fuel reserves. Uranium is a mineral found underground and exploration for it requires drilling holes and taking samples. The United States has been rela- tively well explored, but there are still large uncertainties in the estimates of total uranium deposits. Vast regions in the rest of the world have not been explored at all. In estimating the uranium reserves there, it is common to use "geological analogy." This is a technique in which the rest of the world is assumed to have about the same distribution of uranium as the most-explored region—the United States. Estimates thus arrived at are open to serious question. Further, it is impossible to predict what demands will be made upon future world uranium finds by other nations. Therefore, I shall limit this discussion to U.S. resources and assume that those resources are used in reactors in the United States. This will not necessarily be the case—there is now considerable international trading in uranium— but we need some reasonable assumptions to start from.

Uranium is found in nature as U_3O_8, known as "yellowcake." This oxide of uranium is scattered throughout the Earth's crust in varying degrees. In places where it constitutes a great enough percentage of the crust, it is mined for nuclear fuel. In New Mexico, where much of our reserve is known to be, the concentration of yellowcake in minable deposits is about 0.17 percent. In Canada, small deposits of up to 65 percent yellowcake have been found. If uranium becomes scarce, concentrations that are not economical to mine now may become minable. In 1978 the U.S. Department of Energy estimated the potential resources of this country. Included were known and possible deposits from which the uranium could be recovered for less than $50 per pound of yellowcake. Those estimates are given in Table 14.1

If we count only the proven and probable reserves, there are about 2 million tons of yellowcake in the United States. This is by no means a certain number. For example, the National Research Council set up a study committee to investigate the energy situation (The Committee on Nuclear and Alternative Energy Systems—CONAES). Their estimates of uranium resources, backed up with expert opinion, were only about half as large as the Department of Energy estimates. Other experts think that the DOE estimates are too conservative and that perhaps twice as much uranium exists.

As a happy medium, let's use the DOE estimate that there are about 2 million tons of uranium available. One estimate of the amount of energy this will produce in nuclear reactors is about 1000 quads. This figure is for light-water reactors with no reprocessing of the fuel,

and it may be compared to 5000 or 5500 quads of fossil fuels in the United States, mostly coal.

Rather than approach the figure for the uranium resources as we have the fossil-fuel resources, let's try a different tack. Our "standard" 1000 MWe reactor will need about 4000 to 6000 tons of yellowcake to keep it running for its 30- to 40-yr lifetime, with no reprocessing of fuel. As an average, let's take 5000 tons as the needed amount. Then, a resource of 2 million tons will fuel about 400 reactors. In the period around 1970, most of the experts were predicting a great rate of growth of nuclear power plants, with about 150 predicted for 1980 and around 1000 for the year 2000. Obviously, the supply of fuel would not last very long at that rate. Long before the year 2000 there would be enough reactors to use all of the available fuel, and it would seem foolish to build more after that unless new resources were found in the meantime.

As it seems to be turning out, the situation is not quite as drastic as all that. Instead of the 150 reactors predicted for 1980, there are about 80 now operating. Since 1977 over 70 planned reactors have been canceled or postponed. There are several reasons for this. For one thing, the rate of energy consumption has grown, but not as fast as predicted. The nuclear plants now operating are not running at full capacity. Then, too, it has become very difficult, time-consuming, and expensive to build a nuclear power plant. The average time between when the decision to construct a plant is made and when the plant goes into operation is now about 14 or 15 years. The cost has become almost astronomical. In recent years, no nuclear power plant has been built for the amount of money originally planned, or even close to it. Some 1000 MWe plants have been planned to cost about $300 to $400 million but by the time they are completed they have run well over $1 billion. For example, the two-unit Marble Hill plant under construction in Indiana is expected to be completed in 1988 at a total cost of over $6 billion!

TABLE 14.1 *Uranium Reserves in the United States (Tons)*

Proven reserves	890,000
Probable reserves	1,395,000
Possible reserves	1,515,000
Speculative	565,000
Total	4,365,000

Further, since the cost of electric energy has increased greatly in the past few years, people seem to be doing a better job of conserving. The lower than expected use of energy plus the generally poor state of the economy, coupled with high interest rates, have put the utility companies in a rather bad financial position. A nuclear power plant represents an investment of several billion dollars which will not start paying off for as much as 20 years. (If a loan is taken, the interest might easily total an amount greater than the cost of the plant.) Only the largest utility companies can afford to think about such an investment, whatever the prospect of a long-range payoff might be.

Also, building nuclear power plants has become an uncomfortable proposition. It is almost bound to attract strong criticism and probable demonstrations. It may very well be held up by problems with labor unions. Faulty work or materials during construction could add substantially to the time required and the cost. Changing government safety regulations may force costly changes during construction. (For example, new information on the possibility of earthquakes in the locality has required redesign of one plant after construction was well along.)

Finally, the specter of serious accident hangs over the heads of the decision makers in the power companies. One accident might put a company into bankruptcy. The costly accident at Three Mile Island in Pennsylvania has underscored this possibility.

Nonetheless, let us suppose that the nuclear power program in the United States is not dead. It certainly is not dead in several other countries, which need the power desperately. (For example, Japan is totally dependent upon foreign oil and natural gas and is trying to ease the situation with nuclear power.) You have already seen what some of the consequences are of trying to live off the fossil fuel supplies. If the number of nuclear power plants grows only to one-third of the

original prediction by the year 2000, that will still be more than 300 plants. How much longer will it take to build plant 400? Will it be the year 2020? 2040? Whenever that happens, then no more of the present type of plant can be fueled with known resources. Uranium seems not to be the infinite source of energy it was once thought to be. Used as we now use it, it will run out in the not-so-far-distant future.

There are ways out of this particular dilemma. That is, we cannot increase the amount of uranium that exists in the Earth, but we can possibly increase dramatically the amount of energy to be gained from that uranium. There is also another potential fuel available. As it is sometimes stated, there is a technological "fix" available. The question we all have to face—and soon—is whether the fix is worth the possible social costs.

EXTENDING THE NUCLEAR FUEL

Fuel Reprocessing

Part of the "fix" referred to is the reprocessing of fuel. Even the enriched uranium in light-water reactors is about 97 percent uranium-238. With all of those neutrons flying around in the core, the U^{238} does not just sit there unchanged. Some of the U^{238} nuclei absorb neutrons. Only, instead of fissioning as do U^{235} nuclei, the U^{238} nuclei undergo a transformation to plutonium. The sequence of events, starting with the absorption of a neutron, is given by the following equations:

$$_{92}U^{238} + n \rightarrow {}_{92}U^{239}$$
$$_{92}U^{239} \rightarrow {}_{93}Np^{239} + e^- \text{ (23.1 minutes)}$$
$$_{93}Np^{239} \rightarrow {}_{94}Pu^{239} + e^- \text{ (2.35 days)}$$

The times in parentheses represent the half-lives of U^{239} and Np^{239}. Both of them are short, but the Pu^{239} produced has a half-life 24,360 years. Thus, when the spent fuel is removed from the reactor the U^{239} and Np^{239} will be gone, but Pu^{239} will remain. In LWRs, the rate of plutonium production can be as high as 60 percent of the rate of fission of U^{235}. (That is, for every 100 U^{235} nuclei that fission, about 60 Pu^{239} nuclei are produced.) This plutonium can be removed from the "waste" material in the core after the U^{235} has been exhausted. In fact, even though it requires the handling of highly radioactive materials, the extraction of the Pu^{239} is relatively easy because it is different chemically from the uranium with which it is mixed.

The important thing about plutonium is that it is fissionable. That is, it can sustain a chain reaction with slow neutrons, just as U^{235} can. Thus, a reactor (or a bomb) can be made using plutonium. If the spent fuel from a reactor is reprocessed to recover the plutonium and the remaining U^{235}, then that reclaimed material can be used as fuel. By reprocessing the fuel, the total amount of uranium needed for the lifetime of a reactor could be reduced by at least 15 percent; and possibly much more than that. This could significantly extend the lifetime of the available uranium.

There always seems to be a catch. One problem with reprocessing is that plutonium is a very dangerous substance. As mentioned in Chapter 13, plutonium readily forms an aerosol and thus is easily breathed into the lungs. There the radiation is very damaging. Plutonium can also form radioactive salts which will dissolve in blood and be carried to every part of the body.

The Carter Administration decided that the amount of uranium available is sufficient to carry the nation through to the use of renewable resources, such as solar energy, without the need of reprocessing. That judgment, coupled with the dangerous nature of plutonium, led the administration to place a ban on the reprocessing of the spent fuel. At the time of this writing no reprocessing has occurred for some years. However, it seems that the Reagan Administration disagrees with this judgment, and reprocessing will soon start.

Breeder Reactors

Another possible "fix" for getting more energy from the nuclear fuel available is the *breeder reactor*, mentioned in chapter 13. By surrounding the core of a specially designed reactor with U^{238}, it is possible to produce more plutonium. In fact, it is possible, at least in theory, to produce more fuel than the reactor consumes. That is, for each U^{235} (or Pu^{239}) nucleus which undergoes fission, *more* than one Pu^{239} nucleus is produced from the surrounding uranium. A reactor which does this is called a breeder reactor. The production of Pu^{239} nuclei is called *breeding*, and the number of nuclei converted divided by the number of fissions is called the *breeding ratio*. It is believed that breeding ratios of 1.2 or greater can eventually be achieved. At present, the type of breeder reactor which is favored by the government is called a *liquid-metal fast breeder reactor* (LMFBR). By reprocessing the uranium, somewhat more plutonium can be recovered than is needed to fuel the breeder reactor itself. The time needed to accumulate enough extra fuel to start up an equal size reactor is called the *doubling time*. Practical doubling times of about 19 or 20 years can probably be expected.

In a breeder reactor, the core is fabricated as shown in Figure 14.9. The fuel rods contain fissionable (called *fissile*) material in the centers. The ends of the fuel rods contain *fertile* material, nuclei which can be converted to fissile nuclei. Also, surrounding the fuel rods are similar rods which contain only fertile material. The result is that the active core is surrounded by a *blanket* of fertile material which can absorb neutrons and become fissile material.

The word *fast* in LMFBR describes the neutrons. The LMFBR uses "fast neutrons" rather than "slow," or thermalized, neutrons to produce the fissions. As mentioned earlier, although fissions are not quite as likely to occur with fast neutrons, more neutrons are

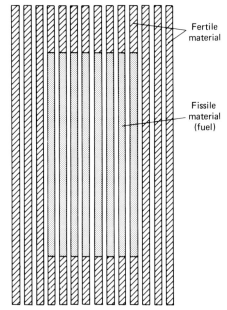

Fertile material

Fissile material (fuel)

Figure 14.9 A "slice" through the core of a breeder reactor. The fuel rods contain fissile material in the centers and fertile material in the ends. This core is surrounded by rods containing only fertile material. The result is a "blanket" of fertile material surrounding the core.

produced per fission caused by them. Therefore, with the same number of fissions, fast neutrons will produce more neutrons in the core. However, since the probability of fission is less, the fuel has to be more highly enriched. The fuel in a LMFBR will contain about 15 percent fissile nuclei.

In a fast-neutron reactor, water cannot be used as the coolant, for it would moderate the neutrons too greatly. That is where the "liquid metal" comes in. The coolant used in a LMFBR is molten sodium. With nuclei much more massive than neutrons, sodium does not slow the neutrons down very much in collisions. Because of its high boiling point (892°C) and good heat conducting properties, sodium can provide the needed cooling at low pressures. Thus, the reactor vessel and the pipes do not have to withstand high pressures.

However, liquid sodium is a highly chemically reactive material! It reacts violently with either air or water, burning so fast as to be explosive. Therefore, it has to be kept away from both air and water.

A further danger of LMBFRs is that a loss of coolant accident has just the opposite effect on the chain reaction as it does in a light-water reactor. You recall that in the light-water reactor, loss of coolant means loss of moderation and the chain reaction stops automatically. If the sodium is lost in the liquid-metal reactor, fewer neutrons are absorbed and the neutron flux becomes greater, thus increasing the rate of the chain reaction. Also, since the neutrons move faster, there is less average time between the time of creation of a neutron and its absorption to cause another fission. Therefore the time available for changing the positions of the control rods is shorter, and the control of the reactor is much more critical in the LMFBR. A prototype LMFBR was built commercially in Michigan during the 1960s. Called Enrico Fermi I, it was operated for a few months in 1966 at its full power of 200 MWe. However, problems developed, and it was shut down. Then, in October of the same year, it was started up again. With everything going well at 10 percent of full power, the operator started to move the control rods out slightly to increase the power. The chain reaction started to "run away," and the reactor was quickly scrammed. Later, it was discovered that there had been some loss of circulation of the coolant and that a partial melting of the fuel occurred. After several years of having problems trying to get the reactor into operation again, it was given up and shut down permanently. The incident points out the fact that the safety factors included in the design of fast breeder reactors must be even more carefully worked out than those for light-water reactors.

The primary sodium coolant in a LMFBR becomes highly radioactive after a period of operation. (Na^{23} is changed to Na^{24} by neutron bombardment, and this product emits high-energy gamma rays with a half-life of 13 hours.) Thus, it is not used directly to produce steam. A secondary loop of liquid sodium is used to isolate the primary loop from the steam generators, as indicated in Figure 14.10. The reactor vessel and the heat exchangers which contain sodium are all surrounded by guard vessels, not shown in the figure, so that large amounts of sodium will not be lost in the event of an accident.

The reactor described here is not the only possible breeder reactor design, but it is the one that has been supported by U.S. Government funding. A European design uses a "pool" of liquid sodium in the reactor vessel, with the first heat exchanger there. This contains the highly radioactive sodium within the reactor vessel. Other possible designs include gas-cooled breeder reactors and reactors that use thorium as the fertile material. There is somewhat more thorium than uranium in the Earth, so the amount of energy available could be doubled if thorium can be used in addition to uranium. In a thorium breeder reactor, the naturally occuring Th^{232} would absorb a neutron to become Th^{233}. This then goes through the following decays:

$$_{90}Th^{233} \rightarrow {}_{91}Pa^{233} + e^- \ (22.1 \ min)$$
$$_{91}Pa^{233} \rightarrow {}_{92}U^{233} + e^- \ (27.4 \ days)$$

The U^{233} produced has a half-life of 1.6×10^5 yr, and it is a fissionable material. In addition to producing more potential fuel, the thorium breeder would have the advantage of being a water-cooled, slow-neutron reactor. That reactor technology is much better understood and developed than is the LMFBR technology. Thus, it would probably be much safer. A disadvantage is that U^{233}, like plutonium, is extremely toxic.

Almost all of the U.S. research on breeder reactors has been concentrated on the LMFBR. The Carter Administration, being convinced that enough uranium existed to provide the needed energy without breeding or reprocessing, substantially slowed down the breeder reactor research program. In fact, except that he was barred from doing so by Congress, President Carter probably would have dismantled the Clinch River demonstration reactor. It appears that President Reagan's administration will emphasize the development of the LMFBR. The Fast Flux Test Facility (FFTF) pictured in Figure 14.11 has been an important part of the Department of En-

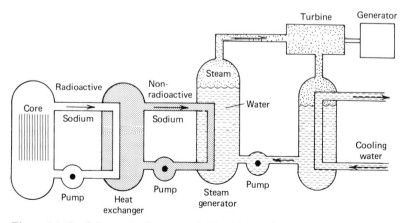

Figure 14.10 Schematic diagram of a liquid-metal fast breeder reactor (LMFBR).

Figure 14.11 The FFTF (Fast Flux Test Facility) at Richland, Wash. (Hanford Engineering Development Laboratory photo.)

ergy's Liquid Metal Fast Breeder Reactor research program.

THE BIOLOGICAL EFFECTS OF RADIATION

The most serious danger of nuclear power is, of course, that somehow humans may be exposed to radioactive materials. Everyone knows that radiation is harmful. However, there is probably more ignorance and fear of this type of danger than there is of any other one can think of, unless it is ghosts and goblins. There should be a healthy respect for radiation, as for any harmful agent, but ignorance breeds an unthinking fear which can distort judgments. In this section, we shall take a look at some of the effects of radiation on humans and, later, we will examine the

extent of the risks that the use of nuclear power could release radiation upon us.

Before discussing the effects of radiation, we need some measure of amounts of radiation. A unit which has come into common use is the *rad* (*radiation absorbed dose*). It is defined as the amount of radiation that, when absorbed by some material, deposits 6.24×10^7 MeV of energy per gram of material. (This energy is equivalent to 0.01 joules per kilogram.) When the material, whether it is human flesh or steel, absorbs the energy, some damage is caused by collisions of the radiation with the electrons and nuclei of the material. This can consist of ionization, of nuclear reactions, or of knocking atoms out of place in a crystal or molecule.

The rad is not quite satisfactory in describing the effects of radiation on people because the different kinds of radiation absorbed in the same amounts are harmful in differing degrees. For example, it is estimated that energy deposited by alpha particles will cause about 20 times as much damage to tissue as the same amount of energy deposited by gamma rays. Therefore, the unit of a *rem* (*roentgen equivalent man*) was invented to take this into account. The radiation dose measured in rems is the same as the number of rads multiplied by a *quality factor*. The approximate quality factors for the different kinds of radiation are given in Table 14.2.

The damaging effects of radiation can be divided into two general categories: *somatic* damage and *genetic* damage. Somatic damage simply means damage to the cells of the body. Genetic damage is damage to a particular group of cells, the genes which determine heredity. Children born to persons with damaged genes are likely to have *mutations*, characteristics different from those they might normally have inherited. Those mutations may be harmless, or they may result in horrible deformation or retardation, stillbirths, and other serious defects. (On the other hand, mutations may result in super genius, but that is very unlikely.) It is very difficult to assess the amount of genetic damage done to a population, for some of the effects may not show up for many generations.

The effects of large doses of radiation are well known. Because of damage to the tissues and organs, persons exposed to large doses of radiation show symptoms of nausea and weakness, high white blood cell counts, and

low red blood cell counts; their hair falls out, the skin may redden or blister, and in severe cases death may follow in a matter of days or weeks. A dose of over 25 rem accumulated in a few minutes or hours can cause radiation sickness. A dose of over 100 rem in a short time period will usually cause radiation sickness, although the person will probably recover. A dose of 400 rem will result in the death of about 50 percent of the persons affected. The other 50 percent will recover. However, of those that recover, a greater than normal percentage will develop cancer or leukemia in later years. A dose of 1000 rem or more is almost certain to cause death within a few weeks. If it is spread out over a period of years, even a large dose of radiation may not produce radiation sickness. However, a person exposed to those doses may have an increased chance of getting cancer or having genetic damage. The effects outlined here are for a *whole body dose*, where the radiation is spread more or less evenly throughout the person's entire body. In some instances certain organs, such as the thyroid gland, will receive doses much more concentrated than those received by the rest of the body. In that case, the figures would be quite different. Unless stated otherwise, the doses mentioned in what follows will be whole body doses.

The effects of exposure to low levels of radiation are much more difficult to determine, and there is not complete agreement among the experts. For example, some think that there is a radiation "threshold," a level of radiation below which there are no harmful effects whatever. Others believe that there is no threshold, and that radiation in any amount, all the way down to zero, is harmful. The International Commission on Radiological Protection (ICRP), which recommends standards for allowed radiation, takes the more conservative view that there is no threshold. The standards they recommend include maximum permissable doses. But they recommend always to keep the exposure to radia-

TABLE 14.2 *The Relative Effects of Various Types of Radiation*

Type of radiation	Quality factor
Gamma rays	1
X-rays	1
Beta particles	1
Slow neutrons	2.5
Fast neutrons	10
Alpha particles	20

tion as low as possible. The Nuclear Regulatory Commission (NRC), which is the U.S. body for setting and policing the regulations, requires that exposures to radiation be kept "as low as is reasonably achievable."

Because the effects may not show up for years or generations and they are masked by natural effects, it is difficult to know how much damage is caused by low-level radiation. For example, there are roughly 360,000 deaths per year from cancer in the United States. How could one then tell if there were a few dozen, or even a few hundred, extra deaths caused by the use of nuclear power? You see the problem. It is even more difficult to judge the genetic effects, since they may not show up in the lifetimes of anyone now living. Let us see if we can arrive at some estimate of the worst possible case concerning cancer.

EXAMPLE

If the conservative theory that there is no threshold for the harmful effects is correct, and if the effect is linear all the way to zero radiation, then the concept of a "person-rem" is useful. That is, one can assume that a given (low) total dose of radiation will eventually cause the same number of deaths whether it is spread out over a few hundred or thousands of people. One frequently used estimate is that an exposure of about 10,000 person-rems will, on the average, cause one cancer death in the following 20 or 30 years. According to the theory that the effect is linear (no threshold), it would not matter whether this dose were 1 rem each to 10,000 persons or 10 rems to each of 1000 people; the result would still be one additional cancer death. (Incidentally, more and more responsible scientists seem to be coming to the conclusion that the effect is not linear and that a threshold exists. It is suggested that a whole body dose of less than 50 rem to a person will not

result in any detectable increase in the risk of cancer. If this is correct, the estimates of probable risk using the linearity theory are much too high. However, we shall be conservative and stick with that for the time being.)

Let's take the worst case we can think of short of a disaster. Suppose that some time in the future, a large enough number of reactors were operating so that the general public was receiving the maximum radiation allowed by present NRC regulations. (We are very much below that limit now. Just how far below, you will see later.) The maximum now allowed for the general public is a total dose of 0.5 rem/yr per person. We now get well over 0.1 rem/yr from sources other than nuclear power plants, so let's assume that 0.4 rem/yr will come from nuclear power.

What is the total dose, in person-rems per year, to the total population of about 250 million people? Assuming one death per 10,000 person-rems, how many cancer deaths per year will the radiation cause? What percentage is this of the total number of deaths from cancer? How does it compare to 50,000 deaths per year, or so, from traffic accidents?

Solution

If each person gets a dose of 0.4 rem, the total number of person-rems is:

$$\text{dose} = 0.4 \text{ rem/yr} \times 2.5 \times 10^8 \text{ persons}$$
$$= 10^8 \text{ person-rem/yr}$$

Assuming one death for every 10,000 person-rems, the number of deaths expected would be:

$$\text{Deaths} = \frac{10^8 \text{ person-rem/yr}}{10,000 \text{ person-rem/death}}$$
$$= 10,000 \text{ deaths/yr}$$

With about 360,000 deaths from cancer per year, the percentage caused by this exposure to radiation would be:

Percentage of cancer deaths

$$= \frac{10,000}{360,000} \times 100\%$$

$$= 2.8\%$$

This is about 20 percent of the number of traffic fatalities per year.

Note well that, if there is a threshold and it is above 0.4 person-rem, then no additional cancer deaths could be attributed to this kind of radiation dose distributed evenly among all of the population.

This exercise gives you a little peek into the potential cost of nuclear power, measured in human lives. The question facing all of us is what price are we willing to pay? In the next section we take a closer look at some of the potential harm from nuclear power.

NUCLEAR REACTOR SAFETY

Concerns about the safety of nuclear reactors break down into four areas:

1. Possible harmful effects of routine operation.
2. Serious accidents.
3. Diversion of plutonium or uranium for nuclear weapons.
4. The handling of nuclear fuels and wastes.

We shall consider each of these, in turn.

Routine Operation

In normal operation, nuclear power plants contain highly radioactive materials. Contrary to popular opinion, the uranium used for fuel is not the source of most of that radioactivity; it is only slightly radioactive and when new fuel is supplied for a reactor, relatively few precautions have to be taken to shield the radiation. However, when the fuel

has been active for a time, highly radioactive fission fragments and by-products are produced. Table 14.3 is a list of some of the fission fragments from the thermal fission of U^{235}, their half-lives, and the approximate percentage each represents of the yield.

A look at this table quickly gives a clue about which of the fission fragments could be troublesome in the environment. Stable isotopes present no radiation problem, for they give off no radioactivity. Isotopes with very short half-lives are very radioactive, but they last only a relatively short time. For example, xenon-135, with a half-life of under 10 hours, decays very rapidly. After 10 half-lives there will only be about one-thousandth of any radioactive substance left. In this case, 10 half-lives is only about four days. A nuclide like cesium-135, on the other hand, with a half-

TABLE 14.3 *Fragments from the Fission of U^{235}*

Fragment	Half-life	Percentage of fragments produced
Krypton-83	Stable	0.6
Krypton-84	Stable	1.0
Krypton-85	10.8 yr	0.3
Krypton-86	stable	2.0
Strontium-89	50.6 days	4.8
Strontium-90	28.8 yr	5.8
Ruthenium-106	1.0 yr	0.4
Tellurium-132	77 hr	4.7
Iodine-129	1.6×10^7 yr	0.8
Iodine-131	8.1 days	3.1
Iodine-133	20.8 hr	6.9
Iodine-135	6.7 hr	6.1
Xenon-132	stable	4.4
Xenon-133	5.3 days	6.6
Xenon-134	stable	8.1
Xenon-135	9.2 hr	6.3
Xenon-136	stable	6.5
Cesium-133	stable	6.6
Cesium-135	2.6×10^6 yr	6.4
Cesium-137	30 yr	6.2
Barium-140	12.8 days	6.4
Cerium-144	285 days	6.0

life of 2.6×10^6 yr, lasts a long time, but it has a relatively low radioactivity.

Two really troublesome fission fragments are strontium-90, with a half-life of 28.8 yr, and cesium-137 with a half-life of 30 yr. They are strongly radioactive and, if they were to escape into the environment, half of their substance would still be there after 30 years. One-fourth of it would remain after 60 years. Further, since they are both very active chemically, they tend to be carried around and into human bodies through the food chain. Strontium is particularly bad because it replaces calcium in chemical reactions. Thus, it collects in milk and in the bones. High concentrations in the bones can cause damage to the bone marrow, the red-blood-cell producing organ. Cesium also tends to become very concentrated in milk.

Iodine is another especially dangerous material, because it gets concentrated in the thyroid gland. There it can cause thyroid cancer. Thus, even though it has a relatively short half-life, radioactive iodine released in large quantities would be particularly harmful. Of the various isotopes of iodine produced, I^{131} is the most harmful; it has both a high level of radioactivity and a longer lifetime than I^{133} or I^{135}.

There is a way for people to avoid much of the harmful effects of I^{131}, in the event some were spilled into the environment. It is to take potassium iodide, in either tablet or liquid form. If this is done, the thyroid gland soon becomes saturated with iodine and will not accept the radioactive iodine. Since the half-life of I^{131} is short, this treatment could be continued until any I^{131} in the environment had decayed away. However, there is not yet any widescale distribution of potassium iodide in this country, and that would be needed in advance of a nuclear accident.

Questions

Suppose there were a major release of radioactive iodine into the environment and that exposed persons were able to get sufficient potassium iodide immediately. Also suppose that people were advised to take the potassium iodide until only about one-thousandth of the original radioactive iodine remained. About how long would that be? (The I^{129} can be ignored; it is only mildly radioactive and will decay slowly for millions of years. Fortunately, it is also a smaller component of the fission products than are the other isotopes.)

Let's go about this one step at a time. First, how much I^{131} would remain after 8 days, its half-life? How many half-lives does this represent for I^{133}? How much of the original I^{133} would remain at the end of the 8 days? How much I^{135} would remain? How much of each would remain after 16 days?

Do you see that after a couple of half-lives of I^{131}, the other two isotopes would have essentially disappeared? If that is the case, then we only have to worry about the I^{131}. Since it represents about 20 percent of the original iodine released, we would want to take the potassium iodide until about 1/200 of the I^{131} remained. About how long would that take?

Fortunately, in the normal operation of a nuclear reactor, most of the highly radioactive nuclei are well contained in the fuel rods. However, some of the radioactive elements diffuse out of the ceramic fuel pellets and through the cladding into the cooling water. That water is continually filtered to remove the radioactive material. Most of it is removed this way, and it is then held in tanks to let it decay. The remaining water can also be held for a time, then slowly mixed with the tremendous flow of cooling water to the outdoors, so that it is very much diluted.

In boiling-water reactors, some of the radioactive gases get into the steam. The important gases are krypton-85 and tritium (hydrogen-3), with half-lives of 10.8 yr and 12.3 yr, respectively. These are removed from the steam and slowly vented to the atmosphere.

Obviously, then, nuclear power plants pollute the environment with radioactive wastes in their routine operation. The question is, how much? To put the answer to this in perspective, look at Table 14.4, which gives the approximate radiation Americans are exposed to from all sources. The dose is *millirems* (mrem) per year. A mrem is 0.001 rem.

This list deserves some comment. First, cosmic rays come to us from outer space and the sun. There is not much we can do to avoid this radiation. The reason for the large variation in cosmic radiation, from 23 to 50 mrem per year, is that the atmosphere shields us to some degree. Thus, there is less cosmic radiation at sea level than at a higher altitude. For example, if you were to move from Dallas to Denver, the extra radiation you would receive annually would be more than 25 times that coming routinely from nuclear power.

There are also radioactive materials in your body (and in your food, in the air you breathe, etc.), and this contributes about 20 mrem a year. The Earth has radioactive materials scattered throughout it. The number given here for radioactivity from the ground and surroundings is an average for the United States. Some regions have much more. Some rocks, such as granite, tend to be somewhat radioactive. One writer has stated (facetiously?) that if Grand Central Station in New York City were a nuclear power plant, it could not be

licensed by the Nuclear Regulatory Commission because its granite produces too much radioactivity. In some parts of India and South America, which have sands that are rich in uranium and thorium, inhabitants receive a dose of about 1500 mrem/yr. Here in the United States, in the West and Southwest, tailings (wastes) from uranium mines have sometimes been used in the construction of buildings as fill dirt and even in cement and mortar. Those buildings are relatively "hot," and people who live or work in them get more than the average radiation from their surroundings. (This practice has now been stopped.)

Medical and dental x-rays contribute their share. The 50 to 75 mrem cited is an average dose, and it could be much larger than that for any given person. A single chest x-ray would be 50 or 55 mrem, and a dental x-ray about 20 mrem. People who undergo radiation treatment for cancer receive doses large enough to produce the symptoms of radiation sickness. The weapons fallout component in the table remains from the days when we were silly enough to test nuclear weapons in the atmosphere. Air travel exposes us to extra cosmic radiation because it takes place above much of the atmosphere; and radioactive items, such as the watch you used in the activity in Chapter 13, also contribute a little bit.

Against all of this background, the small extra radiation received from nuclear power plants does not seem like very much. You can get as much extra radiation in a year by taking a couple of long airplane trips or spending a week or two skiing in the Rockies. Even if the nuclear power plants were all replaced by coal-burning plants, the amount of extra radiation would not decrease. There is enough radioactive material in coal so that the emissions from a coal-powered plant produce more radioactive fallout than the amount produced by a nuclear plant of the same capacity; in addition, coal burning puts millions of tons of pollutants into the air. (It is not clear to me just how much more radioactivity is produced

TABLE 14.4 *Normal Radiation Doses for Americans*

Source of radiation	Dose (mrem/yr)
Cosmic rays	23 to 50
Radioactivity in the body	28
From ground and surroundings	26
Medical and dental X-rays	50 to 75
Air travel, miscellaneous	5 to 6
Weapons testing fallout	4 to 5
Nuclear power	1 (or less)
Total	about 140 to 190

by a coal-fired plant. The highest estimate I have seen is over 400 times as much as an equal size nuclear plant and the lowest is about 10 percent greater.)

To summarize the evidence, when a nuclear power plant is operating *properly and legally*, it contributes very little extra radiation to the already present background. Because of its lower thermal efficiency, it does provide more thermal pollution than that produced by a modern coal-fired plant of the same capacity, but that difference may disappear when the design of reactors becomes more advanced. Some nuclear opponents fear that nuclear plants may purposely release more than the legal amounts of radioactivity to the environment. That practice must be guarded against, as must the release of chemicals and other pollutants by industry. (So far, we seem to be guarding against the release of excess radioactivity far better than against other pollutants, especially chemicals!)

There is one group of persons who are affected more by the use of nuclear power than is the general public. Those are the persons who work in the nuclear power industry: uranium miners, persons working in the refining mills and processing plants, and persons working in the power plants themselves. The standard set by the Nuclear Regulatory Commission for the general public is that the maximum annual exposure from all sources should be no more than 500 mrem (0.5 rem) per year per person. Recently, the Environmental Protection Agency set the tougher standard that no more than 25 mrem per year above the background radiation must come from nuclear power. In addition, the Nuclear Regulatory Commission requires that persons living near a nuclear plant should receive no more than 5 mrem per year whole body dose or 15 mrem per year to the thyroid gland from that plant.

However, for employees exposed to radiation in the workplace, the limit is 5000 mrem (5 rem) per year. Employees are monitored

and, if the dose of radiation they are getting is too high, they are shifted to other parts of the operation, where there is less chance of exposure, or they are furloughed for a time. Studies of the incidence of cancer are rather inconclusive, but it seems quite possible that it is somewhat greater for nuclear workers than for the general public. One way of looking at this is that the nuclear workers may be assuming most of the risk of nuclear power for the rest of us. However, thus far that risk seems to be much less than the risk of injury or death to workers in the fossil-fueled power industry. (It must also be added that not everyone agrees about the level of radiation that is already being received by the public. One study, done in Germany, claims that the NRC estimates of exposure by persons living near nuclear plants are too low because they do not account for all of the pathways by which radioactive materials can enter the body, particularly through food grown in the area. Experts on the other side claim the German report is flawed and invalid. Once again, it seems to be a case of having to choose which group of experts to believe.)

Serious Accidents

Obviously, the routine daily operation of a nuclear power plant is not very harmful. Compared to the presently known alternatives for producing large-scale electric power, nuclear power is very clean and nonpolluting. However, thoughtful critics of nuclear power still have concerns. One of these is of the possibility of a serious accident, which could release large amounts of radioactive materials upon the countryside.

Reduced to simplest terms, the three safeguards against accidents are: quality control, redundancy of safety systems, and operator training. The building and manufacture of the reactor, the containment building, and all of the thousands of components are controlled as carefully as possible. This is what is meant by "quality control," and it is much more stringent for nuclear power plants than conventional quality control is in industry. "Redundancy" of safety systems means that, backing up a system will be a second system, and backing *that* one up may be a third, and so on. Finally, the need for well-trained and highly skilled operators should be obvious.

Nonetheless, the components of nuclear reactors fail regularly, and human errors enter in. The Nuclear Regulatory Commission requires the utility companies to report "incidents or events that involve a variance from the regulations such as personnel overexposures, radioactive material releases above prescribed limits, and malfunctions of safety-related equipment." The NRC then investigates the events reported. In 1979, for example, the NRC received reports of 2300 such incidents from the utilities. The most common incident reported is the failure of a valve. Because of

the redundancy built into the systems, the failure of a component, such as a valve, is not usually serious in itself; there are always one or more backups available. However, as you know, the accident in 1979 at the Three Mile Island plant, shown in Figure 14.12, seriously strained the limits of the safety systems, and it was caused by multiple failures of both equipment and humans.

Rather than going into all of the possible accidents in nuclear plants and their consequences, let us concentrate on the most serious, the dreaded loss of coolant accident (LOCA). Now "small" losses of coolant should be no great problem. "Make-up" pumps could provide water to replace up to 1000 gal/min, if such a leak should occur, until repairs could be made. The real danger comes if a large pipe should suddenly burst, or if a steam turbine should disintegrate explosively. In that case the great pressure in the reactor vessel would

Figure 14.12 The Three Mile Island nuclear power plant, where a serious accident occurred in 1979. (Grant Heilman.)

quickly force the cooling water out, with much of it flashing instantly into steam, in what is called a blowdown. The reactor vessel might lose all or most of its water in less than a minute. Whether any cooling water might remain in the core is not known, and it will depend on the details of the blowdown.

Consider the following imaginary sequence of events. After operating for four and a half years, Number 2 turbine at Unmitigated Edison's Happy Valley plant develops an undetected strain. The Happy Valley reactor is of the boiling-water type. One hot summer afternoon, during the peak demand period, Number 2 suddenly starts to wobble and, before it can be shut down, it shakes itself to pieces. Superheated steam fills the generating room, escaping at a tremendous rate. At this point, a valve between the containment building and the generating room is supposed to close automatically, shutting off the steam, but it fails to do so. A second, backup valve also fails, and the escape of steam continues.

In less than a minute the blowdown is complete, and the reactor core is about two-thirds uncovered. The Emergency Core Cooling System (ECCS) should have gone into action automatically by now, but somehow part of the local power grid was knocked out by the accident. That means that there is no emergency power coming into the plant from the outside, and thus no power for the ECCS pumps. (They need about 6 MW to operate). Automatic flooding of the core with thousands of gallons of water held in reserve has occurred, but it is only partially effective. There are several diesel engine-powered generators to provide emergency power, and they should have started automatically, but something went wrong and they did not. People are dispatched—in a hurry—to start them manually. The ECCS starts pumping great amounts of cooling water through the core within two minutes of the time of the original break.

Meanwhile, before the ECCS starts, the core begins to heat rapidly. This heat is not due to a chain reaction. When the steam pressure started to drop, the computer controlling the reactor automatically scrammed the control rods, and the chain reaction immediately stopped. However, all of the radioactive fission fragments in the fuel rods continue to decay radioactively, releasing energy called the *afterheat*. Immediately after a scram, the afterheat is about 7 percent of the full-power output. In the case of this 1000 MWe reactor, that amounts to 230 MW, a very considerable amount of power. (Do you see why it is 230 MW and not 70 MW?) After about two minutes, because of the very short half-lives of some of the nuclides in the fuel rods, the afterheat drops to 3.5 percent of full power and, after a day or so, it will be less than 1 percent.

Without cooling water to carry it away, the afterheat rapidly raises the temperature of the core. When the fuel rods get to about 1000°C, the zircaloy cladding becomes rather ductile; that is, it bends and stretches easily. When the cooling water was lost, the pressure on the outside of the fuel rods decreased dramatically. Meanwhile, as the fuel rods get hotter, the helium and fission fragment gases (xenon and krypton) in them try to expand, producing higher and higher pressures. This pressure causes the now-ductile zircaloy to swell and buckle, partially blocking the spaces through which the cooling water is supposed to flow. When the water provided by the ECCS does start flowing, it cannot fully cool the center of the core.

Still without sufficient cooling water, the central core continues to get hotter. The hot zircaloy now starts reacting with steam, producing hydrogen in the process. The fuel itself is approaching the melting point. Meanwhile, the hydrogen is collecting at the top of the reactor vessel.

As you probably know, a nuclear explosion is not possible in a reactor of this type. Because of the low concentration of U^{235} in the fuel, an explosive chain reaction just cannot occur. However, chemical explosions or steam explosions are entirely possible. At this point our mythical reactor suffers a hydrogen explosion in the vessel. This bursts many of the now-brittle fuel rods and partially molten fuel starts to escape. Worse, the ECCS is now totally disabled, and only a pool of water remains at the bottom of the reactor vessel. The operators have managed to close off the valves in the steam line leading out of the containment building but, otherwise, the reactor is now completely out of control. A *meltdown* will surely occur. The only hope is that the containment building will stay intact and keep the radioactivity from the environment.

By now several hours have passed since the original accident occurred. The Nuclear Regulatory Commission (NRC) has been informed, and several experts are being rushed from Washington to do what they can to help. Other experts are being gathered together in Washington to discuss the situation and offer long-distance advice. Three miles away, the office of the mayor of the adjoining city of Happyville, population 115,000, has just been informed that there is "a problem" in the nuclear plant. From this point on, the information that reaches the public will be sketchy, confused, and intended more to soothe than to inform. Statements made to the press by the power company will contradict those made by the NRC. As people learn about the "problem" some will panic, others will think it is minor and ignore it. Someone from the NRC thinks about potassium iodide, and frantic calls are started to the major drug companies to see if large amounts can be located.

Meanwhile, the core of the reactor has become several hundred tons of white-hot molten material. It gradually burns its way through the supporting structure and, suddenly, much of it falls into the pool of water in the bottom of the reactor vessel. The resulting enormous steam explosion blows the top right off the reactor vessel, sending a huge piece of it into the containment dome. Somehow, although minor cracks appear on the

outside of the dome, it holds, and the deadly radioactivity now scattered throughout it is contained. The mass of molten material is now melting its way through the bottom of what is left of the reactor vessel.

In the following days, the molten core slowly burns its way through the concrete floor and foundation of the containment building and down into the Earth. There, it is expected to burn its way downward for several hundred feet, in the famous "China syndrome," before coming to rest. (The China syndrome is named for the direction in which the molten core moves, not its final destination.) However, as it moves through the Earth, the core encounters ground water. The resulting steam builds up tremendous pressures, cracking rock and forcing its way around the base of the containment building and into the atmosphere. This releases vast amounts of radioactive materials into the environment. Because of a light steady wind in just the right direction and a temperature inversion which holds the radioactive particles near the ground, much of the radioactivity dumps on nearby Happyville. Thousands of people are exposed to killing doses of radiation. Tens of thousands more receive serious doses and, eventually, millions will be exposed to raised levels of radiation. There are not enough medical facilities, supplies, or personnel available to care for all of the casualties. All over the world, nuclear power plants are shut down in a panic, with the resulting loss of power adding to the suffering.

And so on.

Assessing the Chances   Wow! What a horrible scenario! Is it possible? Well, yes, a very serious accident is possible, though undoubtedly the sequence of events would be quite different from that described above. The final results could be quite different, as well. Some of the events described in this scenario, including the lack of information to the public, are based on actual occurrences at Three Mile

Island. Remember, however, that this scenario is *imaginary*. I confess to having taken considerable liberty with the details in order to dramatize the seriousness of the possible consequences. For example, perhaps the swelling and buckling of the fuel rods described here could never happen. Some responsible nuclear scientists think it is a possibility and others think not. Also, the automatic flooding of the core probably would take place, although in the Three Mile Island accident it did not because the pressure never dropped low enough to trigger it. The ECCS at TMI failed to do the job simply because the operators turned it off. Yet, even after an almost unbelievable combination of equipment and operator failures, the TMI accident could have turned out to be a minor incident. There is reason to believe that as much as an hour and a half, or more, into the accident, had the proper steps been taken, no serious damage to the reactor or the containment would have resulted.

What we need to know, then, is what, really, are the chances that some nuclear catastrophe will happen? There, dear reader, is what the nuclear controversy is all about. No one knows the answer for certain. The two sides in the controversy come up with widely different estimates of the danger. The strongly pronuclear persons point to research, computer simulations, and other studies which indicate that the chances of a meltdown are very, very tiny. The antinuclear people say, "Baloney! Those figures are incorrect, and we can prove it."

In 1972, the United States Atomic Energy Commission (AEC) sponsored a study of nuclear reactor safety. The study, which took three years and about 70 person-years of work was directed by Norman C. Rasmussen, Professor of Nuclear Engineering at the Massachusetts Institute of Technology. It was published by the Nuclear Regulatory Commission, which succeeded the AEC in 1975, and is usu-

ally referred to as the Rasmussen Report. It is about one foot thick. The approach the study group took was to investigate, with the help of computers, thousands of different events and pathways that might lead to a serious accident in a nuclear power plant. They used techniques referred to as *event trees* and *fault trees*. An event tree follows the various possible pathways to a serious accident. The reasoning is something like saying: If A occurs *and* B *and* C *and* . . ., then the results are. . . . An estimate is made of the probability of each occurrence, and an overall probability of a particular pathway is computed. A fault tree works backwards in time, starting with a particular defect and working back to find the various faults that could have caused it. As an example of a pathway in an event tree, we can use our mythical scenario. Suppose the probability of the turbine exploding is 1 chance in 1000: 10^{-3}. (This figure is just pulled out of a hat for the purpose of illustration.) Suppose also that the chance that the blocking valve at the containment wall will fail is 1 in 100: 10^{-2}. Then the chance that the turbine will blow up *and* the valve will fail is the product of the two: 10^{-5}. Going on, we would have to multiply this figure by 10^{-2} to account for the chance that the second valve would also fail, by another small number to allow for the possibility that outside power is lost, another factor for the failure of the diesel engines for emergency power, and so on. Part of this event tree is shown in Figure 14.13.

When the Rasmussen Report was complete, the results of computing thousands of pathways were that the chances of a serious nuclear accident were very slim. The calculations led to the conclusion that non-nuclear events, such as fires, dam failures, air crashes, tornados, earthquakes, and hurricanes, are much more apt to cause fatalities, serious injuries, and property damage than are nuclear accidents. For example, it was estimated that fires were 10,000 times more likely to cause

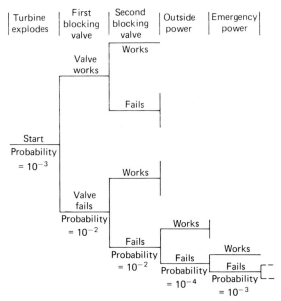

Turbine explodes	First blocking valve	Second blocking valve	Outside power	Emergency power

Figure 14.13 A partial event tree, showing the probability of failure at each step. (The probabilities shown are just guesses.) The total probability of all of these failures occurring to the point shown is 10^{-14}, the product of the individual probabilities.

large numbers of deaths than were accidents in nuclear power plants. This assumed a total of 100 plants operating in the United States.

Naturally, pronuclear people pointed to the Rasmussen Report to support their views that nuclear power is the safest way to produce large amounts of electric power. According to them, the chance of a total core meltdown, with large amounts of radioactivity being released, is laughably small.

On the other hand, antinuclear people claimed that the assumptions made in the report were faulty and the probabilities computed were much too low. In 1977 the NRC organized a Risk Assessment Review Group, led by Harold W. Lewis of the University of California at Santa Barbara. The "Lewis Report" was, in fact, critical of some of the conclusions of the Rasmussen Report, but it praised the approach used and recommended

that it should be used more widely. It also suggested that complete loss of coolant accidents were quite unlikely, but that serious accidents resulting from small failures were more likely and should be studied further. Some nuclear critics pointed out that they had been saying this for years. In 1979 the NRC itself conceded that the Rasmussen Report was flawed. Soon after this the Three Mile Island accident occurred as the result of a series of small faults rather than a sudden LOCA.

One other investigation into the safety of reactors is worthy of mention. There is a LOFT (Loss of Fluid Test) reactor in Idaho which has been built solely for the purpose of experimentally testing accidents that could happen in power reactors. The first planned "accidents" were carried out in late 1978 and 1979, and tests continue. The first tests simulated a major break in the primary coolant circuit, resulting in loss of cooling water to the core. Before the tests, computers had been used to predict the results, as they are used for working reactors. The results of the tests, generally, were that the effects were less serious than predicted by the computers. For example, the temperatures reached in the core were lower than the predicted temperatures. Naturally, the supporters of nuclear power hailed this as a great success. They said that the fact that the core temperatures did not reach predicted levels proved that the computer model was in error on the safe side. The opponents, on the other hand, said that the test proved that the predictions were just plain wrong and that similar predictions should not be trusted. They also said that the LOFT reactor was too small (less than 20 MW) to give any useful information about 1000 MW reactors. The LOFT tests are continuing, and there is intensive effort to improve the computer models.

In one way, the accident at Three Mile Island was a fortunate occurrence. No one was killed as a result of it and, as far as we know, no one was seriously injured. But it gave us an (expensive) opportunity to learn some things about nuclear accidents. The accident involved an almost incredible combination of mechanical failures and human errors. If a probability had been calculated for that particular pathway of faults, it undoubtedly would have been so small as to be negligible. In other words, that particular accident just would not happen. But it did happen. It pointed out at least one deficiency of the fault-tree analysis. The accident was caused, by the closing of two (redundant) valves by a maintenance crew, and failure to reopen them. Then, when the main feedwater pumps, which pump the condensed steam from the turbines back to the steam generator, tripped off temporarily (a fairly common occurrence), the auxiliary pumps that took over were not able to pump the water through the closed valves. This led to the steam generators boiling dry, which then caused the problems that followed. If the fault-tree method estimated that the probability of one valve not being opened was one part in 10,000 (that is, 10^{-4}) then the probability of two valves not being opened would be 10^{-8}. But, those two valves were not closed by separate events; they were both closed and not reopened together. This means that the calculated probability should have been 10,000 times as great as it was. Such oversights might very well have occurred at many points in the event trees and fault trees used for the Rasmussen Report. (Incidentally, those two valves were supposed to be left open at all times. It was permissable, according to regulations, to close only one at a time for brief periods of maintenance.)

Quite characteristically, both sides of the nuclear debate used the Three Mile Island incident as evidence that their own views were correct. The antinukes, naturally, said that the fact that the accident happened is proof that nuclear reactors are not safe, and that similar or worse accidents are just waiting to happen elsewhere, and have been only narrowly averted in the past. The pronukes, on the other hand, said that the accident was proof positive that nuclear power plants are safe. After all, they argued, here was a very serious and unlikely accident, with the core partially uncovered for a long period, and there was not even a meltdown. Perhaps it proved that meltdowns will never occur. Further, no one was killed or hurt seriously, nor was any property damaged. (That does not include the property of the power company. It will probably cost $1 billion or $2 billion to clean the containment building of radioactive materials.)

After the Three Mile Island incident, a group, named The Kemeny Commission after its chairman, John G. Kemeny, was appointed by President Carter to investigate the accident and recommend corrective steps for the future. The most important conclusion reached by that group was that human error is a far greater factor in nuclear safety than had been previously recognized. They also concluded that failures of minor components, along with human error, could lead to serious incidents "much more often" than had been supposed in the past. Foremost among their recommendations was that the training of operators of nuclear power plants must be greatly improved. They also recommended sweeping changes in the operation of the power utilities and of the Nuclear Regulatory Commission itself. Many of the recommended changes have already been made by the nuclear power industry and the NRC.

Permit me a final observation. One would think that the effects of the Three Mile Island accident on the general public would be well known. But that is not the case. The dosage figure published by the NRC was that the nearly 1 million people who live within 50 miles of Three Mile Island received an average dose of 2 mrem as a result of the accident. It seems however, that even officials of the NRC are not sure of the accuracy of this figure, and we will never know for sure how much radiation was received by people living near the reactor. Official government estimates, and the estimates of the Kemeny Commission, are that the accident will cause either no deaths by cancer or so few that they could never be detected. Another expert estimates as many as 50 cancer deaths, and yet another that up to 2500 cancer deaths will result in the next 20 years. It seems likely that the damage was small, but it is a bit distressing that we do not know for certain.

Nuclear Weapons Concerns

One of the serious concerns of our times is that nuclear weapons capability might spread to many nations which do not now have it. Even worse is the fear that some terrorist group might acquire a nuclear weapon. Imagine the results if the Irish Republican Army could demonstrate that it had a nuclear bomb and threatened to set it off in the middle of Lon-

don, or if the Palestine Liberation Organization were to explode one in Jerusalem, or if Iraq were to drop one on Tehran, or Libya on Khartoum in Sudan. The thought is frightening. Under certain conditions, it is even conceivable that a worldwide nuclear war could be triggered (Figure 14.14).

The fear of such consequences is a strong reason for many people to be opposed to the reprocessing of nuclear fuel used in light-water reactors. In a once-through use, that fuel never reaches a condition in which it could be easily adapted for use in a nuclear bomb, for at no point is it rich enough in uranium-235. Enrichment facilities are very expensive and exist only in a few countries in the world. On the other hand, fuel that has been used for a time in a reactor has plutonium-239 in it, produced from the absorption of neutrons by U^{238}. In spent fuel, there will generally be about

Figure 14.14 A nuclear explosion could conceivably be set off by a small nation or even a terrorist group. The explosion shown here was photographed from 50 miles away at an altitude of 12000 ft. (Photo by U.S. Air Force.)

0.8 percent U^{235} remaining and from 0.5 to 1 percent Pu^{239}.

It would take only about 10 kg of Pu^{239}, a piece about the size of a softball, to make a crude nuclear bomb. Such a bomb can be made with relative ease by just a few persons. If preparations were made in advance and enough Pu^{239} was stolen or otherwise obtained, a bomb could be made in just a few days, perhaps less time than it would take to locate the thieves.

Spent fuel, directly from a reactor, is very highly radioactive because of the fission fragments it contains. It also produces substantial amounts of heat as a result of that radioactivity. When the fuel rods are removed from a reactor, they are placed in large tanks of water, with enough separation between them so that they can gradually cool off. Presently, the spent fuel stays in "temporary" storage at the reactor site. No reprocessing occurs. In this form, the spent fuel would be difficult to steal and even more difficult to handle. Thus, the plutonium in it is quite safe.

However, with no reprocessing, the spent fuel rods just accumulate. Further, the valuable U^{235} and Pu^{239} are wasted unless they can be recovered. Those in favor of reprocessing say that we cannot continue to waste a valuable resource, and that reprocessing will reduce operating costs of nuclear power. Those opposed claim that the cost of reprocessing will be such that no money is really saved and that we have enough uranium to provide once-through operation for quite some time to come.

Reprocessing is a somewhat complex and difficult process. First there is a cooling-down period, when the fuel rods are stored in water-filled pools for several months at the power plant. During this period the radioactivity and heat produced will decrease by a factor of about 10,000. One major aim is to get rid of most of the I^{131}, which otherwise would be a problem in reprocessing. After shipment to the reprocessing plant, the fuel rods are chopped into small pieces. The spent fuel is then dissolved into a liquid form, and a rather complex chemical procedure is used to separate out the uranium and plutonium. The process commonly used is called the Purex Process. The separation process can recover more than 95 percent of the plutonium and uranium. The separated uranium–plutonium mixture has about 1 millionth of the radioactivity of the spent fuel. Most of the radioactivity is in the fission fragments, which have now been separated out. This uranium–plutonium mixture would be now sent to an enrichment plant to prepare it to serve as fuel.

The separated material still consists mostly of U^{238}, but it is now much easier to handle than the spent fuel because of the reduced radioactivity. Since uranium and plutonium are alpha emitters, the radiation is not very penetrating, and the material can be handled with light shielding. Furthermore, the plutonium, being chemically different from the uranium, can be separated out by chemical means. This uranium–plutonium mixture, then, could possibly be diverted by thieves and the plutonium separated out with relative ease. If the mixture is processed further to remove the plutonium, then pure plutonium might be stolen in sufficient amounts to make bombs.

You may reasonably ask, how will the fact that the United States is not reprocessing spent fuel prevent other countries from doing so? A good question. In addition to not reprocessing fuel ourselves, the United States has exerted influence on other countries, through treaties and otherwise, to keep the needed technology from spreading. The government has agreed to provide uranium enrichment services for those countries that need it. The International Atomic Energy Agency (IAEA) provides some additional safeguards, mostly by the inspection of nuclear facilities. However, the IAEA can inspect only those nations who allow themselves to be inspected. This consists of signers of the Nuclear Nonproliferation Treaty. Those nations can release themselves from nonproliferation agreements with ease. Further, if the IAEA should find that a nation is diverting nuclear fuel to weapons, it could do nothing more than make this fact public.

Would it be possible for a presently non-nuclear nation to develop nuclear bombs even if the United States continues to refrain from reprocessing fuel? Yes, it would. Let's consider the fictional country of Nonuke. Secretly, those in power in Nonuke decide to produce a nuclear bomb. (Then, they can change the name of the country to Havanuke.) The first step is to get a nuclear reactor, either a power reactor or research reactor. Nonuke is a country poor in fossil fuels, and it needs electric power desperately, so it makes sense for it to get a power reactor. It could try developing the technology and building one itself, but it would be much easier to buy a reactor from one of the nuclear nations. From several countries, it is possible to buy not only the reactor, but supplies of fuel and the technology to operate the reactor. Typically, the country selling the reactor sends people to the purchasing country to get the reactor into operation and to teach the new owners how to run it.

In theory, this should be the end of it. Nonuke now has an operating reactor and knows how to run it. However, the rulers of Nonuke have more than this in mind. They start by purchasing large quantities of U^{238}. The U^{238} may be natural uranium, with about 0.7 percent of U^{235} in it, or it may even be the tailings—the leftovers—from the enrichment plant. Since Nonuke is not after U^{235}, this does not matter. Then, after they have the technology well in hand, Nonuke technicians put a blanket of U^{238} in their reactor, making it something like a breeder reactor. They might do this secretly, and it could be difficult to detect. On the other hand, they might disavow the Nonproliferation Treaty and throw out the IAEA inspectors. Then they could convert U^{238} into Pu^{239}, with less secrecy but

without anyone knowing for sure they were doing so. Meanwhile, they are learning the technology and building a small fuel processing plant. By the time sufficient plutonium has collected, they are ready to purify it, and Nonuke has become Havanuke.

Using a research reactor, this is more or less the procedure that was followed by India to produce their first nuclear bomb, which they exploded in 1974. The reason for the 1981 Israeli bombing of the Iraqi research reactor, built by the French, was that they suspected Iraq was planning to attempt the same thing. (Why did Iraq, an oil-rich nation, have a large stockpile of U^{238}?) The question remains whether refusing to process spent nuclear fuel by the United States, along with other measures, will prevent non-nuclear nations from developing nuclear weapons. There is no agreement among knowledgeable people on this issue.

Radioactive Wastes

The main problem associated with radioactive waste involves the disposal of the highly radioactive spent fuel from the reactors. Before we get to that, however, I would like to mention a related problem that is rarely thought of. In the mining and milling of uranium, there is also waste material, the *tailings*. As has already been mentioned several times, these materials are not highly radioactive, but there is one danger from uncovering large quantities of tailings and leaving them above ground. In the radioactive decay chain, starting with U^{238}, there is radium-226, which decays into radon-222. Radon is a gas, and when it is produced near the surface, it can escape the solid material and get into the atmosphere. Radon-222 has a half-life of 3.8 days, so if it is collected in any quantities, it is quite radioactive. Although it is an alpha emitter, not dangerous on the outside of the body, being a gas, it can be inhaled and cause lung cancer. Leaving huge amounts of tailings on the ground will raise the level on Rn^{222} in the atmosphere. This is not a severe problem, but it is an avoidable one. The simple remedy is to bury the tailings.

A related problem is that uranium miners in underground mines have, in the past, been exposed to increased levels of radon gas. The remedy here is also simple: provide good ventilation in the mines. I do not mean to pass this off as a problem of no consequence at all but, so far, the evidence is that uranium mining is far safer for the workers than coal mining.

As I said, the really serious problem is what to do with the highly radioactive wastes leaving the nuclear reactors. It may seem incredible that the government does not have a policy for the disposal of these wastes, but it is true, 25 years after the start of operation of the first commercial power reactor and after 40 years of military use of nuclear materials. Instead, the wastes that have been produced are being stored. This includes both used fuel rods, which are stored in large pools, and millions of gallons of liquid wastes. The latter, called *high-level wastes*, are stored in large tanks at federal waste repositories. Some of those tanks have been known to leak their deadly contents into the grounds. (One tank lost 115,000 gallons of the stuff!) If spent fuel rods are reprocessed, then high-level liquid wastes will be produced in large amounts.

The high-level wastes contain the fission fragments, with Sr^{90} and Cs^{137} being the most troublesome. They have half-lives of about 30 years and become harmless after about 800 to 1000 years. (The figure most often used is 800 years. What fraction of those two materials remains after 800 years?) A plant that removes these two elements from the wastes is shown in Figure 14.15. The wastes also contain a variety of much longer lived nuclides, including some *transuranic* elements. (That is, elements with an atomic number of more than 92 and that do not appear in nature.) For some of these nuclides to decay to the point at which they are harmless will take thousands of years.

That is the nub of the problem: How do you take care of a material that can be harmful for thousands of years? Our present approach is simply to store it in large tanks, like the one in Figure 14.16.

One thing we should realize, and that our society as a whole tends to ignore, is that this problem is not unique to radioactive materials. As one example, no one seems to worry much that annually we are purifying and throwing into the environment millions of tons of arsenic. As you may know, arsenic is a very poisonous metal used in insecticides and weed killers. What is the half-life of arsenic? Why, it is infinite. Arsenic is stable, radioactively, and never decays. It stays just as harmful forever. The spread of arsenic is just one example of our unceasing efforts to permanently poison the environment. We dump hundreds of harmful substances without a thought to the welfare of future generations. (And present generations, as well. In 1983 the population of an entire town in Missouri had to be moved because of spills of the chemical dioxin.) Even though we do not know everything about the effects of radioactivity, we know even less about the effects of many other substances we dump into the environment. But that is another story, and we do not want to get too far astray.

The problem, then, is to find some way of disposing of high-level wastes in such a way that they will not harm us or future generations. They cannot be burned. Burial might be all right, but they had better be buried securely. The greatest danger in burying wastes is that radioactive materials might get into the ground water and poison water supplies. Thus, any burial that is done must be such that none of the radioactivity escapes for hundreds or thousands of years. In the course of human affairs, this is a very long time. No civilization has ever lasted that long. What happens if we bury the wastes and future civilizations forget they are there? Will they someday find themselves radioactively poisoned, with a large in-

Figure 14.15 The waste separation plant at Hanford, Washington. Here strontium and cesium are removed from high-level wastes from nuclear reactors. After removal, the wastes must go through a five-year "cool-down" period before they can safely be solidified. Then they are "permanently" stored. (DOE photo by Batelle Northwest.)

cidence of cancer and genetic damage? Or can we work out some means of guarding the wastes so that they will never be disturbed? (There is a great science fiction story here, with a mysterious and powerful priesthood growing from the need to guard eternally the radioactive wastes. But I think it has already been written—several times.)

Several possible ways of disposing of the high-level wastes have been suggested. The one that has had the most attention and seems most immediately feasible is to bury them deep underground in salt deposits. Salt deposits are chosen because, where there is salt underground, there has not been much water for millions of years. There is now a method of converting the liquid wastes produced into solid wastes by reprocessing. The solids are

Figure 14.16   Highly radioactive wastes are stored in huge tanks at government facilities. Here, construction is under way on two 1 million-gallon tanks in South Carolina. (Photo courtesy DOE.)

then melted along with silica (sand) to produce a cylinder of glass with the radioactive material a part of it. The cylinder will be about 3 m long and a third of a meter in diameter. Three of them, end to end, would be about the size of a wooden telephone pole. The glass mixture hardens in a stainless steel canister which can then be sealed. The wastes from a 1000 MWe nuclear plant operating for 1 yr will occupy about 10 of these cylinders, not a great volume.

If the steel cans thus produced are to be buried a quarter or half mile below the surface in salt deposits, then there must be a reasonable assurance that the salt deposit will be undisturbed for a sufficiently long time. For example, the salt must not be in an earthquake zone, or near a location where a new volcano could erupt. Any *ground water* (that is, underground water) in the vicinity must have no chance of being diverted into the salt.

Salt, of course, is soluble, and large amounts of water could dissolve it and carry it out of the deposit site. The stainless steel cans will not provide protection forever. Because of the small amount of water present in even the driest salt deposits, they would be gradually eaten away, exposing the radioactive glass. You may not know that glass is soluble, though slightly so. However, if it is in water over a very long period of time, it could conceivably be dissolved away and find its way to surface water.

You will probably not be surprised to learn that there is a considerable difference of opinion about whether deep burial of the wastes is a good idea. Can a guaranteed suitable salt deposit be found, and will the buried waste stay buried for the necessary thousands of years? Some say that not enough is known about geology to predict the required stability. Others claim that there is no problem.

Some nuclear critics do not see how a burial site could be guarded for the necessary number of years. Other persons point out that, if the wastes to be produced from all of the uranium now known to exist in the United States were to be buried, the burial site would be less than 100 square miles in area. That area, they say, could be adequately taken care of by one person who would inspect it regularly and check the water in the vicinity for any signs of escaping radioactivity. One suggestion that has been made is that the longer lasting nuclides in the waste should be removed and "burned" in the reactors. (That is, they would be changed to other, less troublesome nuclides through nuclear reactions.) Then, the remaining wastes would become harmless in "only" about 800 years.

The arguments go on, and the wastes continue to collect. Meanwhile, there have been suggestions for other possible means of waste disposal. Some of them certainly deserve further study. For example, it has been proposed that nuclear wastes be hurled by rockets into the sun, or into solar orbits. This is an attractive idea, for it would get rid of them instantly and permanently. The problem with it is that there is still a chance of a failure of a rocket before it escapes the Earth. What then? Can you imagine a shattered rocket plunging back to Earth and scattering tons of highly radioactive materials?

Another proposal is to bury the canisters deep in ocean trenches. The ocean bottom moves slowly and, if placed in the right location, the canisters would be carried with the movement deep into the inner bowels of the Earth. This would be great, if it could be guaranteed that all of the canisters will, in fact, do so. What if some of them break loose and are carried to other parts of the ocean, where they will gradually corrode?

Another idea is to dispose of the canisters in Antarctica. There, the average temperature has been below freezing for more than a million years. The notion is to put the canisters in the ice and let them melt their way down. The 5 kW or more of heat produced by each canister would then allow it to melt its way down to solid rock, about a mile below. It is estimated that this might take up to five years. The problem with this idea is that it is believed that ice "surges" occur in Antarctica, perhaps once every 100,000 years. Such a surge could possibly move large amounts of ice to the sea, and nobody knows when one might occur. More knowledge of the Antarctic is yet neeed.

Deep burial in drilled holes perhaps 20 km below the surface of the Earth has been proposed. This method has not yet been made feasible, and it would probably be quite expensive.

Finally, there have been suggestions that the long-lived nuclides should be removed and the remaining wastes stored permanently on or near the surface of the Earth. This would require some structure that would resist earthquakes, floods, and other disasters for about 800 years. It would also have to be guarded against intruders for that period of time. On the positive side, wastes stored in this way could be retrieved, whereas this is not the case with other types of disposal. Such a means of storage could buy time for finding other methods of disposal. Then, when a really good method is found, the "permanently" stored wastes could be retrieved and disposed of.

In summary, there are a number of possibilities for safely disposing nuclear wastes. One or more of them may turn out to be feasible. However, thus far, not enough research has been done for us to have complete confidence in any of those methods.

NUCLEAR FUSION

Remember from Chapter 13 that nuclear fusion is the fusing together of two light nuclei to produce a heavier one. When this happens, large amounts of energy are released. This

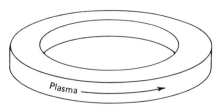

Figure 14.17 In a tokamak-type machine, the plasma moves in a doughnut-shaped space where it is confined by a magnetic field.

approach to producing energy is very attractive. For one thing, if deuterium–deuterium (D–D) fusion can be made practical, the amount of fuel available is almost without limit. It exists in huge quantities in the oceans of the world. Even for the somewhat easier deuterium–tritium (D–T) fusion there is a great deal of fuel available. "Burning" the fuel by means of fusion would be almost a totally clean process, with little or no pollution (except thermal) and only small amounts of ra-

Figure 14.18 The magnetic coils used to confine the plasma can be seen during modification of the "Elmo Bumpy Torus" fusion research apparatus at Holifield National Laboratory. (Union Carbide.)

dioactive materials being produced. Further, it would be a completely safe process. If a fusion machine were turned off, it would shut down completely and instantly; there would be no possibility whatever of a meltdown or any other kind of runaway condition.

So if this fusion power is so wonderful, safe, abundant, clean, and virtuous, why don't we get with it right away? By now you know the answer: there is *always* a catch. In this case, the catch is to provide a container for the fusion. Remember that in order for fusion to occur, a mass of material must be heated to temperatures of about 100 million degrees Celsius. When you have something this hot, what do you keep it in? No materials known, or likely to be known, can withstand temperatures this high, even for very short times.

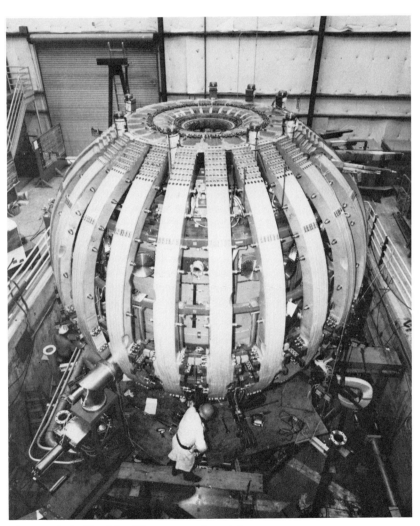

Figure 14.19 Doublet III, an experimental magnetic confinement device built in San Diego. (General Atomic Company.)

The problem, then, is to contain the extremely hot fusing material, but not let it touch anything else.

Physicists and engineers have been working on this problem for over 25 years. The earliest attempts at making a "bottle" that could hold the very hot materials used *magnetic confinement*. When materials are heated to very high temperatures, the electrons are stripped from the nuclei, and the result is a very hot gas of ions and electrons, called a *plasma*. Since a plasma consists of charged particles moving at high speeds, a magnetic field will exert forces on it. The early experiments to produce a "magnetic bottle" used magnetic fields shaped in such a way as to contain the plasma in a given location in a vacuum vessel. (That is, in a container from which the air is pumped out.) As these experiments became more sophisticated, they used superconducting magnets to produce the magnetic fields. You can perhaps imagine the difficulty of producing a region with temperatures of 100 million degrees just a short distance away from a coil which must be held at a temperature just a few degrees above absolute zero!

There has been progress in fusion experiments. A fairly recent development for magnetic confinement was built first in the USSR and is called a *tokamak*. In it the plasma moves in a toroidal magnetic field. That is, it moves in a region shaped something like a doughnut, confined there by a rather complex magnetic field. This is shown crudely in Figure 14.17. You can get an idea of the size and complexity of the equipment needed from Figures 14.18 and 14.19.

Since the late 1960s, experiments have been under way to produce fusion by means of *inertial confinement*. For example, a pellet of deuterium–tritium fuel is dropped into a reaction chamber. When it reaches the center, it is hit simultaneously from all sides by high-energy beams of electrons. One such design, with 36 electron beams, is sketched in Figure 14.20. The electron beams transfer energy to the fuel pellet, causing it to be heated and *imploded*. That is, it is pushed inward from all sides and compressed for a small fraction of time. If fusion can be made to occur, however, that small fraction of time that the fuel is confined to a small space will be enough. The fusion will be so rapid that the fuel material will not have time to scatter before the fusion is complete. Then, perhaps a tenth of a second after the fusion, another fuel pellet will get the same treatment, then another, and so on. In the early inertial confinement experiments, laser beams were used instead of electron beams. Intense beams of light have the same effects of transferring large amounts of energy quickly to the fuel pellets and compressing the pellets. The experimental Shiva apparatus, which uses this approach, is shown in Figure 14.21.

As I said, there has been progress made toward achieving fusion. However, there has still not been a "break-even" experiment,

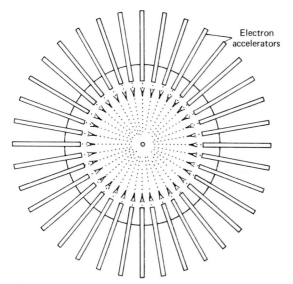

Figure 14.20 An attempt to produce nuclear fusion by smashing electrons from all sides into a fuel pellet. In this figure, 36 electron beams hit a fuel pellet at the center of the reactor chamber.

Electron accelerators

Figure 14.21 Exterior of the Shiva target chamber. Twenty laser beams, half from the top and half from the bottom, enter the chamber simultaneously and are focused on a small sample that is to undergo fusion. A piece of diagnostic equipment will be attached to each of the many ports (shown covered). (Photo courtesy Lawrence Livermore Laboratory.)

where the energy produced is at least as great as the energy put into the machine. Further, it may yet be a long time before nuclear fusion will be commercially feasible. Perhaps by the end of the 1980s we will know whether it can be done or not. However, even if fusion can be made to work—and I certainly hope it can—it will be quite a few years before it is available on a large scale.

THE FISSION DECISION

Well, where does all of this leave us? Are we for or against nuclear power? From this chapter you have seen that the technological aspects of nuclear power are rather complex. Although nuclear power has been studied extensively, there are still many areas that are not well understood. An example of this arose in late September, 1981, when the NRC announced that the pressure vessels of 13 reactors scattered about the country were coming under suspicion. As you know, exposure to intense radiation damages materials. The eight-or-more-inch-thick steel pressure vessels are no exception, but it was thought that they were designed to last the life of the reactors—up to 40 years. Now, it appears that at least some of the pressure vessels are becoming brittle far sooner than expected. This means that a reactor vessel might crack if it is subjected to great strain, such as that from sudden cooling. Replacing them may be impossible; at the least it will be very expensive. This sort of unexpected setback may have a considerable impact on the willingness of power companies to build new reactors.

In fact, the economic aspects of the future of nuclear power are even more complex than the technological ones. The power companies are now in a much more difficult economical position than they have been in the past. Building nuclear plants has become terribly expensive, and unforseen problems could make it more expensive still. For example, in late October 1981, the NRC announced that the El Diablo plant in California, which had just recently been licensed to operate, was improperly constructed and could not be operated. The necessary changes are bound to be expensive. Furthermore, news of incidents like this has a powerful effect on public opinion.

Which brings us to politics, the most complex and unpredictable factor of all. A long discussion of politics would not be appropriate here, nor could I conduct one, but we should recognize that it will be an important factor. Some of the persons exerting political pressures to shape the decisions concerning nuclear power do so with the welfare of the country or the greater society in mind. However, some persons play politics for their own personal gain, or that of the companies or groups they represent. We must somehow "tune them out" and make decisions that will benefit future generations. The choices are not easy; they are of crucial importance, and they must be made by all of us.

What do you say now? Are you pronuke or antinuke?

SUMMARY

Nuclear power is a controversial issue, but an understanding of it is of crucial importance to every citizen. All of us need to look at the facts objectively and decide for ourselves whether to support or not support it. In this chapter, you have had a brief review of how nuclear power is produced and the types of reactors being used and proposed for the near future. Light-water reactors use water both for cooling the reactor core and for moderation, the slowing down of neutrons so they can cause fissions of U^{235}. In the United States, all but one of the presently operating reactors are of this type. Although a nuclear power plant is a complex collection of a reactor and all its parts, miles of plumbing, sophisticated controls, and other components, basically it is simple. The nuclear reactor provides energy in the form of heat to produce high-pressure steam, and the steam runs turbines to turn electric generators.

The fuel used for light-water reactors is generally uranium, enriched to about 3 percent U^{235}. By means of the moderator and the control rods in the core, a controlled chain reaction is sustained at the desired power level. In a boiling-water reactor, the cooling water is allowed to boil at a high pressure and the resulting steam powers the turbines. In a pressurized-water reactor the water is kept under high pressure and not allowed to boil. Steam generators then produce steam in a secondary coolant loop, and that steam runs the turbines. The control rods are automatically moved slightly in and out of the core, as needed to keep the rate of power production constant. Heavy-water and high-temperature gas-cooled reactors have some relative advantages and disadvantages but, with the exception of the HTGR described in this chapter, they are not yet used in this country.

The nuclear fuel is found in the ground as "yellowcake," an oxide of uranium. It must be mined and purified. Then, the most difficult and expensive step, it must be enriched to contain a considerably greater fraction of the fissionable U^{235} than the 0.7 percent that appears in nature. In a once-through fuel cycle using light-water reactors, there is perhaps enough uranium in the United States to produce about 1000 quads of energy, as compared to the 5000 quads, or so, of fossil fuels thought to be available. The uranium fuel may be extended greatly by reprocessing. A breeder reactor produces new fuel by converting U^{238} to plutonium, which is fissionable. If breeder reactors become technically feasible and politically acceptable, they could extend the useful life of the available nuclear fuel many times over a once-through process. Objections to

reprocessing of fuel and to the use of breeder reactors focus largely on the dangers of plutonium. Those dangers include possible accidents, in which the deadly plutonium could be released into the atmosphere, and the acquisition of plutonium for bombs by nations or groups that do not now have nuclear power.

Radiation from radioactive materials is harmful to objects, plants, animals, and, above all, humans. If nuclear power, including breeder reactors, is to be used on a very large scale in the future, we shall have to be very careful to contain the harmful radiation. A properly operating nuclear power plant releases only negligible amounts of radiation into the environment. However, there is some chance that accidents could release radioactive materials in quantity. The many experts are not in good agreement about the magnitude of the risks involved, and the discussion of those risks is likely to be highly emotional. A further important question concerns the disposal of highly radioactive waste materials produced by nuclear reactors. Although the United States has had nuclear power for over 25 years, there is not yet a policy for the disposal of those wastes. There are several possible proposals for that disposal, but there is not yet agreement on a safe way. Long-term storage of the wastes may involve keeping it safe and away from humans for hundreds or thousands of years. Research is continuing, and public concern is growing, so perhaps some policies will be decided upon in the fairly near future.

Research into controlled nuclear fusion continues. If this source of energy should prove possible on a large scale, enormous reservoirs of usable fuel could be tapped. Those resources could be so huge that this type of energy production might be considered a renewable resource. The main obstruction to the development of fusion power is the difficulty of containing materials at temperatures near 10^8 K. Many years of research have not yet resulted in a break-even reaction, with as much energy being produced by the fusion as was needed to cause the fusion to happen. However, it seems worthwhile to continue to try, for the potential pay-off is enormous.

ADDITIONAL QUESTIONS AND PROBLEMS

1. Briefly describe the components of a light-water reactor (LWR).

2. What is the function of the moderator in a reactor? Why does ordinary water make an excellent moderator?

3. What is meant by the term *fuel enrichment*? Why must the fuel be enriched for use in light water reactors?

4. What is the function of the control rods in a reactor? What is a scram?

5. In the event of a loss of coolant accident (LOCA), the reactor would automatically be scrammed, stopping the chain reaction. However, the emergency core cooling system (ECCS) would be needed immediately thereafter. Why? (That is, if the chain reaction is stopped, why is cooling needed?)

6. Light-water reactors (LWRs) are used almost exclusively in the United States for power production, at this time. What are the characteristics of some other types of reactors? Why do you think they are not in use in the United States?

7. Once it was thought that nuclear electric power would be "too cheap to meter," and that the available nuclear fuel is limitless. What are the actual facts about the amount of nuclear fuel available and the cost of nuclear power?

8. If you were president of an electric power company, what arguments could you find in favor of having your company build a nuclear power plant? What arguments would be against that undertaking? Would you build one?

9. What are the dangers associated with fuel reprocessing? What are the advantages of reprocessing?

10. Since $_{94}Pu^{239}$ decays by means of alpha emission, what is the isotope produced? Does this isotope appear in nature?

11. What properties of plutonium make it dangerous to people?

12. Briefly describe the operation of a breeder reactor. Why are liquid metals used as coolants in breeders? Why are there usually two heat exchangers in a breeder reactor?

13. What is meant by the terms *somatic damage* and *genetic damage* when referring to the harmful effects of radiation on humans?

14. What are the four main areas of concern in nuclear reactor safety? Briefly describe the concern in each area.

15. Suppose that the fission fragments are removed from the transuranic elements found in nuclear wastes. It has already been stated that about 800 years are necessary for the cesium-137 and the strontium-90 to decay to the point of being harmless. Using Table 14.3, find the two other fission fragments with half-lives closest to those two. How long will it take each of these to decay to the point that only 1/1024 (10 half-lives) of the original amount remains? Why can long-lived fragments like iodine-129 be ignored, compared to these others?

16. In August, 1982, a fireworks plant in Seabrook, New Hampshire, suddenly exploded, rocking residents for a considerable distance. Some people thought the Seabrook Nuclear Power Station, which was not yet completed at that time, had exploded, and some feared that a nuclear explosion had occurred. If the plant had been in operation, would an explosion of the reactor have been possible? Would a nuclear explosion have been possible? Explain.

17. Describe briefly the technique of fault-tree analysis for determining the probability of a particular accident.

18. What seems to you a reasonable way to dispose of nuclear wastes? Explain.

19. What are the potential advantages of using nuclear fusion for power production? What are the difficulties?

15 *The Future*

Trying to look into the future is a risky business. This has been proved many times in the past as various scientists, social scientists, economists, politicians, and others have peered into their crystal balls. Often, very persuasive cases are made to support claims of what is going to happen. Lately, it has been fashionable to construct complicated, and impressive, computer models which take dozens of factors into account. Yet, the batting average of the prognosticators (predictors), has not been very good. Often the predictions prove to be wildly wrong. Sometimes the claims for the future encourage people in decision-making positions to take courses of action that, later, are seen to be foolish.

There are numerous recent examples of energy predictions, made seriously by "experts," that were very far from reality. For example, around 1960 it was widely touted that nuclear power would produce electricity "too cheap to meter." This has certainly not turned out to be the case—nuclear power has high costs, both economic and social. By 1970 those few voices that were raised to tell us that fossil fuels were becoming scarce and would soon become very expensive were largely ignored. Even the most pessimistic

people in those days did not foresee that the price of imported oil would increase by a factor of 10 in a decade. In 1974, President Nixon announced Project Independence, which would, according to him, make the United States self-sufficient in energy by 1980. (That is, there would be no further oil imports by then.) As we have seen, the amount of oil imported has been increased drastically, not reduced.

So, predictions are very likely to be wrong, even fairly short-term ones. This is partly because our "models" tend to be oversimplified. Often we simply assume that past trends will continue. This is almost necessary to make the models manageable, but then they turn out to be unrealistic. On the other hand, computerized models, which can include dozens of factors, tend to become unstable as they become more complicated. That is, a small difference in the assumptions put into the front end of the model can result in a drastic difference in the results of the output. Thus, neither simple nor complex models turn out to be very reliable. This is further complicated by the unavoidable fact that human desires enter into the models. That is, we tend to predict what we *want* to happen. Thus, the difference between *probable* results and *desired* results is often unclear.

Then what are we to do? A wise person once said, "If you don't know where you are going, any road will get you there." Saying this another way, if there is no *long-range goal*, then events just wander along haphazardly from day to day and year to year. This seems to have been the case with the U.S. energy policy, which has been nearly nonexistent. In the past, we have pumped or dug energy out of the ground as fast as we could and we have used it right away. We have fumbled and stumbled along from one year to the next with no clear idea of what lies ahead. We often seem to think that *somehow* science will come to the rescue in the future. It is almost as if we expect Superman to swoop out of the sky and save us at the last moment.

As you know from studying this book, it is becoming more and more obvious that there are no easy solutions to our problems. Even if vast new reserves of oil and natural gas are found, the time of having ridiculously cheap energy is past. The day when the remaining supply of fossil fuels becomes too precious to burn may be postponed, but it will not be canceled. What can we expect human life to be like at that time?

A goal that seems reasonable is that, at that future time, all people on Earth will have enough energy to live in reasonable comfort. The population of the world will have stopped growing and it will remain at a level that can be supported. Resources will be recycled: iron, copper, glass, paper, and other materials will be used over and over again, with new supplies being used only to make up for materials lost in recycling. Energy supplies will be renewable, including solar energy, wind energy, liquid fuels from wood and other vegetable matter and, possibly, nuclear fusion. Eating habits will be quite different from today's American diet, with much less meat and more grains, fruits, and vegetables. But everyone will have enough to eat. War will

be obsolete, made unnecessary by the fact that everyone has enough of the essential resources. This will not be Utopia, but it will be a world in which people can live comfortably and happily.

Provided that Humankind does not first commit the ultimate stupidity of self-annihilation by means of nuclear war, some version of the future envisioned in the preceding paragraph may well someday come to be. Why would one think so? Well, it is just that humans are too vigorous, ingenious, and filled with a will to live to simply lie down and die out. When will this happen? The *beginnings* may be in 50 years, or 100, or 200. In other words, it will not happen overnight, and much will depend on how well we handle the present energy situation.

If we accept the notion that a comfortable and secure future for the world is possible, then our task is one of *transition*. We must build a bridge from the present situation, which is growing ever more difficult, to the desired future. How we do so is a matter of crucial importance. In the words of former president Jimmy Carter, "We must face the prospect of changing our basic ways of living. This change will either be made on our own initiative in a planned and rational way, or forced on us with chaos and suffering by the inexorable laws of nature." Common sense dictates that we should choose the first path, not the second. How can we do so?

MOTHERHOOD

This section is really about energy conservation. Like motherhood, nobody has anything bad to say about energy conservation. However, if it hurts, even a tiny bit, we usually want the other guy to do it. Yet, in general it costs far less to save a Btu or kWh than it does to produce one. We have seen some encouraging signs in the past few years, with houses being insulated, smaller and more efficient automobiles being used, and other conservation measures being taken. This is resulting in part from a public awareness of the need for energy conservation and in part from the increased price of energy. The question is, is it possible to save *large* amounts of energy without sacrificing the comforts we are accustomed to? I think the answer is yes, although it might mean that we must change some of the ways we live. One encouraging example comes from countries such as Sweden and Canada, where the standard of living is high, but the energy use per person is much less than here.

To examine some of the possibilities, let's take a look at transportation. We have made some progress there, with the 12 mile per gallon gas hogs of a few years ago gradually being traded in for smaller cars that will get 40 miles to the gallon or more on the highway. The 55-mile-an-hour speed limit has helped, also, to conserve gas. (That does not mean that people drive *at* 55, but they tend to drive at 60 or 65, whereas they drove at 75 or 80 when the speed limit was 70.) The lower wind resistance at the lower speeds allows considerable savings in fuel. Also, because of the much higher price of gasoline, people have tended to drive less than before.

However, this is just scratching the surface of what is possible. Generally speaking, there are two ways in which fuel consumption can be reduced. One is the *technical fix*. This means that improvements in design are used to make devices more efficient. It is estimated by knowledgeable people that technical fixes alone could reduce the per capita use of energy in the United States by about half without substantially changing our way of life. In addition, we need to think about ways in which to make changes in the way we use automobiles. This need not lead to a lower standard of living and, indeed, with care and foresight, could lead to an improved quality of life. Some of the things we now do are not only wasteful; they are inconvenient,

time-consuming, and even dangerous. For example, in a city of 100,000 people, the major shopping areas are very likely to be clustered in one or two areas, probably outside the city limits, and miles from most of the population. Any shopping trip means using a car. Going at a convenient time, such as on a Saturday afternoon, usually means a slow, irritating drive in heavy traffic, parking long distances from the stores you want to visit, and fighting crowds of people all afternoon. Is this a modern convenience? What would happen if future cities were designed so that the dozens of stores, many of them competing, now found in a shopping center were scattered around the city so that all of the consumers were near some of them? Going even further with this, imagine a time when every home has a computer linked with a television set. Then, communications can replace travel in many instances. For example, instead of going to a store to shop, why not do it with the computer? Items from different suppliers could be compared at home in much more detail than if a dozen stores were visited. Then a simple entry on the computer would order the item, to be delivered within a day. Payment will automatically be made by your bank. Does this seem as if it could happen only in the very far future? That is not the case. Computers are now becoming very inexpensive and very versatile, and people are becoming accustomed to using them. Soon, everyone will have to know something about using simple computers in order to get along. (For example, this book is being written on a computer.)

Perhaps the idea of replacing automobile travel in our cities can be extended. Why not design cities so that automobiles have no place at all in them? Such cities could not be monsters with populations of millions, but they would be limited to 100,000 or perhaps 200,000 people at the most. A city of 100,000 could be comfortably laid out so that no one is more than three miles from the center. There would be pedestrian and bicycle paths everywhere, and convenient public transporation. This is not just a pipe dream; the cities of Runcorn, England, and Grimaud, France, have already been designed to function without automobiles. Figure 15.1 is one designer's vision of a city of the future. Even more "far out" (if you will excuse the pun) is the notion of a space colony, one version of which is shown in Figure 15.2.

Carrying the idea of changing our modes of transportation a bit farther, we could also eliminate much long-distance automobile, airplane, and truck travel. First, we would need a good system of high-speed rail transportation. That is presently by far the most efficient way to use fuel for overland transportation and, except for the longest trips, it could be the fastest way to travel. We might also simply eliminate some of the present long-distance hauling by airplanes and trucks. For example, with rapid and reliable rail transportation available, air freight shipments could

be eliminated. Also, at present about one-half of the vast amount of truck traffic in the United States hauls food and agricultural products. Most of the food consumed in this country is grown far away from the markets and trucked across the country. If our small cities were spaced so that there was farm land between them, perhaps a sizable fraction of agricultural products could be grown near the local markets, thus cutting down greatly on transportation needs. (Incidentally, to produce 1 calorie of food in the United States these days requires the exenditure of about 15 calories of fossil fuels, used for farm machinery, fertilizers, pesticides, and so on. In addition, transportation, packaging, and distribution require perhaps another 15 calories. Thus, our food chain is heavily dependent on energy and very inefficient in converting that energy into food value.)

This conservation-in-transportation scenario could go on at some length and in some detail, but I think you get the idea.

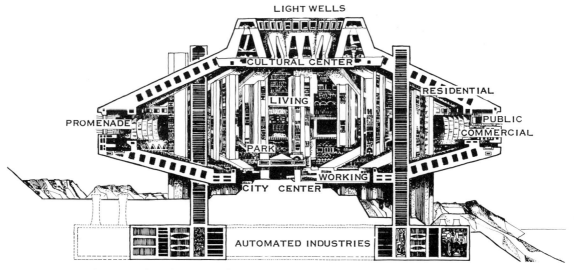

Figure 15.1 Architect Paolo Soleri's plan for a city of the future. Soleri combines architecture and ecology into "Arcology," to plan ecologically sound and energy efficient cities. (Courtesy Cosenti Foundation.)

Figure 15.2 An artist's conception of a future space colony. (Courtesy NASA.)

Overall, it would not be unreasonable to think that 50 years from now the per capita consumption of energy for transportation could be 20 or 25 percent of what it is in the United States now.

Similar observations could be made about the residential, commercial, and industrial uses of energy. In the industrial sector, for example, because of the low cost of energy, we have been very wasteful, particularly in disposing of "waste" heat. It is estimated that industry could save at least one-third of the energy it consumes by conservation practices and production changes, and even more with the cogeneration of electricity. Some of the needed changes are already being made.

Remember, however, that we are talking about a long-range plan. The changes suggested, and others, will not come about overnight. Perhaps in 50 years, if we have the determination and foresight, per capita en-

ergy use could be greatly reduced and, at the same time, the quality of life improved for all.

ENERGY PATHS

Suppose that an energy goal has been selected. The one just sketched so briefly may be one that most people could live with. There still remains the problem of making the transition from now to then. How we go about that may make a world of difference in our lives, our children's lives, and our grandchildren's lives. It is just this process of transition that is the object of so much controversy in this country and in the world.

One extreme view is that there really is no problem. There is plenty of fuel available. What we should do right now is accelerate our efforts to find and use fossil fuels, accelerate the use of nuclear energy, produce alcohol from plants and coal, and continue to *grow*. The future will take care of itself, helped by science and technology. By now, you realize that this "bigger is better" philosophy is highly dangerous and cannot continue indefinitely. The problems are too real and serious to disappear by the application of brute force to them. People who say there is no real energy problem probably do not know what they are talking about. If they do know what they are talking about and they are over 50 years of age, perhaps they are secretly saying to younger people, "There is enough for *me*, so, although there is not enough for *you*, let's not worry about it." We must reject this extreme view and find solutions that are more in line with reality.

At the other extreme, there are those who would shut down all of the nuclear power plants immediately, stop imports of foreign oil, and either shut down or drastically change the operation of coal-burning power plants. This kind of action would probably be beneficial to the environment, but it would wreck the economy and leave us without sufficient power to function. A sudden change of this type would cause untold suffering.

It is rather easy to demonstrate that either approach would be foolish and harmful. Between the two extremes, however, there is a wide spectrum of possibilities, and the choices are not quite so clear-cut. They include both technical advances and changes in life-style. In general, they can be broken down into hard energy paths, and soft energy paths. The term *hard energy path* is used to mean finding ways to increase the available energy through advanced technology and involving large centralized power stations. It is the sort of path that has been most vigorously and consistently supported by the U.S. government. A *soft energy path* would require the efficient use of energy, a rapid shift to renewable energy sources, and a matching of the quality of the energy to the end use. (This means that one should not use high-quality energy, like elec-

tricity, to perform tasks that require only low-quality energy, such as home heating.) Another breakdown has sometimes been termed macropower versus micropower. *Macropower* is power on the large scale, with huge power plants producing energy for large numbers of people and/or industries. *Micropower* is small-scale energy production, such as solar heating for homes or the generation of electric power with a small bank of photovoltaic cells. Note that soft energy is not necessarily primitive; it may make use of advanced technology. Also note that some proposed uses of renewable energy sources belong to the hard energy path, such as a solar power tower or ocean thermal energy conversion.

Choosing between these two paths is not always easy. Many knowledgeable people think the two are mutually exclusive; that is, we cannot do both simultaneously. This has certainly been the case in the past, and the hard energy path is the one that the United States has supported. For example, more than 30 years and billions of dollars have gone into nuclear fission and fusion research. Only in the last few years has any federal money at all been put into solar energy research, and most of that goes to large-scale projects which fit better into the hard energy category. Let us suppose that it is true that only one path is possible. If so, we need to decide soon which one to take. The world, and America in particular, may be thought of as standing at a fork in the road without a map. How do we decide which path to take? Each one of us must make that determination for himself or herself. Let us review for one last time some of the implications of each path.

In a hard energy path, emphasis will be placed upon increasing our energy supplies, particularly those that come from within the United States. This means that efforts will be increased to find new sources of oil and natural gas. Coal will have to be exploited more fully than it is now. We will have to pursue the production of liquid fuels from organic materials, from coal, and even from shale. It may be necessary to relax—perhaps temporarily, perhaps not—many of the air pollution and water pollution standards. Power plants will remain large, with many of the new ones being nuclear. Energy storage in a form that can produce large-scale electric power will have to be developed. New technologies will have to be pursued very hard, with emphasis on breeder reactors and nuclear fusion. If nuclear fusion turns out not to be possible on a practical scale, we must be ready to find substitutes. If breeder reactors turn out not to be safe, then they must be made safe.

The soft energy path would emphasize conservation to the point of changing our way of life in some manner. We would have to find much more efficient ways to deliver energy to the users. Instead of large, centralized power plants and massive distribution systems, the emphasis would be on small units placed near the users. This would include solar heating and hot water systems, photovoltaic systems, wind power, cogeneration of power for industry, and others. Electricity would be used mainly for those end uses which require electricity, such as lighting and electronics, and not for heating.

Painted in broad strokes, those are the options. We, the American people, need to realize, as some of the extreme advocates on both sides seem not to, that it may not be necessary to follow a hard energy path or a soft energy path exclusively. Some combination of the two is probably more sensible than either alone. We need to start making *conscious*, and perhaps tough, decisions about the path we ought to follow. Thus far, we have just wandered along allowing events to control us rather than vice versa, and this must not continue.

A FINAL WORD

In a democracy, it is up to the people to let their will be known. Members of the U.S. Congress and state legislators tend to support those causes they *think* the voters favor. You must let them know, frequently and consistently, what you favor in terms of energy. In the past, concerns about energy have been shoved to the rear and ignored, and this must not continue. If you have studied this book conscientiously, you now know far more than the average legislator about energy, about the situation we are in, and about the alternatives available. No doubt, you will forget three quarters of the material you learned here. Much of the factual material will be obsolete in a very short time. However, you should have gained a way of looking at the situation and an ability to learn more about it. When you read or hear information on energy, you should be able to analyze it and judge if it is reasonable. It is amazing how often a little simple arithmetic, similar to what you have learned to do here, will show up false claims and arguments. Further reading is open to you, and I hope you will pursue it. To get you started, I have included some references in Appendix E that present several different viewpoints.

Remember that what *you* do about energy in the future *is* important.

Basic Algebra

VARIABLES AND EQUATIONS

Algebra is a system whereby quantities are manipulated and calculated by means of symbols, rather than being limited to numbers, as is arithemtic. The symbols used may be anything, such as ☺, but usually letters of the alphabet are used for convenience. If we use a symbol, such as x, it may assume any numerical value or values we wish to give it for a particular problem. For example, I may say "x is equal to 5." In that case, x keeps the value of 5 until I change it explicitly. This is called a *constant*, a quantity whose value does not change during the course of a particular problem. Of course, for the next problem, the same symbol x probably will have a different value entirely.

If the only use we made of algebraic symbols was as constants, they would not be very helpful. It is when a symbol is allowed to be a *variable* that it is really useful. A variable changes in a predictable way in the course of a problem. For example, suppose you are walking along a road and you would like to have a symbol to represent the distance you have walked from the starting point. Since that distance keeps changing as you walk,

you need a variable to represent it. We might as well use d, for distance. (Or, we could have used h for how far, or f for farness, or s for space covered, or whatever. It really does not matter, as long as we keep track of it.) Anyway, suppose we use d, and you walk at the rate of 4 miles per hour. Then, at the end of 1 hour d is 4 miles, at the end of 2 hours d is 8 miles, at the end of 3 hours, 12 miles, and so on. If we let the symbol t stand for time, another variable, then we can say that the distance covered in miles is 4 times the time in hours. Egad! We have just written an algebraic equation:

$$d = 4t$$

An *algebraic equation* is simply a relationship between two or more variables, and that is what we have here. This equation says exactly what was just said in quite a few words. That is, if $t = 1$, $d = 4$; if $t = 2$, $d = 8$, and so forth. If we wanted to get just a bit fancier, we could have written $d = Ct$, where C is the constant 4. One major virtue of the equation is that it is a much shorter way to write down the same thing it takes quite a few words to say.

TABLES

The relationship between *d* and *t* can also be expressed in a couple of other ways: with a table or with a graph. To make a table showing this, just substitute into the equation the various values of *t*. For example, if $t = 0$, then $4t = 0$ and $d = 0$. That is the place you start the walk from. If $t = 1$ then $d = 4$; if $t = 2$, $d = 8$, and so forth. The table is made, then, by placing corresponding pairs of values side by side, as illustrated.

t *(hours)*	d *(miles)*
0	0
1	4
2	8
3	12
4	16
5	20

This table contains precisely the same information as the equation, but the equation is more concise. Another advantage of the equation is that it gives the answer for any time whatever. You can find, for example, that you walked 10 miles in 2.5 hours, or 3 miles in 45 minutes (.75 hours), or 24 miles in 6 hours without having to fill in every possible value in a table.

GRAPHS

To show the same information in a graph, first draw two perpendicular lines, called *axes* (singular, *axis*), and label them. Either axis may be chosen for *t*, and the other for *d*, but it is customary to put the variable on the right side of the equation on the horizontal axis and the one on the left side on the vertical axis, as shown in Figure A.1. Also on Figure A.1 there are distances marked off representing hours on the *t-axis* and miles on the *d-axis*. Those distances are arbitrary and are just chosen for

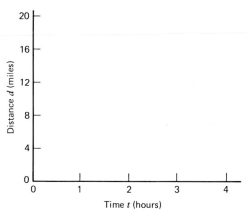

Figure A.1   Labeled axes for a graph.

convenience, so that they fit nicely on the paper. Graph paper is already marked into equal segments, and is handy for this.

Next, the information from the table must be put on the graph. This is called *plotting the points*, and is shown in Figure A.2. The point where both *d* and *t* are zero, where the two axes meet, is called the *origin* of the graph. To plot the point where *t* is 1 hour and *d* is 4 miles, move the pencil horizontally to the mark for 1 hour, then straight up to the mark for 4 miles, and that is the location of the point. Plot the other points in the same way.

Finally, connect the points with a line. In this case the line is a straight line, but in many

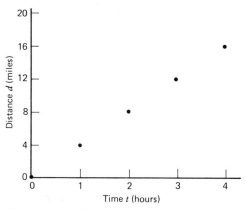

Figure A.2   Plotting the points for the graph.

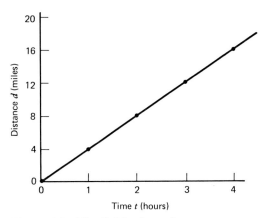

Figure A.3 The finished graph.

cases it will be a curved one. The finished graph is shown in Figure A.3.

Questions

You can buy coffee for $3.00 a pound. Using *c* for cost and *p* for pounds, write an equation that gives the cost of a given number of pounds of coffee. Using that equation, make a table and graph showing the same information.

You rent a car for $20.00 per day plus $0.20 per mile driven. Write an equation that will give you the total cost of the car. (*Hint*: Since the charge for zero miles driven is $20.00, you have to add this amount in on the right side of the equation.) Again, make a table and a graph. How does this graph differ from the preceding one?

MANIPULATING EQUATIONS

The basic rule for manipulating equations is very simple: Whatever you do to the left side, do to the right side also. This will ensure that the equation remains an equation. The one exception to this is that you can *never* divide by zero. Any number divided by zero is undefined and without meaning. Even your

electronic calculator knows this: try it and see. Suppose we have the simple equation

$$x = 4$$

Then, multiply both sides by 5:

$$5x = 20$$

Now subtract 7 from both sides:

$$5x - 7 = 13$$

And this equation is still correct. To see that it is, substitute the original value of 4 for *x* and see if the left side is 13. It works, doesn't it?

The real value of this kind of manipulation is to go in the other direction, from a more complex to a simpler equation. For example, suppose you have the equation

$$7y + 9 = 3y + 33$$

What is the value of *y*? To find out, start by subtracting 9 from both sides:

$$7y + 9 - 9 = 3y + 33 - 9$$
$$7y = 3y + 24$$

Now subtract 3y from both sides:

$$7y - 3y = 3y - 3y + 24$$
$$4y = 24$$

Finally, divide both sides by 4:

$$\frac{4y}{4} = \frac{24}{4}$$
$$y = 6$$

In this case, *y* is a constant. To see that the answer is correct, just substitute 6 for *y* everywhere it appears in the original equation. You should get $51 = 51$. Now let's try the same thing when y is not a constant:

$$5y - 3x + 4 = 2y + 3x - 5$$

First subtract 4 from each side:

$$5y - 3x + 4 - 4 = 2y + 3x - 5 - 4$$
$$5y - 3x = 2y + 3x - 9$$

Then add $3x$ to each side:

$$5y - 3x + 3x = 2y + 3x + 3x - 9$$
$$5y = 2y + 6x - 9$$

Then subtract $2y$ from each side:

$$5y - 2y = 2y - 2y + 6x - 9$$
$$3y = 6x - 9$$

Finally, divide both sides by 3:

$$\frac{3y}{3} = \frac{6x}{3} - \frac{9}{3}$$

Or:

$$y = 2x - 3$$

Notice that *both* terms of the right side were divided by 3. Now we have an expression relating y to x for which we could make a table and draw a graph. In this case y is not a constant but it is a variable whose value depends on the value of x, which is also a variable.

One important comment: In this last equation, if x is 5 oranges, then y is 7 oranges, *not* apples. As the fella says, you can't add apples and oranges.

Questions

In each of the following equations, solve for y, either as a constant or as a variable depending on x.

$$5y - 14 = y + 2$$
$$7y + 9 = 4y + 3$$
$$5y - x = 3x - 9 + y$$
$$14 + 7x = 4 + 2x - 5y$$
$$y - 2x = 5x^2 + 11$$

In this last equation did you try to add $5x^2$ and $2x$ to get 7 something? Don't! Those cannot be added that way; their sum is $5x^2 + 2x$ and is not equal to $7x^2$ *or to* $7x$.

MULTIPLYING BINOMIALS

A *binomial* is an algebraic expression involving the sum or difference of two quantities. For example $a + b$, $x - y$, $z + 5$, and $17 - w$ are all binominals. When multiplying one binomial by another, the rule is to multiply each term of one with each term of the other and take the sum. Some examples will show this better than words:

$$(a + b) \times (c + d) = ac + ad + bc + bd$$
$$(x + 3) \times (y - 4) = xy - 4x + 3y - 12$$
$$(x - 3) \times (x + 5) = x^2 + 5x - 3x - 15$$
$$= x^2 + 2x - 15$$
$$(5y + 7) \times (3y - 4) = 15y^2 - 20y + 21y - 28$$
$$= 15y^2 + y - 28$$
$$(2x + 5y) \times (7x - y) = 14x^2 - 2xy + 35xy - 5y^2$$
$$= 14x^2 + 33xy - 5y^2$$

(Note that $xy = yx$.)

$$(x + y)^2 = (x + y) \times (x + y)$$
$$= x^2 + xy + xy + y^2$$
$$= x^2 + 2xy + y^2$$
$$(2a - 3b)^2 = 4a^2 - 12ab + 9b^2$$

Questions

Why don't you try some?

$(3t + 5) \times (w - 2) =$

$(5z + 6) \times (2z + 4) =$

$(3x + y)^2 =$

$(2a - 5b)^2 =$

$(x - y) \times (3x + 5) =$

$(a - 2b) \times (4a - 3b) =$

To check these results, let $a = 1$, $b = 2$, $t = 3$, $w = 4$, $x = 5$, $y = 6$, $z = 7$. Now, into each equation, substitute the appropriate values. Does the number on the left side equal the number on the right side in each case? If not, look for an error.

B Numerical Manipulations

There is not a great deal of mathematics used in this book. The basic algebra presented in Appendix A will suffice. However, the application of physics to energy concerns will entail considerable numerical calculation, at times. I strongly recommend that you buy an inexpensive electronic calculator you can take to class every day. This appendix will help you to use your calculator wisely.

INSIGNIFICANT DIGITS

The electronic calculator is a boon to people, like myself, who never cared all that much about memorizing multiplication tables. It also relieves us of the agony of long division. Fortunately, calculators are now so inexpensive and readily available that you should be able to have one to help you through this course. The electronic calculator does, however, insidiously promote one kind of nonsense, which might be called the production of "insignificant digits." For example, suppose three of us are hanging around and would like to go to a movie, but you are the only one with money. Being a nice person you are willing to share it equally. I ask you how much you have and, without counting, you say, "around

ten dollars." I whip out my handy-dandy pocket calculator, divide 10 by 3 and announce that we can each have $3.3333333. Ridiculous! In the first place, there is no way to subdivide pennies, and the final five digits make no sense at all. Worse than that, the 33 cents may also be misleading. If, without counting, you know that you have more than $9 and less than $12, a reasonable answer might be, "about $3 each" or, "between $3 and $4." If you are certain that it is $10 and change, "around $3.50 apiece" would do and, if you count the money and find that it is $10.38, then the exact answer of $3.46 can be confidently given.

So don't be seduced by your calculator into quoting meaningless digits or, worse, putting them down on paper. Learn to round off, to approximate. For example, if a table top has sides of *about* 4.5 by 5.5 ft, its area can be computed as:

$$4.5 \text{ ft} \times 5.5 \text{ ft} = 24.75 \text{ square feet}$$

But this is misleading. It implies that you know the area to the nearest hundredth of a square foot. A better answer would be 24.8 ft². (Often *ft²* is used as a shorthand symbol for *square feet*). To show why, suppose you are measuring roughly and you are not certain of the measurements to any better than 0.05 ft (about one-half inch). Then, the area may be anywhere between 4.45 ft × 5.45 ft ≈ 24.25 ft² and 4.55 ft × 5.55 ft ≈ 25.25 ft², so rounding to 24.8 ft² does make sense as a best estimate. (Incidentally, the symbol ≈ means "approximately equal to" or "about." It appears fairly frequently in this book.)

As another example, suppose you are an engineer and you must calculate the air-conditioning load for a building that is approximately 45 by 65 ft by 25 ft tall. You need to know the volume, which is

$$V = 45 \text{ ft} \times 65 \text{ ft} \times 25 \text{ ft}$$
$$= 73{,}125 \text{ cubic ft}$$

Once again, this is misleading, and an answer $V \approx 73{,}000$ cubic ft would be more appropriate, since a small error in each of the length measurements would produce a rather large difference in the computed volume. The zeros on the right side of a whole number like 73,000 are not considered to be significant digits.

There exist formal rules for rounding off the results of calculations. However, rather than learn a complex set of rules for rounding off and approximating to fit the various circumstances, I suggest you simply use good judgement (common sense), with a consideration of the particular circumstances for each case. As a very rough rule of thumb, decimal answers probably should not have more decimal places than the least precise number used for the calculation. In a number like 2.4000, the trailing zeros are significant and should not be used unless the number is exactly 2.4. Whole number answers probably should not have more significant digits than the least precise number used for the calculation; thus the two significant digits of 73,000 used earlier.

Questions

In the volume calculation just done, suppose one knows that 45 ft, 65 ft, and 25 ft, are not off by more than 0.1 ft each. What is the range, from lowest to highest, of volumes that result with this much uncertainty? Is the approximation 73,000 ft³ a reasonable one?

DIMENSIONS AND UNITS

In the real world, physical quantities and objects are never measured in pure numbers; they have dimensions and are measured in some units. For example, the distance from one point to another has the dimension of length. The length may be measured in a number of different units, such as feet, meters, inches, kilometers, miles, cubits, centimeters, furlongs, fathoms, light-years, or yards. Each and every unit of measurement is arbitrary, and is used only because somebody, sometime, found it convenient. The unit of measurement is not important, as long as people can agree on it, and two persons measuring the same thing can get the same answer. However, whatever unit is being used, it is *essential* to write it down with the number each time.

Another example is time, which is extremely difficult to define, but easy to measure in seconds, hours, days, weeks, months, years, or centuries.

In 1974 a General Conference of Weights and Measures recommended an international system of units, called *SI* for *Le Système International d'Unités*. It has been widely adopted, but is certainly not universally used, especially in this country. In this book, the SI system is used whenever possible. (However, in many instances, the book leaves the SI system in order to use the units commonly used in American engineering practice.) In the SI system, the basic units are the *meter* (m) for measuring length, the *second* (s) for measuring time, and the *kilogram)* (kg) for measuring *mass*. The unit of energy used in this system, the *joule* (J), is derived from these three.

You are urged to write down the units for the quantities involved in any problems you attempt. Worked out examples in the text illustrate the manipulation of units.

POWERS OF TEN

In dealing with energy concepts, we often will need to use very large numbers, and sometimes very small ones. A number like 2,500,000,000,000 is difficult to write and even more cumbersome to use in arithmetic manipulations. It is far easier to write 2.5×10^{12}, which reads, "2.5 times 10 to the twelfth power." (The power to which 10 is raised is called the *exponent*.) The number 2.5×10^{12} represents 10 multiplied by itself 12 times, and that result multiplied by 2.5. that is,

$$2.5 \times 10^{12} = 2.5 \times 10 \times 10 \times 10 \times 10 \times$$
$$10 \times 10 \times 10 \times 10$$
$$\times 10 \times 10 \times 10 \times 10$$
$$= 2.5 \times (1,000,000,000,000)$$
$$= 2,500,000,000,000$$

Using the powers of 10 (also called Scientific Notation) does not seem like much of an advantage yet, but consider the following multiplication:

$$2,500,000 \times 6,000,000,000$$
$$= 15,000,000,000,000,000$$

Already it becomes cumbersome, and an easier way to do the same thing is

$$(2.5 \times 10^6) \times (6 \times 10^9) = (2.5 \times 6) \times (10^6 \times 10^9)$$
$$= 15 \times 10^{15}$$
$$= 1.5 \times 10^{16}$$

There are a few things to note about this process:

1. The multiplication can be done separately for 2.5×6 and for the exponents.

2. When multiplying, the exponents simply add. That is, $1,000,000 \times 1,000,000,000 = 1,000,000,000,000,000$ is exactly the same as $10^6 \times 10^9 = 10^{15}$. The power to which 10 is raised is equal to the number of zeros when the number is written out for each of these three numbers.

3. Multiplying a number, such as 2.5, by the sixth power of 10, 10^6, is equivalent to moving the decimal point six places to the right; that is, 2.5×10^6 becomes 2,500,000.

Division is similar, except that the exponents are subtracted:

$$\frac{6 \times 10^9}{3 \times 10^5} = \left(\frac{6}{3}\right) \times \left(\frac{10^9}{10^5}\right)$$
$$= 2 \times 10^4$$

That is, the exponent in the denominator, 5, is subtracted from that in the numerator, 9, to yield 10^4.

Suppose it had been the other way around, with the larger number in the denominator. Then the process is the same:

$$\frac{3 \times 10^5}{6 \times 10^9} = \left(\frac{3}{6}\right) \times \left(\frac{10^5}{10^9}\right)$$
$$= 0.5 \times 10^{-4}$$
$$= 5 \times 10^{-5}$$

The meaning of the negative exponent is shown by

$$10^{-4} = \frac{1}{10^4}$$

thus

$$0.5 \times 10^{-4} = \frac{0.5 \times 1}{10,000} = 0.00005$$

so the result of the exponent, -4, is to move the decimal place four places to the *left*.

The only remaining rule to remember is that to add or subtract two numbers, they must have the *same* power of 10. (That is, they must be of the same *order of magnitude*). For example,

$$3 \times 10^3 + 2 \times 10^4 \text{ is } not \ 5 \times 10^3 \text{ or } 5 \times 10^4$$

It is

$$3000 + 20,000 = 23,000$$

We can do this easily by writing it as

$$3 \times 10^3 + 20 \times 10^3 = 23 \times 10^3$$
$$= 2.3 \times 10^4$$

Or, an equally good way to do the same thing is

$$0.3 \times 10^4 + 2 \times 10^4 = 2.3 \times 10^4$$

You might practice some arithmentic using powers of 10 by answering the following questions.

Questions

Try the following arithmetic operations to see if you get the answers indicated:

$2.4 \times 10^5 \times 3.1 \times 10^3$ $\qquad (7.4 \times 10^8)$

$\dfrac{5.3 \times 10^2}{200}$ $\qquad (2.7)$

$7 \times 10^4 + 2.2 \times 10^5$ $\qquad (2.9 \times 10^5)$

$\dfrac{5,000}{3 \times 10^6}$ $\qquad (1.7 \times 10^{-3})$

$1.4 \times 10^{17} - 4 \times 10^{16}$ $\qquad (1.0 \times 10^{17})$

$6 \times 10^{-11} \times 5 \times 10^3$ $\qquad (3 \times 10^{-7})$

$\dfrac{4 \times 10^4}{3 \times 10^{-2}}$ $\qquad (1.3 \times 10^6)$

$5 \times 10^4 + 7 \times 10^{-1}$ $\qquad (5 \times 10^4)$

$\dfrac{2 \times 10^5 \times 3 \times 10^{-2}}{5 \times 10^3}$ $\qquad (1.2)$

$\dfrac{3.16 \times 10^6}{3 \times 10^4} - 15$ $\qquad (105)$

C

Graphs

It is said that a picture is worth a thousand words. For the prepared person, a good graph is worth at least a hundred pictures, which I suppose makes it worth 1×10^5 words. Figure C.1 and Table C.1 provide the same information, the daily rate of consumption of oil in the United States for the years 1969 through 1981.

Although the basic information can be gleaned from a close inspection of the table, the trends leap out at you from the graph. First, there is a very noticeable overall increase from 1969 to 1978. This includes a rather dramatic decline in oil consumption following the Arab oil embargo of 1973, but with the pattern of growth resuming shortly thereafter. Finally, there is the encouraging fact of a substantial decrease from 1978 through 1981.

However, before one gets too encouraged, note that it is risky to draw conclusions about the future shape of the graph from insufficient information. An optimist might look at the decline since 1978 and conclude that it is the beginning of a new trend. A pessimist might decide that this was just a temporary dip in a steady climb, as in 1974 and 75, and continue the graph on its upward way. A fence-sitter might opt for a future with constant use of oil. The three alternatives are shown in Figure C.2. This points out the danger of extending graphs into the future—of *extrapolating*. Any extrapolations should be done thoughtfully and with a clear description of the assumptions being made. Even then extrapolation is risky.

Learning to interpret graphs easily is largely a matter of practice; the more you do the easier it gets. To start getting some prac-

TABLE C.1 *The Rate of Consumption of Oil in the United States*

Year	Daily consumption (thousands of barrels)
1969	13,815
1970	14,350
1971	14,845
1972	15,990
1973	16,870
1974	16,150
1975	15,875
1976	16,980
1977	17,925
1978	18,255
1979	17,910
1980	16,470
1981	15,480

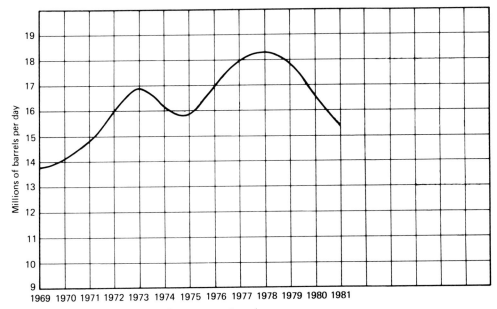

Figure C.1 United States oil consumption from 1969 to 1981.

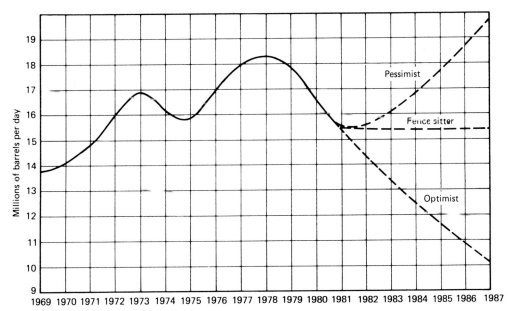

Figure C.2 Three possible extrapolations of the graph for oil consumption.

TABLE C.2   *A Listing of x versus y*

x	y
0	5
1	6
2	7
3	8
4	9
5	10

tice, let's take a look at some features of graphs. For example, look at a graph of x versus y, from Table C.2.

Figures C.3(*a*) and C.3(*b*) show two graphs that present this same information. The only difference is a difference in *scale* on the vertical axis. We need to keep in mind that two graphs that look different may really be the same. Graphs of this sort are called *linear*, meaning that the graph is a straight line rather than a curved one. It also said that y is a *linear function* of x when the result is a linear graph.

Some features of this type of graph are worth looking at. Figure C.4 shows three different lines drawn on the same graph. Each represents a somewhat different relationship between x and y. *A* and *B* start at different points on the *y*-axis, but climb at the same rate, parallel to one another. *C* starts at the

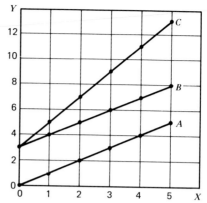

Figure C.4   Three linear graphs.

same point as *B* but climbs at a greater rate. From the graph, let us make a table (Table C.3), showing x versus y in each case.

Using these three sets of data, we can write equations relating x and y in each case:

$$A \quad y = x$$
$$B \quad y = x + 3$$
$$C \quad y = 2x + 3$$

You can easily check this out by substituting in various values for x. For example, if in case C $x = 4$, then $y = 2 \times 4 + 3 = 11$, which matches the value in the table. Try it for some

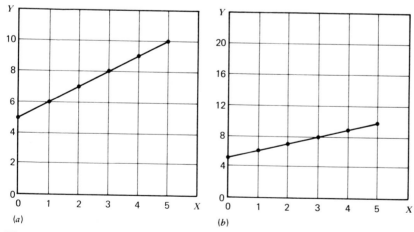

(a)   (b)

Figure C.3   The same graph drawn to two different scales.

TABLE C.3.   *Three Different Straight Lines*

A		B		C	
x	y	x	y	x	y
0	0	0	3	0	3
1	1	1	4	1	5
2	2	2	5	2	7
3	3	3	6	3	9
4	4	4	7	4	11
5	5	5	8	5	13

x	y
0	−2
1	1
2	4
3	7
4	10

Finally, draw the graph, as in Figure C.5.

other values of x.

In fact, if we were to write down equations for linear (straight-line) graphs of all imaginable appearances, they would all look a lot like C. We can generalize this by writing the equation

$$y = ax + b$$

Where a and b are any constants, whatever. Because the graph of such an equation is always a straight line, this is called a *linear equation*.

EXAMPLE

Suppose, for a linear equation, that $a = 3$ and $b = -2$. Make a graph of the equation.

Solution

First, substitute into the general linear equation the values of a and b:

$$y = ax + b$$

becomes

$$y = 3x - 2$$

Then, make a table of values, picking convenient numbers for x and finding y in each case from the equation.

Questions

For each combination of a and b listed here, make a table of x versus y and a graph. Use the equation $y = ax + b$ for each case.

1. $a = 4$, $b = 0$
2. $a = 0$, $b = 4$
3. $a = 3$, $b = 9$
4. $a = 0$, $b = 9$
5. $a = -3$, $b = 9$

In each case, b is the *y-intercept*, the place where the line crosses the y-axis. That is, it is the value of y when $x = 0$. The last three examples show that a determines how steeply the line climbs (falls, in the case of negative a). This is called the *slope*.

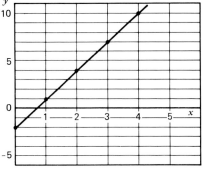

Figure C.5   The graph of $y = 3x - 2$.

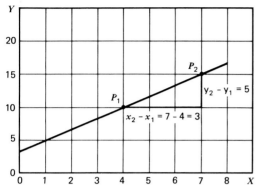

Figure C.6 Finding the slope of a straight line.

The way to find the slope from a graph is indicated in Figure C.6. Pick any two points, P_1 and P_2, on the line and draw a right triangle, as shown. Then the slope of the line is defined as the vertical leg of the right triangle divided by the horizontal leg. In this case $a = 5/3 = 1.7$. (If you know trigonometry, you can see that this is the tangent of the angle between the line and the x-axis.) The greater the slope, the steeper the line. As you just discovered in the Question, a negative slope means that the line falls rather than rises. Also, since the line strikes the y-axis at 3, $b = 3$. Thus, the equation for the line graphed in Figure C.6, $y = ax + b$,

becomes

$$y = 1.7x + 3$$

Questions

For each of the three graphs of Figure C.7, find the slope and the y-intercept and write the equation of the line.

Note that Figures C.7(a) and C.7(c) have the same slope, even though they do not look the same. The different appearance is caused by a different choice of scale.

Of course, not all graphs are straight lines; they may consist of *curves*, which may be simple or complex in nature. A simple curve is a graph of $y = x^2$. It is shown in Table C.4 and in Figure C.8.

A question one might ask is, What is the slope of the line? The answer is more difficult

TABLE C.4. *Values of x versus y when $y = x^2$*

x	y
0	0
1	1
2	4
3	9
4	16
5	25
6	36
7	49
8	64
9	81
10	100

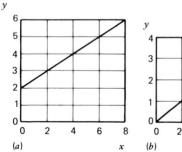

(a)

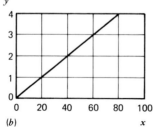

(b)

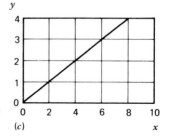

(c)

Figure C.7 Three linear graphs.

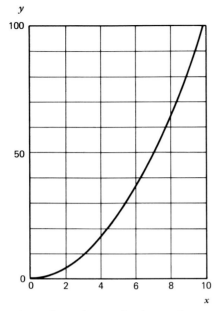

Figure C.8 The graph of $y = x^2$.

than before because here the slope is changing, getting steeper and steeper as x increases. In this case, we have to determine the slope at each point on the curve. One way to do so approximately is shown in Figure C.9(a). We are finding the slope at $x = 6$ by picking points P_1 and P_2 nearby on either side of the desired point and drawing the triangle. The answer is that the slope is about 12. Another method is to draw a straight line that touches the curve at only one point (we say it is *tangent* to the curve), as shown in Figure C.9(b). Then the slope of that straight line is defined to be the slope of the curve at that point. Again, it comes out to about 12 in the example we are using.

Questions

Make a table and draw a graph of $y = 2x^2$ and, by both methods illustrated in Figure C.9, find the slope at $x = 2, 4, 8,$ and 10.

In general any graph, however complex, will consist of various combinations of straight

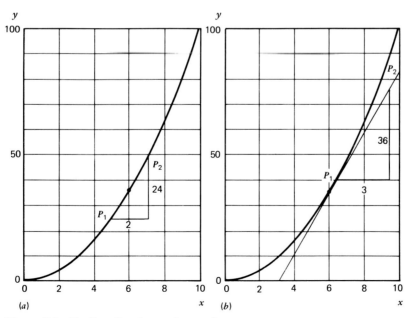

(a) (b)

Figure C.9 Finding the slope of $y = x^2$.

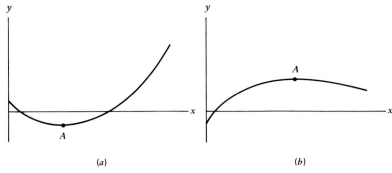

Figure C.10   (*a*) An upward turning curve. (*b*) A
downward turning curve.

lines, upward turning curves, as in Figure
C.10(*a*), and downward turning curves, as in
Figure C.10(*b*). The upward turning curve of
Figure C.10(*a*) has the characteristic that its
slope is always increasing as *x* increases. To
see this, put a ruler down on it, tangent to
the curve. In this case, the slope is negative
for small values of *x*; it becomes less negative
(increases), as we move to the right, then finally becomes positive, still increasing.

On the other hand, the downward turning curve of Figure C.10(*b*) has a slope that
starts out as positive, but decreases with increasing *x*, finally going negative.

Questions

What is the slope at the point labeled A on
Figures C.10(*a*) and C.10(*b*)?

D *Energy Units*

As you read through this book, you will see a number of different energy units used. This is made necessary in part because, as science has developed, different forms of energy such as heat and kinetic energy were not at first recognized as being related. Also, there have grown up several different systems of units for measuring things. In the British Engineering System, mechanical energy was measured in foot-pounds, and thermal energy was measured in British thermal units. The SI units for the same quantities are joules and calories. Unfortunately, persons who presently write and present statistics about the energy situation do so in more than a dozen different units. Even worse, different writers are not consistent with one another (and sometimes even not with themselves!). If you do some reading in the energy field, particularly statistics, it is easy to get confused about the units. To help you to sort it out, I present here some of the more common units. Inside the back cover is a table relating them.

Joule (J).   The amount of work done by a force of one newton acting through a distance of one meter.

Calorie (cal).   The amount of heat required to raise the temperature of one gram

of water one degree, Celsius. (From 14.5°C to 15.5°C.) The modern definition of the calorie is that it is 4.184 joules.

Kilocalorie (kcal).   Also called a large calorie. 1000 calories.

Kilowatt-hour (kWh).   Used for electric energy, 1000 watts (joules per second) flowing for one hour. Thus, it will be 3.6 million joules.

British thermal unit (Btu).   The amount of heat required to raise one pound of water one degree, Fahrenheit.

Million British thermal units (MBtu).   This large unit of energy is used in describing energy resources, as are the following units.

Quad (quad). 10^{15} Btu.

Q (Q). 10^{18} Btu.

Milli-Q (mQ).   One-thousandth of a Q. That is, 10^{15} Btu, which is the same as a quad.

Tonnes of oil equivalent (mtoe).   This is the amount of any fuel that will produce the same heat, when burned, as will a tonne of oil. A tonne is a metric ton, 1000 kg. A tonne of oil will produce about 4.11×10^7 Btu when burned. The abbreviation *mtoe* refers to "metric tons of oil equivalent."

Some references use 1 million tonnes of oil as the standard.

Tonnes of coal equivalent (mtce).   Some authors prefer to use coal, rather than oil, as the standard. This unit is equal to about 2.8×10^7 Btu. The abbreviation *mtce* stands for "metric tons of coal equivalent." Some books report energy resources on the basis of millions of tonnes of coal.

Barrels of oil equivalent (boe).   Then again, instead of using tonnes of oil, some authors use barrels of oil. A barrel of oil is 42 gallons, and it has a heating value of approximately 5.9×10^6 Btu.

Trillion cubic feet (TCF).   This is a unit of volume for measuring natural gas. It is 1 trillion (10^{12}) cubic feet, measured at standard temperature (0°C) and standard atmospheric pressure. Its heating value is about 1.1×10^{15} Btu, and it is thus equal to 1.1 quads.

Teracalorie (Tcal).   One trillion (10^{12}) cal. This unit is not used very often, but you will run across it occasionally. You will also see terajoules (10^{12} J) once in a while.

Table D.1 on the back inside cover gives the relationships between all of these units. Of course, the heating value of a tonne of coal or a barrel of oil varies considerably, so some sort of average had to be used. Where I could find them, I have used United Nations standards for average heating values. I confess that, in some instances, I had to just make estimates, but the values should be in the right ballpark. The following examples will show how to convert from one unit to another.

EXAMPLE

Suppose a coal field is reported to have a total of 4×10^7 tonnes of coal in it. How many barrels of oil would be needed to provide the same heating value as that much coal? How many Btus can be provided by burning the coal.

Solution

First find mtce in the left column of the table. Move in that row to the column under boe. The number in the intersection is 4.76, which is the number of barrels of oil that are equivalent to 1 tonne of coal. That is, there are 4.76 boe per mtce. Then

$$4 \times 10^7 \text{ mcte} \times 4.76 \frac{\text{boe}}{\text{mtce}} = 1.9 \times 10^8 \text{ boe}$$

To find the heating value of the coal in Btus, note from the table that 1 mtce = 2.8×10^7 Btu. Then:

$$4 \times 10^7 \text{ mtce} \times 2.8 \times 10^7 \frac{\text{Btu}}{\text{mtce}} = 1.1 \times 10^{15} \text{ Btu}$$

Or, alternatively, using the heating value of oil found in the table:

$$1.9 \times 10^8 \text{ boe} \times 5.9 \times 10^6 \frac{\text{Btu}}{\text{boe}} = 1.1 \times 10^{15} \text{ Btu}$$

EXAMPLE

Imports of oil to the United States have been running about 20 quads/yr for the past few years. How many barrels of oil is that?

Solution

$$20 \text{ quads} \times 1.7 \times 10^8 \frac{\text{boe}}{\text{quad}} = 3.4 \times 10^9 \text{ barrels}$$

(over 3 billion!)

Unfortunately, there is a similar wide range of units for reporting power, the rate of using or producing energy. All of them are derived by dividing one of the energy units by a unit of time. Some examples are watts (J/s), kilowatts (1000 J/s), megawatts (10^6 W), gigawatts (10^9 W), barrels per day, tonnes per year (either oil equivalent or coal equivalent), teracalories per year, and quads per year. In the text, when discussing large-scale energy production or consumption, I shall stick to quads for energy and quads per year for the rate of energy use or production wherever possible.

E *A Reading List*

The following books represent a wide range of views and approaches to the energy problem.

Perspectives on Energy, Third Edition, Lon C. Ruedisili and Morris W. Firebaugh, Eds, Oxford University Press, New York. 1982.

McGraw-Hill Encyclopedia of Energy, Second Edition, Sybil P. Parker, Ed, McGraw-Hill Book Company, New York, 1981.

Energy and the Environment, John M. Fowler, McGraw-Hill Book Company, New York, 1975.

Energy Environment Source Book, John M. Fowler, National Science Teachers Association, Washington, D.C., 1978.

Annual Review of Energy, Annual Reviews, Inc., Palo Alto, Cal., (One volume each year.)

America's Energy Famine, Ruth Sheldon Knowles, University of Oklahoma Press, Norman, 1980.

Soft Energy Paths: Toward a Durable Peace, Amory B. Lovins, Friends of the Earth, International, Ballinger Publishing Company, Cambridge, Mass., 1977.

The Last Chance Energy Book, Owen Phillips, The Johns Hopkins University Press, Baltimore, Md., 1979.

The Politics of Energy, Barry Commoner, Alfred A. Knopf, New York, 1979.

Energy: Created Crisis, Anthony C. Sutton, Books in Focus, Inc., New York, 1979.

Energy Handbook, Robert C. Loftness, Van Nostrand Reinhold Co., New York, 1978.

World Energy Survey, Ruth Leger Sivard, Published under the auspices of the Rockefeller Foundation, New York, 1979.

BP Statistical Review of the World Oil Industry, Published annually by The British Petroleum Company Limited, London.

Energy in Transition 1985–2010, Final Report of the Committee on Nuclear and Alternative Energy Systems, The National Research Council, W.H. Freeman & Company, San Francisco, 1980.

Energy and Man: Technical and Social Aspects of Energy, M. Granger Morgan, Ed, IEEE Press, New York, 1975.

Practical Physics: The Production and Conservation of Energy, Joseph F. Mulligan, McGraw-Hill Book Company, New York, 1980.

Energy: An Introduction to Physics, Robert H. Romer, W.H. Freeman & Company, San Francisco, 1976.

Energy: The Next Twenty Years, Hans Landsberg, Ed, Ballinger Press, Cambridge, Mass., 1979.

Energy-Efficient Community Planning, James Ridgeway, J. G. Press, Emmaus, Pa., 1978.

Energy Future: Report of the Energy Project of the Harvard Business School, Robert Stobaugh and Daniel Yergin, Eds, Boston, Mass., 1979.

The Passive Solar Energy Handbook, Edward Mazria, Rodale Press, Emmaus, Pa., 1979.

Introduction to Appropriate Technology, R. J. Congdon, Ed, Rodale Press, Emmaus, Pa., 1977.

Solar Electricity: An Economic Approach to Solar Energy, Wolfgang Palz, Butterworth Publishing Company, Woburn, Mass., 1978.

Wind Power: Recent Developments, D. J. DeRenzo, Ed, Noyes Data Corporation, Park Ridge, N.J., 1979.

The Poverty of Power, Barry Commoner, Bantam Books, New York, 1976.

Energy Conservation and Public Policy, John Sawhill, Ed, Prentice-Hall, Englewood Cliffs, 1979.

The Nuclear Power Debate: Issues and Choices, Scott Fenn, Praeger Publishers, New York, 1981.

A Guidebook to Nuclear Reactors, Anthony V. Nero, Jr., University of California Press, Berkeley, 1979.

The Health Hazards of Not Going Nuclear, Petr Beckmann, The Golem Press, Boulder, Colo., 1976.

Nuclear Reactor Safety, F. R. Farmer, Ed, Academic Press, New York, 1977.

A Guide to Radiation Protection, J. Craig Robertson, John Wiley & Sons, New York, 1976.

The Nuclear Fuel Cycle, The Union of Concerned Scientists, MIT Press, Cambridge, Mass., 1975.

The Atom Besieged, Dorothy Nelkin and Michael Pollak, MIT Press, Cambridge, Mass., 1981.

The Accident Hazards of Nuclear Power Plants, Richard E. Webb, University of Massachusetts Press, Amherst, 1976.

An Irreverent, Illustrated View of Nuclear Power, John W. Gofman, Committee for Nuclear Responsibility, 1979.

One of the best sources of informative articles on energy is *Scientific American*. Here are a few recent articles.

The entire September, 1971, issue is devoted to energy.

Tides and the Earth-Moon System, April, 1972, page 42.

Clean Power from Dirty Fuels, October, 1972, page 26.

The Hydrogen Economy, January, 1973, page 13.

Energy Policy in the U.S., January, 1974, page 20.

The Gasification of Coal, March, 1974, page 19.

The Disposal of Waste in the Ocean, August, 1974, page 16.

The Rise of Coal Technology, August, 1974, page 92.

The Fuel Consumption of Automobiles, January, 1975, page 34.

The Necessity of Fission Power, January, 1976, page 21.

Oil and Gas from Coal, May, 1976, page 24.

The Reprocessing of Nuclear Fuels, December, 1976, page 30.

The Importation of Liquified Natural Gas, April, 1977, page 22.

The Disposal of Radioactive Wastes from Fission Reactors, June, 1977, page 21.

Passive Cooling Systems in Iranian Architecture, February, 1978, page 54.

World Oil Production, March, 1978, page 42.

Nuclear Power, Nuclear Weapons, and International Stability, April, 1978, page 45.

Fusion Power with Particle Beams, November, 1978, page 50.

Progress Toward a Tokamak Fusion Reactor, August, 1979, page 50.

The Safety of Fission Reactors, March, 1980, page 53.

A Ban on the Production of Fissionable Materials for Weapons, July, 1980, page 43.

Oil Mining, October, 1980, page 182.

The Fuel Economy of Light Vehicles, May, 1981, page 48.

Advanced Offshore Oil Platforms, April, 1982, page 39.

Carbon Dioxide and World Climate, August, 1982, page 35.

Index

TABLE D.1 *Conversion Table for Various Units of Energy*

	Joule (J)	Calorie (cal)	Kilocalorie (kcal)	Kilowatt-hour (kwh)	Megawatt-year (MW-yr)	British thermal unit (Btu)	Million British thermal unit (MBtu)
Joule (J)	1	0.239	2.39×10^{-4}	2.78×10^{-7}	3.17×10^{-14}	9.49×10^{-4}	9.49×10^{-10}
Calorie (cal)	4.184	1	10^{-3}	1.16×10^{-6}	1.33×10^{-13}	3.97×10^{-3}	3.97×10^{-9}
Kilocalorie (kcal)	4184	1,000	1	1.16×10^{-3}	1.33×10^{-10}	3.97	3.97×10^{-6}
Kilowatt hour (kWh)	3.6×10^{6}	8.6×10^{5}	860	1	1.14×10^{-7}	3,413	3.41×10^{-3}
Megawatt-year (MW-yr)	3.16×10^{13}	7.54×10^{12}	7.54×10^{9}	8.77×10^{6}	1	2.99×10^{10}	2.99×10^{4}
British thermal unit (Btu)	1,054	252	0.252	2.93×10^{-4}	3.34×10^{-11}	1	10^{-6}
Million British Thermalunit (MBtu)	1.05×10^{9}	2.52×10^{8}	2.52×10^{5}	293	3.34×10^{-5}	10^{6}	1
Quad (quad)	1.05×10^{18}	2.52×10^{17}	2.52×10^{14}	2.93×10^{11}	3.34×10^{4}	10^{15}	10^{9}
Q	1.05×10^{21}	2.52×10^{20}	2.52×10^{17}	2.93×10^{14}	3.34×10^{7}	10^{18}	10^{12}
Milli-Q (mQ)	1.05×10^{18}	2.52×10^{17}	2.52×10^{14}	2.93×10^{11}	3.34×10^{4}	10^{15}	10^{9}
Metric tons of oil equivalent (mtoe)	4.3×10^{10}	1.0×10^{10}	1.0×10^{7}	1.2×10^{4}	1.4×10^{-3}	4.1×10^{7}	41
Metric tons of coal equivalent (mtce)	3.0×10^{10}	7.1×10^{9}	7.1×10^{6}	8,100	9.4×10^{-4}	2.8×10^{7}	28
Barrels of oil equivalent (boe)	6.2×10^{9}	1.5×10^{9}	1.5×10^{6}	1,700	2.0×10^{-4}	5.9×10^{6}	5.9
Trillion cubic feet (TCF)	1.2×10^{18}	2.8×10^{17}	2.8×10^{14}	3.2×10^{11}	3.7×10^{4}	1.1×10^{15}	1.1×10^{9}
Teracalorie (Tcal)	4.184×10^{12}	10^{12}	10^{9}	1.2×10^{6}	0.13	4.0×10^{9}	4.0×10^{3}